T0282374

Medical and Scientific Publishing

Medical and Scientific Publishing

Author, Editor, and Reviewer Perspectives

Edited by

Jasna Markovac
Publishing Officer, Health Information Technology & Services,
University of Michigan, Ann Arbor, MI, United States

Molly Kleinman
Program Manager for Clinical Education & Basic Science,
Health Information Technology & Services,
University of Michigan, Ann Arbor, MI, United States

Michael Englesbe
Cyrenus G. Darling Sr. & Cyrenus G. Darling Jr. Professor of Surgery,
University of Michigan, Ann Arbor, MI, United States

Academic Press is an imprint of Elsevier
125 London Wall, London EC2Y 5AS, United Kingdom
525 B Street, Suite 1800, San Diego, CA 92101-4495, United States
50 Hampshire Street, 5th Floor, Cambridge, MA 02139, United States
The Boulevard, Langford Lane, Kidlington, Oxford OX5 1GB, United Kingdom

Library of Congress Cataloging-in-Publication Data
A catalog record for this book is available from the Library of Congress

British Library Cataloguing-in-Publication Data
A catalogue record for this book is available from the British Library

ISBN: 978-0-12-809969-8

For information on all Academic Press publications visit our website at https://www.elsevier.com/books-and-journals

Publisher: Mica Haley
Acquisition Editor: Tari Broderick
Editorial Project Manager: Pat Gonzalez
Production Project Manager: Mohana Natarajan
Designer: Vicky Pearson

Cover Legend: Front cover by Marc R. Stephens, MAED (marque@med.umich.edu).

Typeset by TNQ Books and Journals

To our students, past, present,
and future—thank you for your
enthusiasm, energy, and excitement
for publishing and the importance of
communicating important research

Contents

Part IV
Expanding Access and Increasing Impact

List of Contributors

Ellen R. Abramson, Medical Development, University of Michigan, Ann Arbor, MI, United States

Christina N. Bennett, PhD, American Physiological Society, Bethesda, MD, United States

John P. Bilezikian, MD, Division of Endocrinology, Columbia University, New York, NY, United States

Robert M. Cermak, MM, Medical Student Leadership Program, University of Michigan, Ann Arbor, MI, United States

Chris Chapman, MA, Health Information Technology and Services,University of Michigan, Ann Arbor, MI, United States

Jason Colman, MS, Michigan Publishing, University Library, University of Michigan, Ann Arbor, MI, United States

Marisa L. Conte, MLIS, Taubman Health Sciences Library, University of Michigan, Ann Arbor, MI, United States

Shannon Cramm, MD, Department of Surgery, Massachusetts General Hospital, Boston, MA, United States

David C. Cron, University of Michigan Medical School, Ann Arbor, MI, United States

Sagar Deshpande, University of Michigan Medical School, Ann Arbor, MI, United States; John F. Kennedy School of Government, Harvard University, Cambridge, MA, United States

Justin B. Dimick, MD, MPH, Department of Surgery, Center for Healthcare Outcomes and Policy, University of Michigan, Ann Arbor, MI, United States

Michael Englesbe, MD, University of Michigan Medical School, Ann Arbor, MI, United States; Section of Transplantation Surgery, Department of Surgery, University of Michigan, Ann Arbor, MI, United States

Jason Engling, MA, Health Information Technology and Services,University of Michigan, Ann Arbor, MI, United States

David Fessell, MD, Department of Radiology, Medical Student Leadership Program, University of Michigan, Ann Arbor, MI, United States

Hunter Heath III, MD, Department of Medicine, Indiana University School of Medicine, Indianapolis, IN, United States

Nathan Houchens, MD, Department of Internal Medicine, University of Michigan, Ann Arbor, MI, United States; Department of Medicine, Inpatient Care, Veterans Affairs Ann Arbor Healthcare System, Ann Arbor, MI, United States

Andrew M. Ibrahim, MD, MSc, Department of Surgery, University of Michigan, Ann Arbor, MI, United States

Todd A. Jaffe, University of Michigan Medical School, Ann Arbor, MI, United States

Molly Kleinman, MSI, MA, Health Information Technology and Services, University of Michigan, Ann Arbor, MI, United States

Karen Kost, Health Information Technology and Services, University of Michigan, Ann Arbor, MI, United States

Melissa Levine, JD, Copyrights Office, University Library, University of Michigan, Ann Arbor, MI, United States

Spencer Lewis, University of Michigan Medical School, Ann Arbor, MI, United States

Joseph R. Linzey, University of Michigan Medical School, Ann Arbor, MI, United States

Alisha Lussiez, MD, Department of Surgery, University of Michigan, Ann Arbor, MI, United States

Preeti N. Malani, MD, Department of Internal Medicine, University of Michigan, Ann Arbor, MI, United States

Dave Malicke, MA, Office of Academic Innovation University of Michigan, Ann Arbor, MI, United States

Adam Marcus, MA, Retraction Watch and The Center for Scientific Integrity, New York, NY, United States

Robert Marcus, MD, Department of Medicine, Stanford University, Stanford, CA, United States

Jasna Markovac, PhD, Health Information Technology and Services, University of Michigan, Ann Arbor, MI, United States

Michael W. Mulholland, MD, Department of Surgery, University of Michigan, Ann Arbor, MI, United States

Vahagn C. Nikolian, MD, Department of Surgery, University of Michigan, Ann Arbor, MI, United States

Tyler Nix, MLIS, Taubman Health Sciences Library, University of Michigan, Ann Arbor, MI, United States

Susan E. Old, PhD, National Institutes of Health, Bethesda, MD, United States

Ivan Oransky, MD, Retraction Watch and The Center for Scientific Integrity, New York, NY, United States

Laura Ostapenko, MD, John F. Kennedy School of Government, Harvard University, Cambridge, MA, United States; Department of Surgery, Brigham & Women's Hospital, Boston, MA, United States

Evan Oxner, Freelance Medical Illustrator, New York, NY, United States

Jennifer L. Pesanelli, Federation of American Societies for Experimental Biology, Bethesda, MD, United States

Hanna Saltzman, Medical Student Leadership Program, University of Michigan, Ann Arbor, MI, United States

Judith Smith, MLIS, Taubman Health Sciences Library, University of Michigan, Ann Arbor, MI, United States

Jean Song, MSI, Taubman Health Sciences Library, University of Michigan, Ann Arbor, MI, United States

Paula T. Ross, PhD, Office of Medical Student Education, University of Michigan, Ann Arbor, MI, United States

Meng H. Tan, MD, Division of Metabolism, Endocrinology and Diabetes, Department of Internal Medicine, University of Michigan, Ann Arbor, MI, United States

Neil Thakur, PhD, National Institutes of Health, Bethesda, MD, United States

Aki Yao, Health Information Technology and Services,University of Michigan, Ann Arbor, MI, United States

Foreword

Developing communication skills has long been recognized as a fundamental part of a medical student's education. The focus traditionally has been on communicating with patients, their families, and professional colleagues mutually involved in the care of one's patients. Likewise, involvement in research is recognized as important to a medical student's professional development. However, teaching how to convey the results of one's research findings is generally not a part of the formal medical school curriculum.

As part of a curriculum transformation at the University of Michigan medical school, the scope of communication skills development was broadened to include the breadth of skills that future physicians will employ. Writing and publishing research papers and presenting scholarly work are communication skills specifically included in the curriculum. While recognizing that not all students will undertake scientific studies or pursue academic careers, the faculty felt that experiencing the process would enhance the students' understanding of the importance of open communication of scientific findings and develop skills that are widely applicable.

In *Medical and Scientific Publishing*, the editors and authors provide insight and advice based on their experience in developing and implementing a curriculum designed to develop skills and understanding of scientific publication. Multiple options for publishing, the process involved in not only writing but also the review of articles, and legal and ethical issues are but some of the topics covered. Very importantly, heeding the wisdom of Albert Einstein, "Learning is experience. Everything else is just information," the student-led *Michigan Journal of Medicine* was launched to provide opportunities for students to practice what they have been taught and develop their skills. Sharing their approach to learning, the authors provide educators interested in developing their students' scientific and medical communication skills a practical guide that can be readily applied at other institutions.

James O. Woolliscroft, MD, MACP
Lyle C. Roll Professor of Medicine
Professor of Internal Medicine and Learning Health Sciences
University of Michigan Medical School

Preface

The ability to effectively communicate and disseminate research is an integral part of any academic's professional responsibilities. Unfortunately, many scientists and most physicians do not learn in any organized and systematic way how to write and publish their work. As part of its integrated medical school curriculum, in 2015 the University of Michigan introduced a course on Medical Writing and Editing, concomitant with the launch of a student-run medical journal, the *Michigan Journal of Medicine*, which would in essence serve as a "lab" for the class. This book was initially conceived based on this course. Based on student feedback, the breadth of the course evolved and expanded since the first time it was taught, as did ultimately the scope of our book.

While various texts cover some aspects of scientific or medical writing, there has—until now—been no single guide devoted to helping clinicians and academics understand how to navigate the publishing world, while also preparing them for other related roles, such as author, editor, and peer reviewer. We hope that *Medical and Scientific Publishing: Author, Editor, and Reviewer Perspectives* will fill that crucial gap.

Many of our contributors are guest lecturers in the course, some are (or were) our students, and others are our publishing colleagues. We chose to leave the chapters in the authors' individual voices rather attempting to standardize across the entire contents. We feel that the book benefits from many diverse perspectives—including those of physicians, medical students, writers, editors, and publishers—to provide a comprehensive, start-to-finish overview of the complexity (and opportunity) that exists within the medical and scientific publishing landscape.

In addition to covering "standard" scientific and medical writing, the book explores many other publishing and editorial topics that are essential to helping academic physicians and scientists gain critical knowledge and perspective, including Distinguishing the many different types of publishing; Explaining different kinds of editing and their importance; describing how peer review plays a central role in academic publishing; highlighting legal and ethical issues such as copyrights, contracts, and publishing ethics—including scientific misconduct and conflict of interest; and other relevant topics including open access and funder mandates; the importance of effective communication, publishing case studies and clinical trials, measuring impact, and more.

This book also includes several chapters based on the *Michigan Journal of Medicine* by the University of Michigan medical students, staff, and faculty who worked together to establish and publish the new journal.

We hope that *Medical and Scientific Publishing* will become an essential resource for any undergrad, grad student, medical student, future or current scientist, academic professional—or anyone else—who is interested in gaining a better understanding of academic publishing and wishes to use this knowledge to facilitate broad dissemination of important research and scholarship.

Jasna Markovac, Molly Kleinman and Michael Englesbe

Acknowledgments

The editors would like to thank all the authors for their insightful and amazingly timely contributions to this book. This could not have happened without you. We really appreciate your support, both of the book and of our publishing schedule. We are also very grateful to our colleagues, Karen Kost and Marc Stephens in Health Information Technology and Services at the University of Michigan for their assistance in developing the various elements of the book that needed to come together in a very short timeframe. Karen was instrumental in content development and Marc kindly allowed us to use his photographs in the book and took new pictures whenever we needed to fill any gaps. In addition, we thank Nancy North, our freelance writer in Ann Arbor, who assisted with the lecture transcripts and who encouraged us to expand our table of contents to include topics we had not previously considered. Finally, we acknowledge the hard work and infinite patience of our publishing editor, Tari Broderick, our editorial project manager, Pat Gonzalez and our production project manager Mohana Natarajan, at Elsevier. Without Tari's perseverance on all levels from start to finish and without Pat's and Mohana's gentle but persistent (and much deserved) reminders and their very efficient editorial and production work, this book would not exist.

Thank you to everyone who made it possible.

PERSONAL ACKNOWLEDGMENTS

Many thanks to Gary for his patience, support, and love—J. Markovac

All my thanks to Piet, Helena, and Renske, for everything—M. Kleinman

I would like to thank Audrey, Mia, Ava, and William for their support—M. Englesbe

Chapter 1

Teaching Publishing in Medical Education—An Overview

Michael Englesbe, MD[1,3], Jasna Markovac, PhD[2]
[1]*Section of Transplantation Surgery, Department of Surgery, University of Michigan, Ann Arbor, MI, United States;* [2]*Health Information Technology and Services, University of Michigan, Ann Arbor, MI, United States;* [3]*University of Michigan Medical School, Ann Arbor, MI, United States*

INTRODUCTION

Academic institutions agree on the importance of encouraging the principles of open science and open scholarly communication throughout the research life cycle. Yet, while many stakeholders have recognized the need, the best route toward achieving transparency remains unclear. The University of Michigan Medical School believes that the most sustainable approach for creating an open science culture is to endow that ethos within our students. Incorporating ethical research conduct, open science principles, and effective communication techniques in early research experiences will prompt and encourage students to both demand and embody openness throughout their careers as clinician scientists. With the support of key stakeholders, including administrators, principal investigators, and Library partners, we have begun to incorporate the necessary tools and services into our curriculum.

MOTIVATION

Medical education lacks established curricula in which students can engage with the dissemination of their work to their peers, the broader scientific community, and the lay audience. Traditionally, any discussion of communication practices is predicated on having results deemed worthy of dissemination. Medical students, who are often exposed to concentrated and not longitudinal research experience, have few opportunities to engage in the world of publishing and can become easily discouraged with the process. We believe that there are essential skills to be learned, independent of the success of the project or the quality of the mentor. To ensure that all medical students have these foundational skills, they must have structured opportunities. Under the banner of professional development, we are working toward the development of a student-led and faculty-driven research community that allows for the sharing, criticism, and publication of data with the establishment of large-scale

Medical and Scientific Publishing. https://doi.org/10.1016/B978-0-12-809969-8.00001-2

program for open scholarly communications. We are actively working toward this vision through the growth and refinement of our medical school curriculum and specific student-led curricular initiatives [1].

UNIVERSITY OF MICHIGAN MEDICAL SCHOOL CURRICULUM

In 2015 we launched a course designed to teach medical students about medical publishing, Medical Writing and Editing, and a new journal, the Michigan Journal of Medicine (MJM) [2] (Fig. 1.1). MJM, now in its third year, includes a team of student editors and reviewers and serves as a practical lab component for the course. As students are responsible for the production of the journal, from initial article review to final publishing decisions, involvement with the journal provides several leadership and educational opportunities. In addition to providing students the chance to practice structured scientific criticism, the journal serves as an outlet for student research that is open to the public, transparent, and free from publication bias. Student editorial team operates under the guidance of faculty and staff at the University of Michigan Medical School and the University Library.

The course has expanded into Academic Communications and now includes the Michigan Medical Student Communication Collaborative (CC) [3]. The CC began as a student-run class, founded with the goal of promoting clear communication skills and providing forums for students to practice public speaking. The

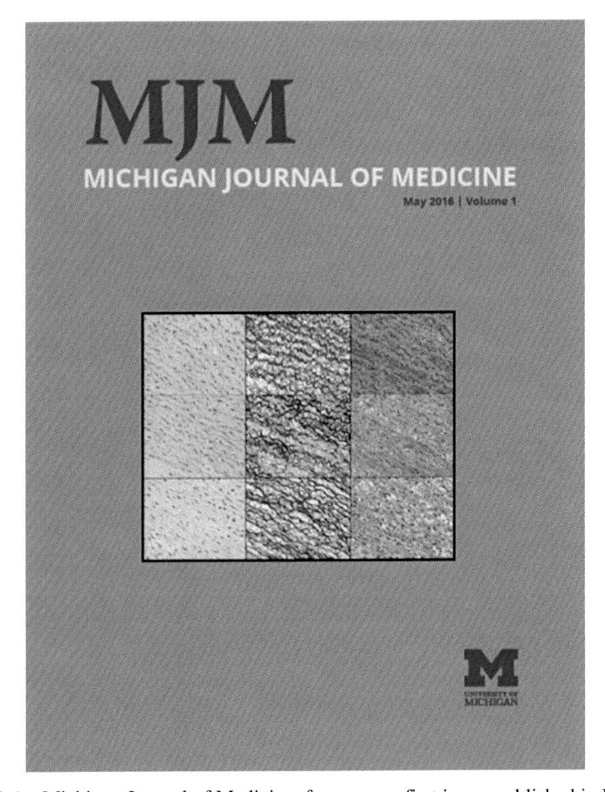

FIGURE 1.1 Michigan Journal of Medicine, front cover, first issue, published in May, 2016.

medical students are responsible for organizing a speaker series, "Medical Student Grand Rounds," [4] (Fig. 1.2) and coaching speakers for the delivery of refined oral presentations. Coaching involves an initial brainstorming session, slide deck critique, and in-person presentation practice. Final presentations are recorded and available for viewing by the medical community. In addition the material presented by the students can be developed into manuscript submissions to MJM.

These initiatives provide rare educational opportunities for students, enabling them to see multiple sides of the publication process and allowing them to learn about the end-to-end workflow involved in the dissemination of information. In addition, we work closely with our colleagues at the Library and elsewhere on campus to provide training for students, fellows, and junior faculty in publishing, giving presentations, and funding compliance.

WORKING WITH TOGETHER

As an academic institution, the University of Michigan is a staunch supporter of open dissemination of scholarship. We strive to teach our students about the many content delivery modes that are available to the medical and scientific communities and encourage them to explore various options and make informed decisions about how best to publish their research.

In addition to curricular training, students have opportunities to work with faculty and staff who are involved in projects with professional publishing companies and other publishing organizations. This includes journal editorial work (Fig. 1.3),

FIGURE 1.2 Example of a lecture, part of Medical Student Grand Rounds.

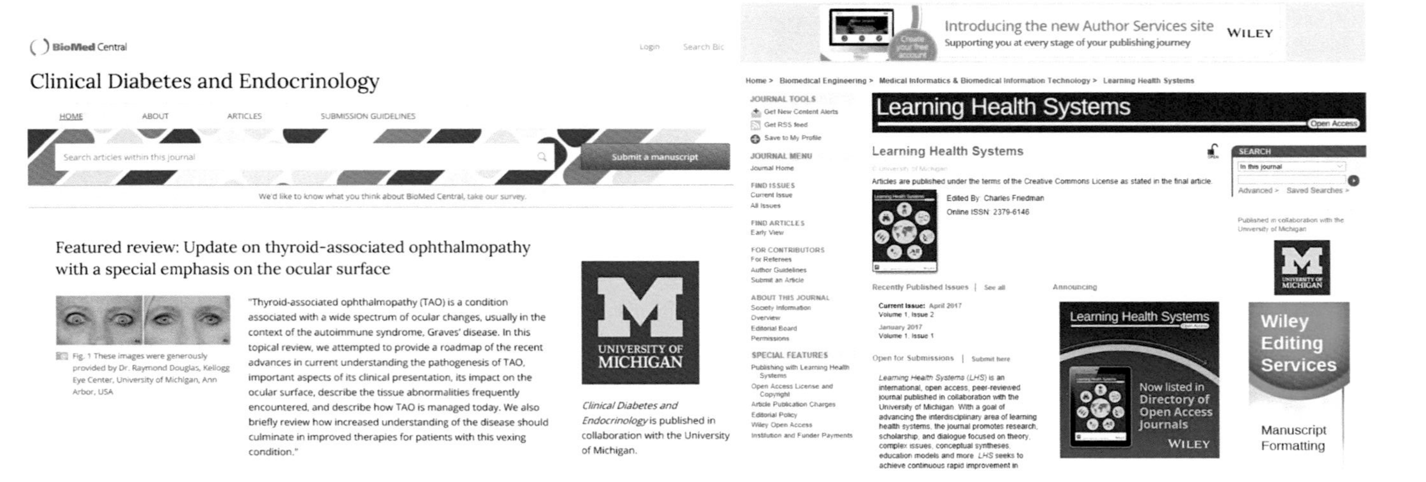

FIGURE 1.3 Journals published by the University of Michigan, in collaboration with publishing partners. *(From http://clindiabetesendo.biomedcentral.com/ and http://onlinelibrary.wiley.com/journal/10.1002/(ISSN)2379-6146/.)*

book development (Fig. 1.4), content app development (Fig. 1.5), and collaborations with departments that provide publishing services across campus [6]. We encourage students to discuss publishing opportunities at all levels, from how to publish an article to how to navigate an editorial career path, with our faculty and staff as well as with professional publishers.

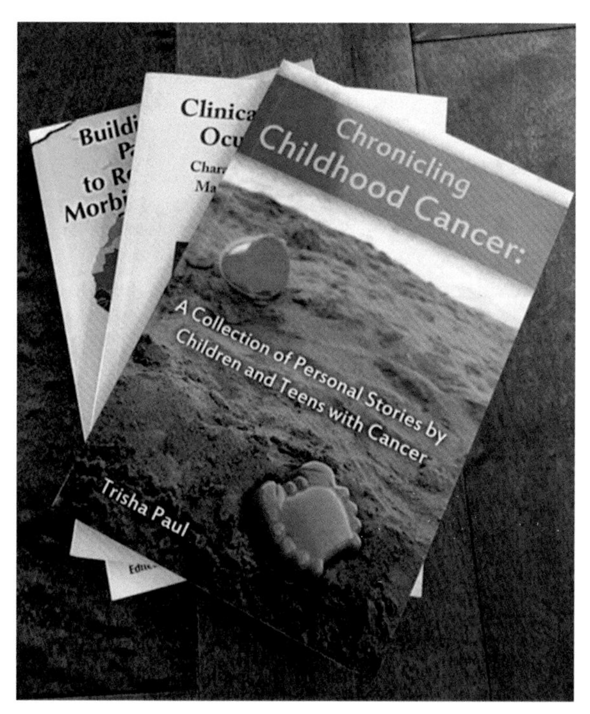

FIGURE 1.4 Examples of books published by the Medical School, in partnership with the Library's Michigan Publishing [5]. The top title, *Chronicling Childhood Cancers*, was developed by a then student working at C.S. Mott Children's Hospital.

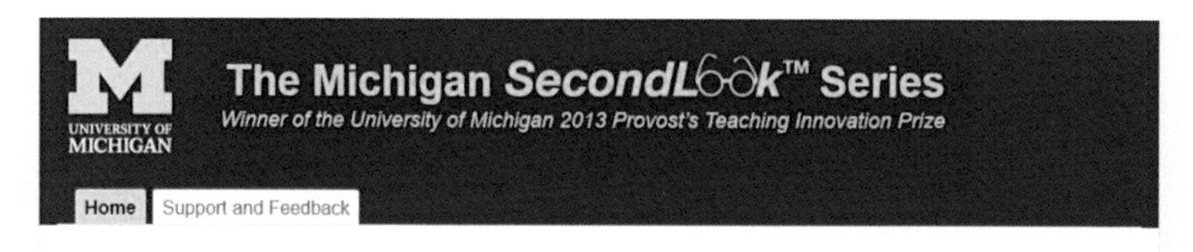

FIGURE 1.5 Home page for SecondLook mobile applications, self-review study tools. http://secondlook.med.umich.edu/home.

This book was conceived through discussions with a publisher about our course and its related initiatives. Our initial plan was to use our course syllabus for the book outline. But, as we consulted with our students and our colleagues, it became clear that our scope needed to expand beyond what we had been teaching. In addition, we decided to broaden our author pool to include contributors beyond the University of Michigan and outside of academia. As we started having discussions about our book plans with potential authors, we also talked about our course and what we were aiming to achieve. As a result, we were able to generate considerable enthusiasm for our teaching efforts among our authors and even managed to recruit a few guest lecturers.

CONCLUSION

We are seeking to promote and cultivate a culture of open science by establishing a program that will provide medical students with the knowledge and experience to critique, communicate, and publish their research, regardless of result, within a community of their peers. We are implementing our new medical school curriculum with initiatives, tools, and opportunities for students that allow them to publish their research without the external pressures of the broader scientific community as well as to prepare them for the publishing responsibilities that come with a professional academic career. As our initiatives grow to reach a broad population, we are hopeful that our curriculum will be easily replicable at other institutions and that our students will pass along open science principles, skills, and behaviors to those that they will eventually train.

REFERENCES

[1] https://medicine.umich.edu/medschool/education/md-program/curriculum.
[2] http://www.michjmed.org/.
[3] https://ummscommunicationcollaborative.com/.
[4] https://maizepages.umich.edu/organization/MSGR.
[5] https://www.publishing.umich.edu/services/.
[6] http://hits.medicine.umich.edu/.

Part I

Publishing and Editing

Chapter 2

Open Access Journal Publishing

Dave Malicke, MA[1], Jasna Markovac, PhD[2]
[1]Office of Academic Innovation University of Michigan, Ann Arbor, MI, United States; [2]Health Information Technology and Services, University of Michigan, Ann Arbor, MI, United States

INTRODUCTION

Open access (OA) scholarship is available online and can be read for free. It is often available for use and sharing without copyright and licensing restrictions commonly placed on published works. By making their work available in this manner, authors ensure that the broadest possible audience can read and use it, without limiting it to only those who are affiliated with research libraries or who can afford journal subscriptions. OA is not defined by a particular business model or type of content. Rather, OA is an approach to sharing one's work with the wider world [1]. This introductory section on OA was adapted from the University of Michigan Library's research guide on OA, as an excellent example of the usefulness of high quality freely available content (Source: http://guides.lib.umich.edu/openaccess; © The Regents of The University of Michigan, published under a CC BY license).

WHAT IS OPEN ACCESS?

The term "Open Access" has at least three primary definitions. One of the original principles, known as the Budapest OA Initiative, requires that OA content be freely available and that all uses, including commercial uses, be allowed so long as the authors are "properly acknowledged and cited" [2]. Others, such as the Bethesda [3] and Berlin [4] statements, say that OA materials should allow others to "copy, use, distribute, transmit and display the work publicly and to make and distribute derivative works, in any digital medium for any responsible purpose, subject to proper attribution…." It is important to note here the key phrase "responsible purpose."

Some OA supporters feel that to be truly OA, commercial uses must be allowed, which is in accordance with the Budapest definition. Others believe that allowing all commercial uses may be irresponsible and so interpret the Bethesda and Berlin statements to mean that OA materials should be made freely available for all noncommercial purposes, such as for education and

Medical and Scientific Publishing. https://doi.org/10.1016/B978-0-12-809969-8.00002-4

research, but should not be permitted to be monetized by others. Articles that are made freely available but do not permit any reuses without direct permission from the copyright holder are often referred to as "publicly available" or "free to view," but these are not considered to be OA.

HISTORY OF THE OPEN ACCESS MOVEMENT

Initially intended for primary research journal articles, OA has expanded over the years to include other types of materials (e.g., review articles, book chapters, teaching materials). The "modern" OA movement is thought to have started in the 1950s, but it did not become prominent until the 1990s when the Internet started to take hold and as a response to journal publishers' high subscription prices. In life and biomedical sciences, OA gained momentum during the National Institutes of Health (NIH) leadership of Harold Varmus. As Director of the NIH, Varmus proposed a new "journal" to serve as a platform for preprints as well as peer-reviewed articles that would eventually become PubMed Central (PMC) [5] (Fig. 2.1).

In 2000 Varmus, then at Memorial Sloan Kettering Cancer Center, Pat Brown from Stanford, and Michael Eisen from Berkeley started a petition among the scientific community calling for a ban on publishing in any journal that did not make its content freely available online either on publication or at most after a 6-month embargo. While tens of thousands of scientists signed the petition, most did not follow through, and in 2001 Eisen and Brown announced that they would start their own publishing company that would provide free online access to all articles. This publishing company, The Public Library of Science (PLOS), became one of the first major OA journal publishers when it launched in 2002 with the support of the Gordon and Betty Moore Foundation. Others soon followed suit. Some OA publishers are commercial companies, for example, BioMed Central (a division of Springer Nature), while others are not-for-profit (e.g., PLOS).

According to a recent report from the media and publishing intelligence firm, Simba Information, OA articles (whether freely accessible on publication or after an embargo) now represent one-third of all research articles published, and the number of OA papers is increasing at twice the annual rate for all articles. Simba projects that the total number of articles published as OA will reach 3 million worldwide by 2020 [6]. However, some top OA journals have shown decreases in numbers of published articles. Notably, PLOS One, the largest, by far, of all the PLOS journals, has seen a drop of nearly 30% since its peak in 2013, due primarily to a decrease in submissions [7]. The submission dip may be due to an increase in the number of broad-based, interdisciplinary OA research journals, some with higher impact factors, or perhaps to recent increases in author fees for PLOS One. Nevertheless, PLOS One continues to publish over 22,000 articles per year, nearly 90% of the total output of the PLOS journals.

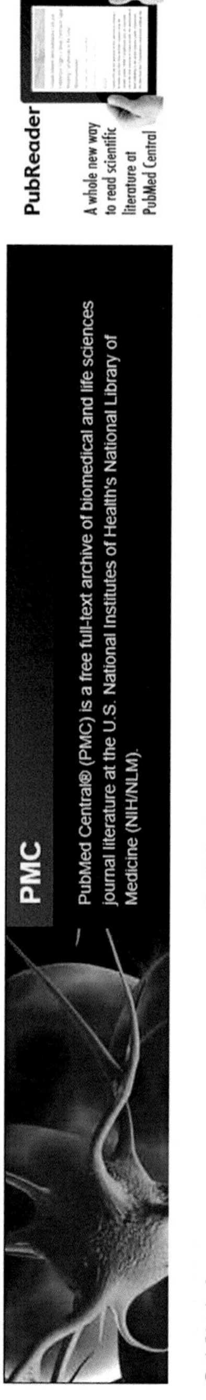

FIGURE 2.1 Screenshot of PubMed Central website (June 2017) https://www.ncbi.nlm.nih.gov/pmc/. *(National Center for Biotechnology Information, U.S. National Library of Medicine.)*

OPEN ACCESS, COPYRIGHT, AND LICENSES

Copyright is an important factor in OA publishing. Authors who publish in OA journals or choose to make their articles available openly via an OA option generally retain copyright to their work. This is in contrast to traditional subscription-based journals, which may require that authors assign copyright either to the publisher or to the learned society that owns the journal. Because authors keep copyright for OA content, they must assign a license to their work that provides guidance for the journal and for the reader regarding reuse of the copyright-protected material. There are many different licenses, most created by third parties, including publishers, other industry organizations, and not-for-profits such as Creative Commons (CC). The CC licenses are the ones most frequently used by OA journals. There are several types of CC licenses that range in level of reuse restriction (Table 2.1).

Most OA journals use either the CC BY or CC BY-NC license, some offering a choice to authors. Several funding agencies have mandated assignment of one of the least restrictive licenses, CC BY, to articles resulting from research funded by those bodies. For more information on copyright and terms of use, please see Chapter 17.

In 2014, the International Association of STM Publishers (STM) introduced a set of model licenses that would complement the CC licenses. The STM model licenses provide additional options that enable publishers and authors the ability to better accommodate several factors, including those related to text data mining, translations, and the overall globalization of research [8]. Because CC licenses were designed for a very broad creative community and not specifically for academic and/or medical publishing, the additional clarity made by the STM model licenses may be helpful regarding terms of use, particularly commercial reuse.

TYPES OF OPEN ACCESS

There are two types of OA: green and gold. Green OA refers to deposition of articles into a publicly accessible subject repository or an institutional repository. These can be preprints, authors' final peer-reviewed manuscripts, or final journal articles. Since green OA articles published in traditional subscription-based journals, authors do not pay any fees for this type of OA. The cost of publication is borne by the subscribers. These articles are generally under copyright and may or may not carry a CC license. If a CC license is used, then it is usually the most restrictive type (CC BY-NC-ND). There may also be an embargo until the article can be made publicly accessible. That embargo can be between 6 and 24 months after the official publication of the article, but this is highly variable, depending on the journal and the discipline. Because of the restrictions, there are those who maintain that green OA is not true OA.

Gold OA refers to OA content published in (peer-reviewed) journals. The journal can be subscription based and offer an option to authors who are willing to pay a fee to have their articles made openly available on publication (hybrid

TABLE 2.1 Types of CC Licenses ("Creative Commons Licenses" by Creative Commons Is Licensed Under CC BY 4.0)

 **Attribution CC BY** This license lets others distribute, remix, tweak, and build upon your work, even commercially, as long as they credit you for the original creation. This is the most accommodating of licenses offered. Recommended for maximum dissemination and use of licensed materials.	 **Attribution-ShareAlike CC BY-SA** This license lets others remix, tweak, and build upon your work even for commercial purposes, as long as they credit you and license their new creations under the identical terms. This license is often compared to "copyleft" free and open source software licenses. All new works based on yours will carry the same license, so any derivatives will also allow commercial use. This is the license used by Wikipedia and is recommended for materials that would benefit from incorporating content from Wikipedia and similarly licensed projects.	 **Attribution-NonCommercial CC BY-NC** This license lets others remix, tweak, and build upon your work non-commercially, and although their new works must also acknowledge you and be non- commercial, they don't have to license their derivative works on the same terms.
 **Attribution-NonCommercial-ShareAlike CC BY-NC-SA** This license lets others remix, tweak, and build upon your work non-commercially, as long as they credit you and license their new creations under the identical terms.	 **Attribution-NoDerivs CC BY-ND** This license allows for redistribution, commercial and non-commercial, as long as it is passed along unchanged and in whole, with credit to you.	 **Attribution-NonCommercial-NoDerivs CC BY-NC-ND** This license is the most restrictive of the six main licenses, only allowing others to download your works and share them with others as long as they credit you, but they can't change them in any way or use them commercially.

option). Alternatively the journal can be fully OA (no subscriptions) and may charge fees to authors (or their institutions, if applicable) for accepted papers. All gold OA articles are freely available for all readers with Internet access.

There are nearly 10,000 gold OA journals in the Directory of OA Journals (DOAJ) and many more that do not qualify for the directory due to questionable editorial and publishing practices. The latter are commonly known as "predatory journals" or "predatory publishers." Until recently, there was a publicly available list of these journals and publishers (maintained by a librarian at the University of Colorado), but this list has been removed. No official explanation has been provided. Nevertheless, we strongly encourage caution for those who may receive email solicitations to submit papers to journals with titles remarkably similar to established, legitimate publications. Authors beware: there are many OA journals, but quality and reputation vary widely.

ARTICLE PROCESSING CHARGES

Whether or not a journal charges Article Processing Charges (APCs) varies widely among disciplines. OA journals in some fields, such as the arts and humanities, tend to have no fees, while others, especially in the "hard sciences," are more likely to charge APCs. In 2013, Koziak and Hartley published a study of publication fees among OA journals across disciplines [9]. Interestingly, that paper is not OA and could only be obtained with a subscription to the Journal of the Association for Information Science and Technology or by pay-per-view.

Many reputable OA journals in life and biomedical sciences and in medicine require authors to pay OA fees. APCs can range from a few hundred dollars to several thousand dollars, depending on the journal. Journals that attract many submissions and have very high rejection rates tend to charge higher APCs since the cost of the rejections (editorial workflow, peer-review, and administrative resources) essentially needs to be absorbed into the cost of the content that is ultimately accepted and published. Since its launch in 2012, a notable exception to charging APCs was eLife, an OA journal established as a joint venture between the Howard Hughes Medical Institute (HHMI), the Wellcome Trust, and the Max Planck Society. But starting in 2017, eLife began charging $2500 for all accepted articles [10].

For many journals, especially society publications, APCs are often similar in cost to other publication fees (e.g., page charges, submission charges). These standard publication fees are assessed to authors for subscription-based journals even where authors do not retain copyright and the articles are not publicly available or reusable on publication.

IMPACT OF OPEN ACCESS

Over the past decade, a great deal of research has been done on the impact of OA on citations, with mixed results. Some studies seem to show that OA does lead to more citations [11,12], while others suggest that there is no statistical difference [13,14]. Self-selection may be a factor as is the particular discipline or field of study [13]. A recent report from Thomson ISI suggests that OA in itself does not necessarily mean more (or fewer) citations and that increasing the article's (or the journal's) potential readership does not change the basic relevance and value of a particular paper to the research community [15]. Furthermore, the authors suggest that as long as the publication is available to its target readership, then other components (e.g., article quality, prominence of the editorial board) will continue to be more important than who pays the production and access costs. As OA publications continue to grow and expand, further research will be required to determine what the impacts are on various metrics and how significant they may be in the long run.

Many funding agencies and an increasing number of universities and countries have established mandates for grant recipients and university researchers to make their scientific articles available publicly, usually after a set embargo period postpublication. Furthermore, most of the funders' regulations include mandatory deposit of the article to PMC or an equivalent repository or in university repositories. Specific requirements vary by institution, funding agency, and country. In some cases, the granting agency will provide funds especially designated to pay OA fees and will stipulate that the final journal article be deposited and made freely available on publication without an embargo. Wellcome Trust is an example. HHMI, on the other hand, strongly advised that authors make their articles be publicly available as soon as is possible and reasonable and no more than 12 months after publication. Other funders, such as the NIH in the United States, require that the final accepted author manuscript (after all revisions but prior to copyediting and coding) be deposited to PMC and made freely available no more than 12 month after official publication. For more information about US government funding and publishing, please see Chapter 23.

A survey conducted by Taylor and Francis on researcher OA behaviors indicates that OA as a way to publish scholarly content will continue to have a high profile and that the overall positive attitude toward this publishing model will grow. In addition, the report suggested that authors may not fully understand the implications of the various OA licenses and tend to be unsure of the various rules and regulations regarding deposits into institutional (and other) repositories [16]. There is a clear opportunity for universities, professional societies, and publishers to become proactive and more engaged in educating authors about the many choices that are offered to them as they decide how to publish their scholarly works.

ACKNOWLEDGMENTS

Portions of this chapter are adapted from an article on Open Access, published by the authors in 2015 in the International Journal of Neuropsychopharmacology by Oxford University Press on behalf of the CINP [17]. The journal had just switched its business model from subscription-based to fully OA, our article was published under a CC BY license, and we, the authors, are the copyright holders.

REFERENCES

[1] http://guides.lib.umich.edu/openaccess.

[2] Budapest OA Initiative. http://budapestopenaccessinitiative.org/.

[3] Berlin Declaration on OA. https://openaccess.mpg.de/Berlin-Declaration.

[4] Bethesda Statement on OA. http://legacy.earlham.edu/~peters/fos/bethesda.htm.

[5] Varmus H. E-BIOMED: a proposal for electronic publications in the biomedical sciences. 1999. Retrieved from: NIH.gov.

[6] http://www.simbainformation.com/Open-Access-Journal-10338054/.

[7] Davis P. PLOS one output drops again. The scholarly kitchen. January 5, 2017. https://scholarlykitchen.sspnet.org/2017/01/05/plos-one-output-drops-again-in-2016/.

[8] http://www.stm-assoc.org/copyright-legal-affairs/licensing/open-access-licensing/.

[9] Kozak M, Hartley J. Publication fees for OA journals: different disciplines—different methods. J Am Soc Inf Sci Technol 2013;64:2591–4.

[10] Butler D. Open-access journal eLife to start charging fees. Nature. http://dx.doi.org/10.1038/nature.2016.20700.

[11] Gargouri Y, Hajjem C, Lariviere V, Gingras Y, Brody T, Carr L, Harnad S. Self-selected or mandated, OA increases citation impact for higher quality research. Futrelle RP, editor. PLoS One 2010;5:e13636.

[12] Jiang X, Tse K, Wang S, Doan S, Kim H, Ohno-Machado L. Recent trends in biomedical informatics: a study based on JAMIA articles. J Am Med Inform Assoc 2013;20:e198–205.

[13] Davis P, Lewenstein BV, Simon DH, Booth JG, Connolly MJL. OA publishing, article downloads, and citations: randomised controlled trial. BMJ 2008;337:a568.

[14] Davis PM. OA, readership, citations: a randomized controlled trial of scientific journal publishing. FASEB J 2011;25:2129–34.

[15] Pringle J. Do open access journals have impact?. Nature 2017. http://www.nature.com/nature/focus/accessdebate/19.html (referencing The Impact of Open Access Journals: A Citation Study from Thomson ISI).

[16] http://www.tandf.co.uk/journals/explore/open-access-survey-june2014.pdf.

[17] Markovac J, Malicke D. Making sense of OA. Int J Neuropsychopharmacol February 2015;18(3). http://dx.doi.org/10.1093/ijnp/pyu108. pyu108. Published online February 16, 2015.

Chapter 3

Society Journal Publishing

Jennifer L. Pesanelli
Federation of American Societies for Experimental Biology, Bethesda, MD, United States

VALUE OF A SOCIETY JOURNAL PUBLISHER

There are numerous types of scholarly publishers including academic institutions, libraries, university presses, commercial publishers, independent publishers, and society publishers. Kent Anderson, former editor-in-chief and founder of *The Scholarly Kitchen*, a blog aimed at advancing and supporting scholarly publishing and communication, wrote a seminal post on the "96 Things Publishers Do (2016 Edition)" extolling the value publishers contribute to the publication process. While the list is too long to include here, among the 96 are peer review, providing infrastructure, editing, conveying prestige, format migrations, search engine optimization, standards, archiving, analytics, and compliance [1]. Society journal publishers do all of these things and more, contributing greatly to the scholarly publishing ecosystem.

THE SCHOLARLY MISSION

A scholarly or learned society is a group that is formed to promote an academic discipline or profession. A group's agreed on purpose and organizational commitments, or mission, is the cornerstone of a scholarly society and is the basis from which all activities of the organization stem. The Royal Society, founded in 1660, promotes the mission: "*The Society's fundamental purpose, reflected in its founding Charters of the 1660s, is to recognise, promote, and support excellence in science and to encourage the development and use of science for the benefit of humanity*" [2].

In the context of promotion, development, and use, the Royal Society published the first issue of *Philosophical Transactions* (Fig. 3.1) in 1665, which "established the important concepts of scientific priority and peer review, is now the oldest continuously-published science journal in the world" [3]. A scan across the mission statements of academic societies covering disciplines from humanities through sciences frequently include terms such as advance, develop, disseminate, educate, promote, and understand. For more than 350 years, since

Medical and Scientific Publishing. https://doi.org/10.1016/B978-0-12-809969-8.00003-6

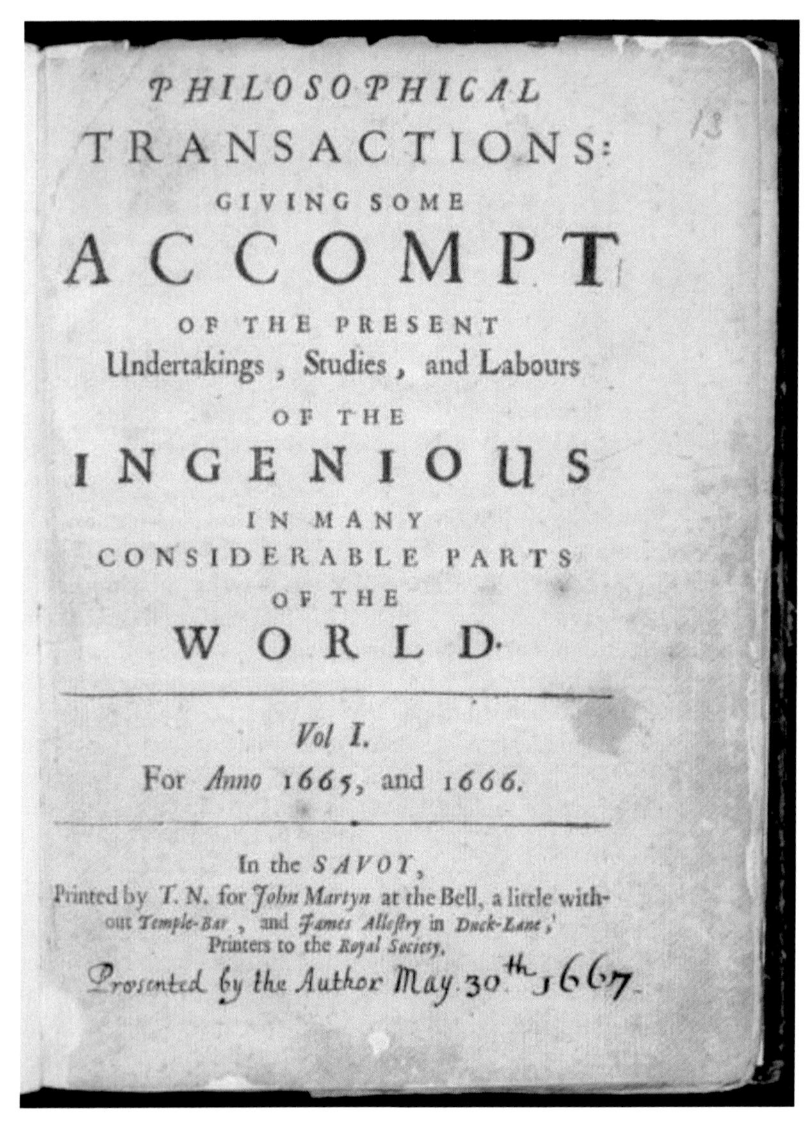

PHILOSOPHICAL
TRANSACTIONS:
GIVING SOME
ACCOMPT
OF THE PRESENT
Undertakings, Studies, and Labours
OF THE
INGENIOUS
IN MANY
CONSIDERABLE PARTS
OF THE
WORLD.

Vol I.
For *Anno* 1665, and 1666.

In the *SAVOY*,
Printed by *T. N.* for *John Martyn* at the Bell, a little with-
out *Temple-Bar*, and *James Allestry* in *Duck-Lane*,
Printers to the *Royal Society*.

Presented by the Author May 30th 1667.

FIGURE 3.1 Volume 1 of The Royal Society's Philosophical Transactions from 1665 [4].

the publication of *Philosophical Transactions*, the scholarly journal has been a mechanism for furthering the mission of scholarly societies.

STRATEGIC ROLE OF SOCIETY JOURNAL PUBLISHING

A journal publishing program should support a society's strategic plan which, inherently, should support the mission. It may be journal publishing is a part of a larger publishing program for a society, and journals are just one component of a multifaceted program, which may include books, databases, magazines, newsletters, etc., but regardless, the strongest programs have a clearly defined purpose and objectives.

In addition to supporting the scholarly mission directly through dissemination, education, and the like, journals often provide significant financial resources for societies which are used to further the mission. Revenues from journals can directly fund education programs, travel grants, scholarly achievement awards, advocacy efforts, and, in many cases, help keep membership dues lower than they might otherwise need to be. In many cases, journals are the largest source of nondues revenue for a society, and, without that revenue, other work of the organization would not possible.

Not all society publishing programs are profitable endeavors, however, but publication continues because the journals have a strategic role in supporting the mission. For some of these organizations, a portion of membership dues or revenues from other programs are used to offset the costs of publishing, and still other journals are published at a financial loss to the society.

PUBLISHING OPTIONS

There are many approaches societies use to publish journals, and a society may take advantage of multiple approaches depending on its resources as well as the size and needs of its publishing program. One approach is for a society to self-publish whereby all aspects of the program, including development and deployment of technology, are handled within the organization. This scenario is most likely to occur within a resource-rich society where in-house development is a priority for the organization.

A variation on this approach would be self-publishing where the organization supports the majority of the process internally but relies on external vendors for certain aspects, quite often technology related. For example, a society may employ staff to coordinate the peer review process but contract with a third-party vendor for software to enable online manuscript submission and peer review. Similarly, a society may handle article production internally but host the articles on a third-party online platform rather than develop its own. In working with these external organizations, societies can benefit from the experience and work of others that typically result in a higher quality product for a lower cost.

For other organizations, working with a university press or other not-for-profit press may be a more viable option. These business arrangements typically leave editorial control with the societies but move some or all aspects of the production process, online hosting, and even sales and marketing to an external organization.

Working with a commercial publisher is yet another option available to societies, and there is a wide variety of contractual relationships in this arena. In some instances, the commercial publisher may wholly own a journal title, including copyright, and contract with a society to brand the journal with the society's name, usually paying the society a fee or royalty to do so. This type of arrangement may be extended to allow the society a degree of editorial oversight of journal content. A society may agree to this type of arrangement because they do not have the resources to publish on their own, and the aim and scope of the publication support their mission. A society does not have much leverage in this type of arrangement, and dissatisfaction with the outcomes may result in a society removing their name from the journal. Similarly, the commercial publisher can also decide it is not in their best interest to retain the society involvement with the journal and discontinue the relationship.

Other types of commercial publishing arrangements include societies contracting with commercial publishers to handle all aspects of journal production and distribution while the societies retain copyright and editorial control of the journal. These are typically limited contractual arrangements (3–10 years) that can periodically be renegotiated to address changes in the industry or concerns that either party might have regarding execution of the prior agreement.

BUSINESS MODELS

Regardless of the publishing approach, the financial success of each journal relies heavily on an effective business model. These models are not unique to society publishing programs but represent alternatives for journal publishing programs whether independently or commercially published.

A traditional business model relies heavily on revenue from subscriptions. Quite often, the primary subscriber type is an academic institution. Depending on the subject matter, other types of organizations such as government agencies or for-profit companies may also be subscribers as well as individuals who may or may not be members of the society. If societies self-publish, they may rely on direct subscription purchases from institutions, academic, or otherwise. Societies may also work in groups and/or with external partners to aid in licensing their content to institutions. If they work with a university press or commercial publishing partner, those organizations typically sell packages, or bundles, of journals to institutions rather than license them title by title.

A common component of a traditional business model may also include revenue from author fees such as submission charges, page charges, color charges, supplemental data charges, and even author alternation charges. Whether or not these fees are charged and at what rate differ from journal to journal. Some societies waive some or all of the author fees for member authors or have reduced rates as an incentive for membership. Color charges have fallen by the wayside for a number of journals that have ceased print publication in favor of an online-only distribution. Other societies have created a combined page and color charge to cover the cost of generating an article page regardless of final format.

Another business model used by society journal publishers is an open access model. In this model, no subscription is necessary to access the journal content, and, in lieu of revenues from subscriptions to support the publishing process, authors pay article processing charges (APCs). APCs vary from journal to journal but are typically a flat rate per article. In many instances, an author's institution or funding agency will pay the APCs. APCs may be charged in addition to author fees such as submission and page charges, but other APCs may be all inclusive. Open access journal publishing is discussed in more detail in Chapter 2 on Open Access Journal Publishing.

A journal may offer a hybrid approach where some articles are not open access and therefore available only to subscribers, and other articles are covered by APCs and made immediately available to all who wish to read it.

Beyond just a hybrid journal, a society may use a mixed strategy in its journal publishing program and publish journals using both traditional and open access business models. Content, audience, and objectives are central factors in determining which business model is appropriate for a given journal.

A CASE IN POINT

To illustrate the strategic role of journal publishing for a society as well as to provide an example of publishing options and business models, consider *The FASEB Journal*. *The FASEB Journal* is published by the Federation of American Societies for Experimental Biology (FASEB), a federation of scholarly societies, whose mission is to promote progress and education in biological and biomedical sciences. The journal directly supports the mission by disseminating multidisciplinary original, high quality, peer-reviewed research in the life sciences. Net revenue from the journal plays a critical role in supporting FASEB's public affairs, science policy, and other mission-based activities. Fig. 3.2 shows the April, 2017 issue of *The FASEB Journal*.

The journal is self-published by the Federation, which owns the copyright, and FASEB uses a mix of in-house staff and external vendors to support the process. Peer review, editorial, and production management are handled internally while FASEB contracts with outside companies for the peer-review system, copyediting, composition, printing (yes, in 2017, *The FASEB Journal* is still available in print), and online hosting.

The business model for *The FASEB Journal* is based on a hybrid of subscription, author, and open access fees. The majority of revenue for the journal comes from institutional subscriptions. However, if authors or their funders desire to publish open access, they can pay a fee to have their article made openly available online immediately at publication. If the authors chose the open access option, they may choose a creative commons copyright license type of either CC-BY or CC-BY-NC. For more information on licenses, please see Chapter 2 on Open Access Journal Publishing and Chapter 17 on Copyright and Contracts.

In response to the growth of open access and authors' demands for different publishing models, FASEB is looking to expand its publications program with the launch of a fully open access online-only journal that will complement *The FASEB Journal*. The new journal will aim to serve the global biology

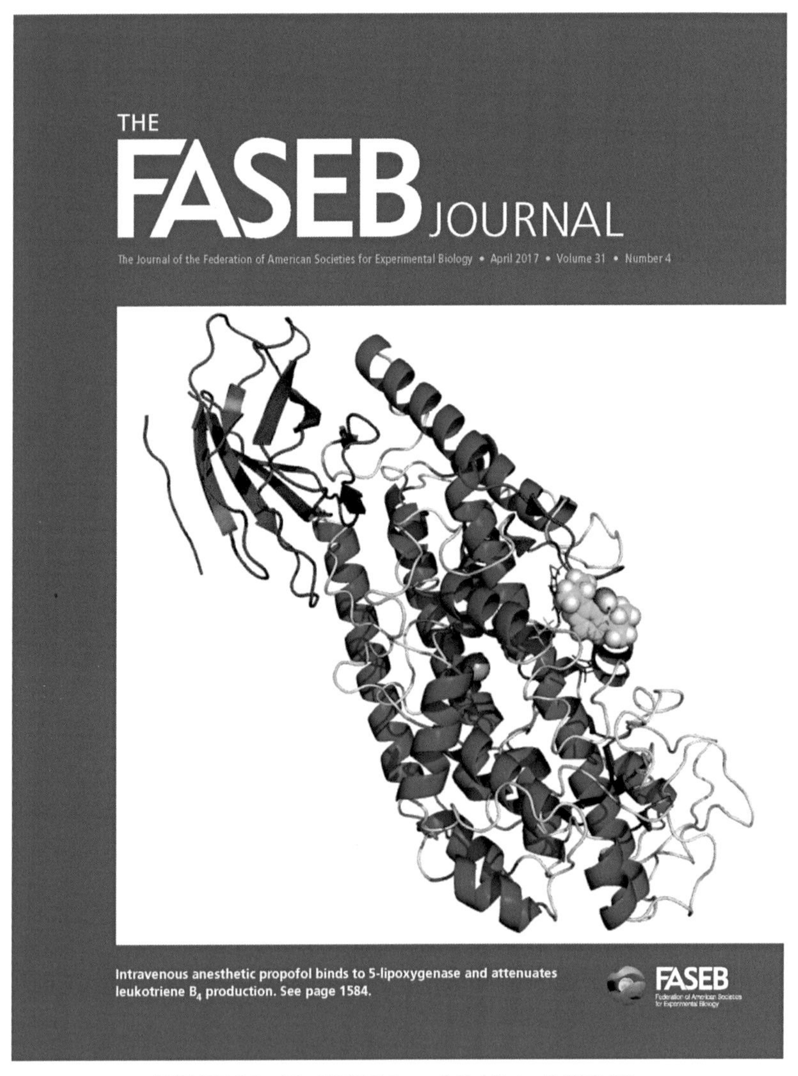

FIGURE 3.2 The FASEB Journal 31 [4], April 2017 [5].

community by providing a vehicle for scientists to publish well-designed and executed research papers, and the scope will include original, multi/transdisciplinary biomedical research from the international scientific community. The aim and scope support FASEB's mission to promote progress in biological and biomedical sciences. FASEB will retain ownership of the journal though it will work with a commercial publishing partner to produce the journal, and the articles will be published under a CC-BY license.

THREATS TO SOCIETY PUBLISHING PROGRAMS

Funding is a broad issue that poses challenges to society journal publishing at many levels. At the most fundamental level, author (researcher) funding has decreased across many disciplines, which has a ripple effect on an author's ability to perform research and publish research findings. Reduced funding may influence where an author chooses to submit and publish research as some society journals may be cost prohibitive, and authors may elect to publish in lower cost, nonsociety-related journals.

Along with decreased funding for authors, academic library subscription budgets are being outpaced by subscription price increases as well an ever increasing number of journals. As a result, libraries are often forced to make cancellation decisions. For a small society with one or two niche journals, cancellations can be financially devastating and force societies to discontinue their publishing program or seek alternative arrangements, such as shifting the journal to a commercial publishing arrangement, to maintain the title.

In addition to threats from decreases in funding, society journals can face threats from within the community based on their business model. While subscription and open access models were presented above as a choice for the society journal publisher to make, depending on the academic discipline, some societies are subject to pressure from their communities to provide partial or, increasingly, all open access journals. These pressures can materialize as author preference or even funder mandates. For example, some funding agencies will pay author publication charges only if the research output is open access. Other agencies will allow research to be published in a subscription-based journal but under the condition that the article is made freely available after a defined period of time, often referred to an embargo period. Perhaps the hybrid model will satisfy the funder requirements in some instances, but not all. And, as the market will only bear a certain level of APCs, it may not be financially viable for a society to transition their journal from subscription to open access, and, again, the society publisher may be forced to discontinue their program or make other arrangements to sustain the journal.

Beyond funding and business models, society journal programs are also threatened by activities such as predatory publishing and piracy. In predatory publishing, organizations misrepresent themselves as actually being or being aligned with a particular society or publisher or as publishing peer-reviewed journals (often using the same or similar name as a reputable publisher) and deceive authors into paying money to publish with them. Although, sometimes authors knowingly and intentionally publish in these journals [6], this diverts authorship and publication dollars from society journals. With piracy, individuals or organizations abuse copyright laws which, depending on the scale, can be detrimental to a society publishing program. For example, the search engine Sci-Hub illegally provides access to tens of thousands of research articles by circumventing paywalls or subscription requirements, which would otherwise return revenues for content use to the publisher.

While the threats described above may be the most obvious or the most critical for some society journal publishers, it is likely that dozens more could be identified and detailed.

CONCLUSION

Despite present and increasing threats to society publishing programs, journal publishing remains a mission critical activity for many scholarly societies. Journals can be produced by a variety of options and under various business models depending on the resources, needs, and goals of each society.

REFERENCES

[1] https://scholarlykitchen.sspnet.org/2016/02/01/guest-post-kent-anderson-updated-96-things-publishers-do-2016-edition/.
[2] https://royalsociety.org/about-us/mission-priorities/.
[3] https://royalsociety.org/about-us/history/#timeline.
[4] http://rstl.royalsocietypublishing.org/.
[5] http://www.fasebj.org/content/31/4.toc.
[6] https://scholarlykitchen.sspnet.org/2017/02/28/predatory-publishing-rational-response-poorly-governed-academic-incentives/.

Chapter 4

Library Journal Publishing

Jason Colman, MS[1], Jasna Markovac, PhD[2]
[1]*Michigan Publishing, University Library, University of Michigan, Ann Arbor, MI, United States;*
[2]*Health Information Technology and Services, University of Michigan, Ann Arbor, MI, United States*

INTRODUCTION

Library publishing is defined by the Library Publishing Coalition as the set of activities led by college and university libraries to support the creation, dissemination, and curation of scholarly, creative, and/or educational works. Library publishing generally requires a production process and disseminates original work that has not previously been published. In addition, the library can provide validation, either by association of the university brand or through peer review [1]. Library publishing plays an important role in scholarly communications by fostering new and innovative ways to distribute information as widely and as openly as possible. Publishing services are made available to the institution's faculty, staff, and students, often free of charge or at cost recovery.

Michigan Publishing is the hub of scholarly publishing at the University of Michigan and is a part of the University Library, publishing scholarly and educational materials in a range of formats for wide dissemination and permanent preservation. Michigan Publishing provides publishing services to the University of Michigan community and beyond and advocates for the broadest possible access to scholarship everywhere [2]. Fig. 4.1 presents the home page of Michigan Publishing, showing the three components—University of Michigan Press, Michigan Publishing Services (MPS), and Deep Blue, Michigan's institutional repository. The Journals program at Michigan Publishing is part of MPS.

MPS focuses on three primary missions:

- Enabling new publishing endeavors consistent with the University of Michigan's mission
- Enhancing existing publishing programs at University of Michigan
- Providing a space for experimentation in scholarly communication, especially open access

This chapter presents the University of Michigan's library journal publishing program and discusses the role of Michigan Publishing in the development and

Medical and Scientific Publishing. https://doi.org/10.1016/B978-0-12-809969-8.00004-8

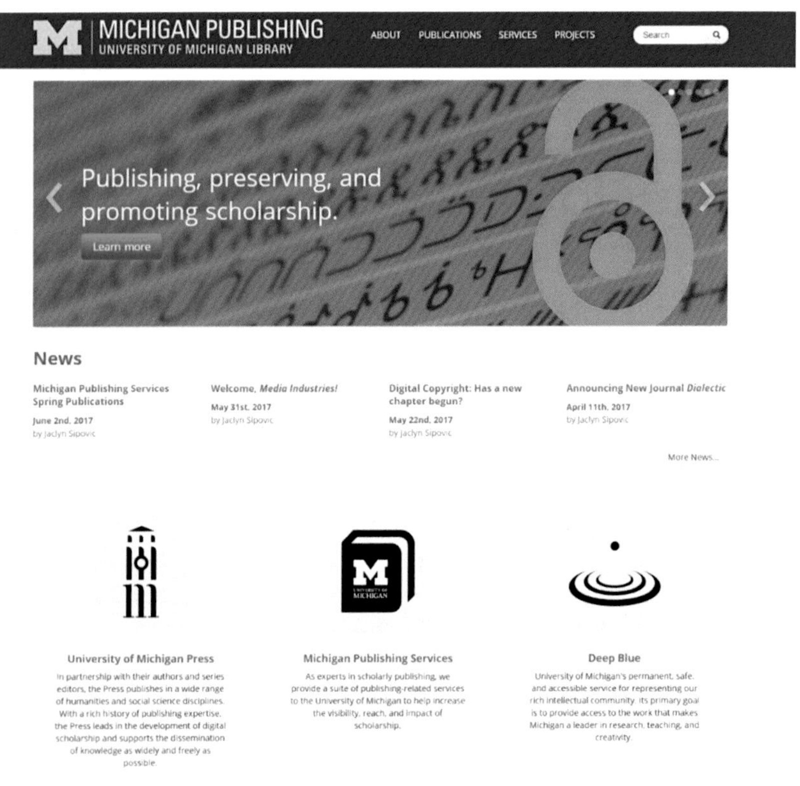

FIGURE 4.1 Michigan Publishing home page.

launch of the Michigan Journal of Medicine (MJM), a student-run journal used for medical education. Chapters 26–28 present additional information about MJM.

JOURNAL PUBLISHING

MPS provides an affordable, efficient publishing solution for important journals in niche markets for publishing partners who are interested in broad, open dissemination and access of scholarly content. While there is an emphasis on open access, MPS publishes subscription-based journals with a strong mission alignment. Although the business model for journal publishing services is evolving, the cost to the publishing partner continues to be minimal. The early business model (before 2015) is presented in Fig. 4.2. The business model currently in place is outlined in Fig. 4.3.

Because MPS charges its publishing partners very low prices for journal publishing services, journals are often able to operate on alternative funding models. These include donations and author fees, institutional support, and mixed publishing models (free plus monetized).

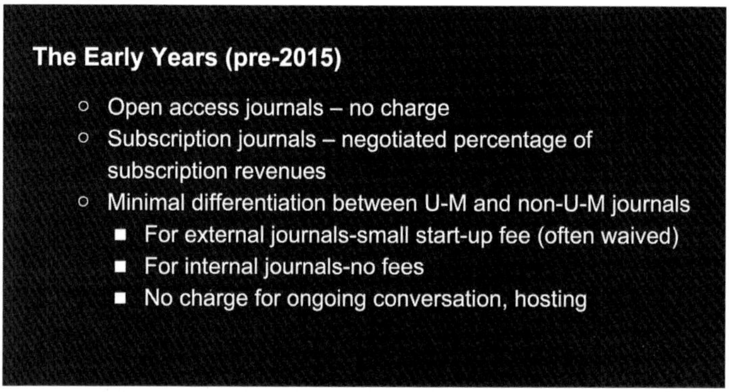

The Early Years (pre-2015)

- Open access journals – no charge
- Subscription journals – negotiated percentage of subscription revenues
- Minimal differentiation between U-M and non-U-M journals
  - For external journals-small start-up fee (often waived)
  - For internal journals-no fees
  - No charge for ongoing conversation, hosting

FIGURE 4.2 Michigan Publishing Services pre-2015 business model.

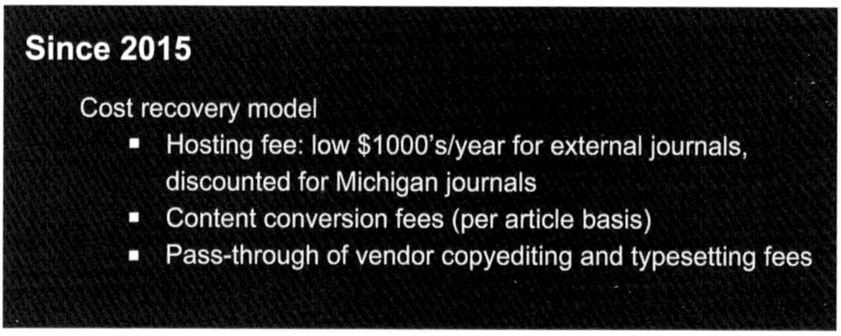

Since 2015

Cost recovery model

- Hosting fee: low $1000's/year for external journals, discounted for Michigan journals
- Content conversion fees (per article basis)
- Pass-through of vendor copyediting and typesetting fees

FIGURE 4.3 Michigan Publishing Services current cost recovery model.

To explore funder support options, Michigan Publishing will work with the journal editors' university to identify the strategic direction of the institution and to articulate the alignment of the journal with this higher-level strategy, particularly engagement with the public sphere. Performance metrics including Google Analytics, alternative metrics (Altmetric), and Google Scholar Alerts, as well as Annual Reports, are presented as a way to demonstrate impact to journals' home institutions.

In addition, a combination of free access and monetized print-on-demand options are being implemented for select publications. The journals themselves are published open access, but single issues are made available in print and e-book formats and are for sales through various channels. Fig. 4.4 shows an example of single issue sales. The pros and cons of this model are illustrated in Fig. 4.5.

MICHIGAN JOURNAL OF MEDICINE

When the idea for MJM was conceived by the University of Michigan medical students, they, along with their faculty advisor, began exploring options for publication (Chapter 26). Since the University Library offered very practical and affordable publishing services, the students and their advisor developed a proposal for Michigan Publishing using the guide provided in Fig. 4.6. The students

FIGURE 4.4 Example of an Open Access journal that makes single issues available for sale.

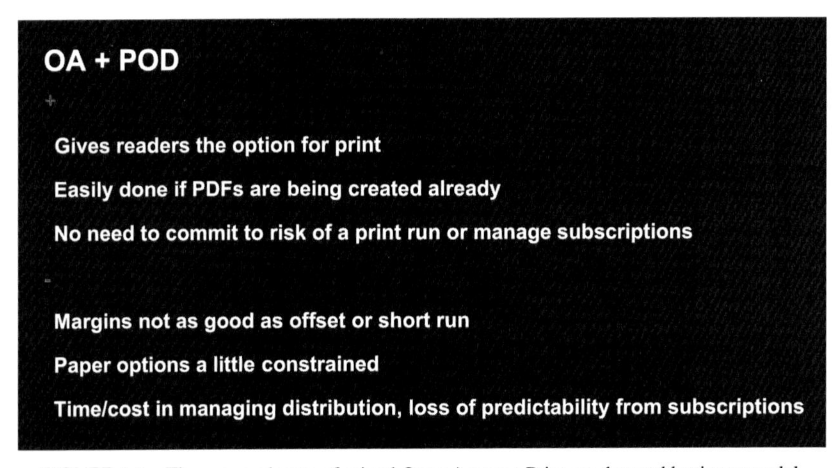

FIGURE 4.5 The pros and cons of mixed Open Access + Print-on-demand business model.

established an editorial workflow (Fig. 4.7) for the journal as well as an editorial organizational chart (Fig. 4.8). After careful internal review, Michigan Publishing approved the proposal and plans for publication moved forward.

A publishing contract was issued for MJM that presented the roles and responsibilities of the editors and the publisher. This became an important teaching tool for the students. For many of them, it was the first publishing agreement they had seen, and it provided an opportunity to learn more about the legal aspects of the publishing process.

As part of the agreement with Michigan Publishing, the editors (and faculty advisors) of MJM agreed to the following:

- To establish an editorial board and editorial staff, including a succession plan so that the journal's activity may be sustained without interruption.

Plan and Launch a Scholarly Journal

The table below is an overview of the major steps to plan and launch an academic journal, whether or not you partner with Michigan Publishing.

Step	Michigan Publishing's role
1. Identify a clear need in the field	
2. Recruit editors and editorial board. In consultation with them:	
2a. Establish policies for journal management (editors elected or appointed? Term limits? Who has authority to make decisions?)	
2b. Establish responsibilities and expectations for editors and board members (Scheduled meetings or phone calls? Time commitment? Commitment to review? To write?)	
2c. What responsibilities remain, and who will handle these? Need to hire a managing editor or administrative assistant?	
3. In consultation with editorial board develop:	Michigan Publishing can consult
3a. Journal title	
3b. Journal scope	
3c. Frequency of publication	
3d. Submission policies/guidelines	
3e. Peer review policies/guidelines	
3f. Access model (Subscription? Embargo? Open access?)	
3g. Copyright and licensing (Author retains rights? Journal acquires rights? Creative Commons?)	
3h. Institutional home (is the journal permanently tied to any particular university, organization, or society?)	
3i. Funding model. What startup and ongoing costs will you have? How will they be covered?	
4. Investigate options for publisher and/or platform (if working with a publisher, they will shape what platform and format choices are available to you)	Michigan Publishing can consult. If our services seem like a good match for your needs, now's the time to submit a proposal to partner with us.
4a. Online only? Open Journal Systems (), Blogging platform, publisher platform, other?	
4b. XML/HTML, PDF, both?	
4c. Ebook formats?	
4d. Print?	*If your proposal to work with Michigan Publishing has been accepted:*
5. Establish presence online	
5a. Website	Michigan Publishing will handle
5b. Blog? (If so, how will it be distinct from the journal itself?)	
5c. Social media?	
5d. Email address(es)	
6. Apply for ISSN (need to have a website and expected publication date)	Michigan Publishing will handle
7. Write & publicize a call for papers and deadline; begin soliciting content	Michigan Publishing will post your CFP to the journal website
8. Monitor incoming submissions and assess them for rejection or review	Michigan Publishing can provide access to software to help manage this
9. Assign reviewers and manage peer review process	Michigan Publishing can provide access to software to help manage this
10. Manage author revisions	Michigan Publishing can provide access to software to help manage this
11. Copyediting	Michigan Publishing can provide access to software to help manage this
12. Get author approval on final copy	Michigan Publishing can provide access to software to help manage this
13. Deliver content to publisher/host/platform	
14. Prepare content for publication online (e.g. conversion to HTML/XML, typesetting)	Michigan Publishing will handle
15. Preview/proof journal site and first issue content	Michigan Publishing makes corrections and changes per your comments
16. Register article-level DOIs	Michigan Publishing will handle
17. Launch!	Michigan Publishing will handle
18. Spread the word about the journal	Michigan Publishing will work together with you
18a. Press release	
18b. Social Media	
18c. Listservs	
18d. Conferences	
19. Register the journal's existence	Michigan Publishing will work with you to determine appropriate venues, and then will handle
19a. Directory of Open Access Journals (DOAJ) if applicable	
19b. Web of Science, if applicable.(See selection criteria.)	
19c. Ulrich's (http://ulrichsweb.serialssolutions.com/login), Cabell's (http://www.cabells.com/index.aspx)	
19d. MedLine, if applicable. (See selection criteria)	
19e. RSS feed and OAI-PMH feed	
20. Repeat steps 7-17 for each issue	

FIGURE 4.6 Journal orientation guideline provided by Michigan Publishing Services for Michigan Journal of Medicine.

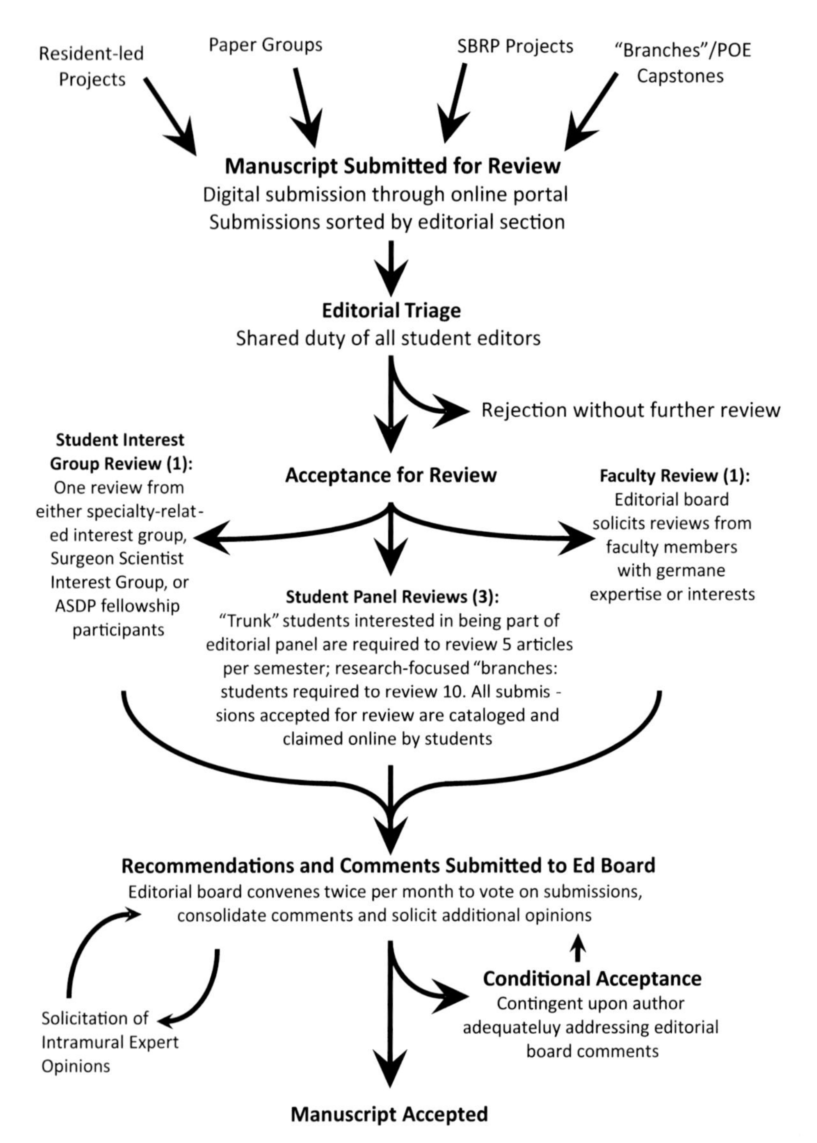

FIGURE 4.7 Proposed editorial workflow for Michigan Journal of Medicine.

- To produce and edit the content of the journal, including recruiting submissions, soliciting reviews, managing review and revisions, and obtaining final author approval.
- To ensure that authors sign the author agreement and obtain permissions for any embedded materials (images, multimedia, etc.) for which they do not own the copyright.

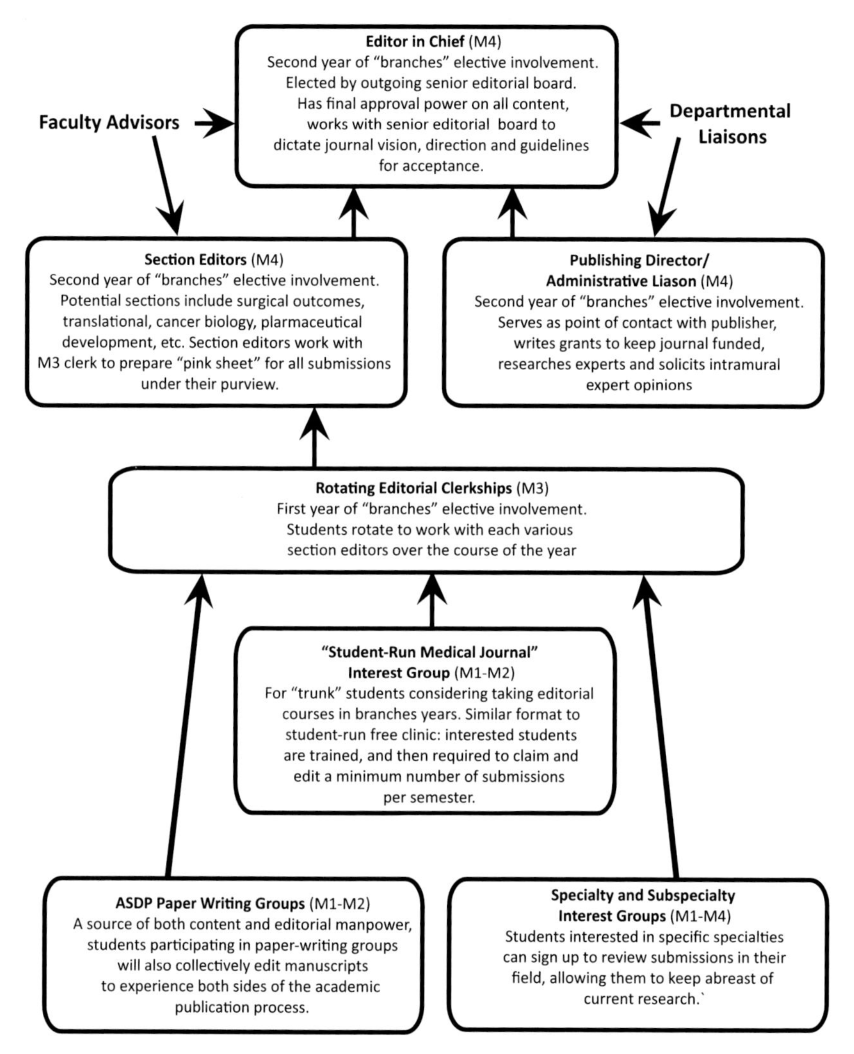

FIGURE 4.8 Proposed editorial organizational chart for Michigan Journal of Medicine.

- To ensure that all content adheres to our own stylistic guidelines before delivering it to Michigan Publishing. (Michigan Publishing is not responsible for finding and making such changes to the content.)
- To deliver all content in its complete and final form to Michigan Publishing according to an agreed on attached schedule.
- To preview the publication website before new content is published, apprising Michigan Publishing of any changes that need to be made and granting approval to publish.

- To monitor the publication and alert Michigan Publishing immediately of any problems with the site or the content.
- To acknowledge that each author accepted to The Michigan Journal of Medicine retains copyright in their work, granting Michigan Publishing permission to publish the work under a Creative Commons license.
- To retain copyright in any intellectual property inherent to The Michigan Journal of Medicine and grant Michigan Publishing permission to publish the journal under a Creative Commons license.

Michigan Publishing agreed to the following rights and responsibilities pertaining to the publication of MJM:

- To ingest MJM content into the Library's publishing platform and deliver it via the web.
- To endeavor to keep the MJM site online 24h a day, 7days a week, 365days a year and to alert the editors of any known outages, including when services are restored.
- To provide the editors with access to usage statistics.
- To continue to preserve and host content that has been published on the Library platform, even if the partnership is dissolved or MJM ceases to publish.
- To take responsibility for the following on behalf of MJM:
  - ISSN registration
  - DOI registration
  - Submitting the journal for consideration for indexing by third parties (e.g., Web of Knowledge, Directory of Open Access Journals, Ulrich's, etc.)
  - Granting permission for the journal to be indexed by third parties such as those listed above
  - Changing the functionality, structure, or interface of the journal website when our publication platform, external policies (e.g., ADA compliance, COPE or DOAJ guidelines, etc.), or web standards require it.
- To reserve the right to make the final decision about whether to develop or incorporate a new feature, function, customization, or enhancement to the publishing platform.
- To ensure that the design and branding of the journal site comply with web and accessibility standards and University of Michigan policies.

The MJM editors and Michigan Publishing agreed to the following:

- To communicate openly and consistently as needed throughout the year. At least once a year to meet (in person if possible, by telephone call or video conference if not) for a PUBLICATION CHECK-UP to check in on the overall health of the journal, look ahead to the coming year, and discuss any changes that affect the contractual agreement or the partnership.
- To consult together about any new feature requests, customization, or enhancement, with Michigan Publishing's Technology Group to evaluate feasibility, desirability, and cost.

- If any royalties are generated by indexing the content in a third-party product, these will go directly The Michigan Journal of Medicine. Michigan Publishing recommends (but does not dictate) that the Editors to use any such income to sustain the journal.
- To work together to publish the journal according to the stated schedule. If The Michigan Journal of Medicine falls more than 6 months behind its stated schedule, to work with the editors to develop a revised schedule to get back on track. If The Michigan Journal of Medicine falls more than 1 year behind without sign of improvement, Michigan Publishing reserves the right to "archive" the publication and deactivate the partnership. The Editors would need to submit a new proposal, with a new schedule and sustainability plan, to revive the partnership.

After the contractual formalities were completed, the MPS staff then worked closely with the student editors to establish a publication workflow and a realistic schedule. Michigan Publishing's Director of Strategic Integration and Partnerships participated in the Medical Editing course, part of the medical school curriculum, and worked with the students on the various aspects of the publishing process. MJM became the "lab" for the course, and the students were able to do hands on work on the journal as part of their curriculum.

The biggest challenge for the editors and the publisher was to ensure that there were enough manuscripts to publish the first issue and to generate sufficient interest and engagement in MJM among the medical students to have a sustainable pipeline for future issues. MPS and the medical school provided local marketing support for the journal and the student editors rallied their colleagues; papers were submitted, reviewed, revised, and accepted. The first editor-in-chief of MJM presents her story in Chapter 27.

Michigan Publishing produced the journal and the first issue of MJM published less than a year after the contract was signed. On publication, the editors and the sponsors organized a launch party to thank everyone for their hard work.

Thanks to the thriving partnership between Michigan Publishing, the Medical School, and the student editors, MJM has been able to maintain a healthy pipeline of manuscripts and published its second issue 10 months after the publication of the first. The second publication year offered opportunities for MPS and the editors to make improvements to publishing processes, journal marketing, and web presence. Chapter 28 discusses the second year editorial experiences of that issues EiC and Chapter 29 presents the web redesign approach used for the MJM site.

REFERENCES

[1] https://librarypublishing.org/about-us.

[2] https://www.publishing.umich.edu/.

Chapter 5

Combining Medicine and Writing: My Journey to Becoming an Editor

Preeti N. Malani, MD
Department of Internal Medicine, University of Michigan, Ann Arbor, MI, United States

THE EARLY YEARS

My path to becoming an associate editor at *The Journal of the American Medical Association* (JAMA) was rather serendipitous. I am a Michigan native. My journey starts back in junior high school. I wanted to take two math classes in 10th grade, but my math teacher would not sign the waiver because he thought that taking algebra and geometry at the same time would be too challenging.

This left an open block in my tenth-grade schedule. So, on a whim I decided to take journalism. This changed the entire trajectory of my life. I loved journalism! It made sense to me.

Journalism became a way for me to learn how to write. I became the editor of my high school newspaper, and I also wrote a weekly column for my local paper.

THE UNIVERSITY OF MICHIGAN

When I came to the University of Michigan (U-M) as an undergraduate student, I was interested in medicine and in journalism, but I did not know how to combine the two. Today it seems like there are a million ways to make that combination, but in 1987 it was not apparent. The usual path was to go to college, become a biology major, then in 4 years go to medical school. There were no gap years—students just went straight through. I was not sure that medicine was the right career for me. There are no other physicians in my family. I did not know many women in medicine, but I was interested, so I pursued that path.

Although I was focused on medicine, I continued to enjoy writing and looked for ways to combine these interests. Michigan did not offer an undergraduate degree in journalism, but I wrote for the Michigan Daily and, at the same time,

Medical and Scientific Publishing. https://doi.org/10.1016/B978-0-12-809969-8.00005-X

I designed an independent major in "medical journalism." I also did a second major in communications. I was not following that straight line.

The way it worked out, I finished a year early, but I was not ready to apply to medical school yet because I had not done all the prerequisites. I do not quite know how, but I decided to apply to Northwestern University's School of Journalism and I completed my master's degree there in Reporting and Writing. I look back at that as one of the most important but somewhat serendipitous decisions I ever made. It opened my eyes in so many ways and it gave me rigorous training in writing.

BECOMING A JOURNALIST AND A PHYSICIAN

During the summer before I started at Northwestern, I worked at the *Dayton Daily News* in Ohio. I was sent out to cover a story in the summer of 1990. It was about a family whose son had been killed in a motorcycle accident. The family donated his organs. That would not be a news story today, but it was back then (Fig. 5.1).

So, I found this family and I went to see his grieving parents and grandparents. It felt really uncomfortable. Honestly, I probably was not emotionally prepared to do this. I had just turned 21 and I had not seen much of life. I grew up in the suburbs and went to school and suddenly I was thrust into this situation. I remember thinking, "I really don't want to do this - I want to be harvesting the organs." I wanted to be on the other end of this story.

At the time, applications to medical school were typewritten—everything was on paper in those days. I went home to my little apartment after that organ donation story and started writing my essay for my medical school application. I was accepted to medical school while at Northwestern.

At Northwestern, I took a seminar on science writing and a magazine writing course. As part of that course I got to spend some time at the American Medical News and the JAMA. I did an internship at the *American Medical News*. That publication is now defunct but it was a newspaperish magazine. One of the pieces I wrote there is probably the best thing I have ever written. It was a piece on women in surgery. I had to figure out how to find women in surgery and then interview them about their experiences. It was a pretty big deal at that point to be a woman in surgery, although today one does not think twice about a surgeon who happens to be a woman (Fig. 5.2).

When I started medical school at Wayne State University, I was interested in urban health and all types of social justice issues. Journalism really awoke this interest in me. I was going to be a doctor in the city. I was going to take care of pregnant teens and help prevent repeat pregnancies. That is what one of my major papers from the journalism master's was about, which is about as far away from what I actually do today as possible. But, at 22 it was what I wanted to do and it was a good thing. During medical school I became involved with the student section of the JAMA and became first an assistant editor, then deputy editor, and finally editor-in-chief.

As part of this work, I arranged to spend a month at JAMA as a fourth-year medical student. I slept on my friend's couch and took the train in from

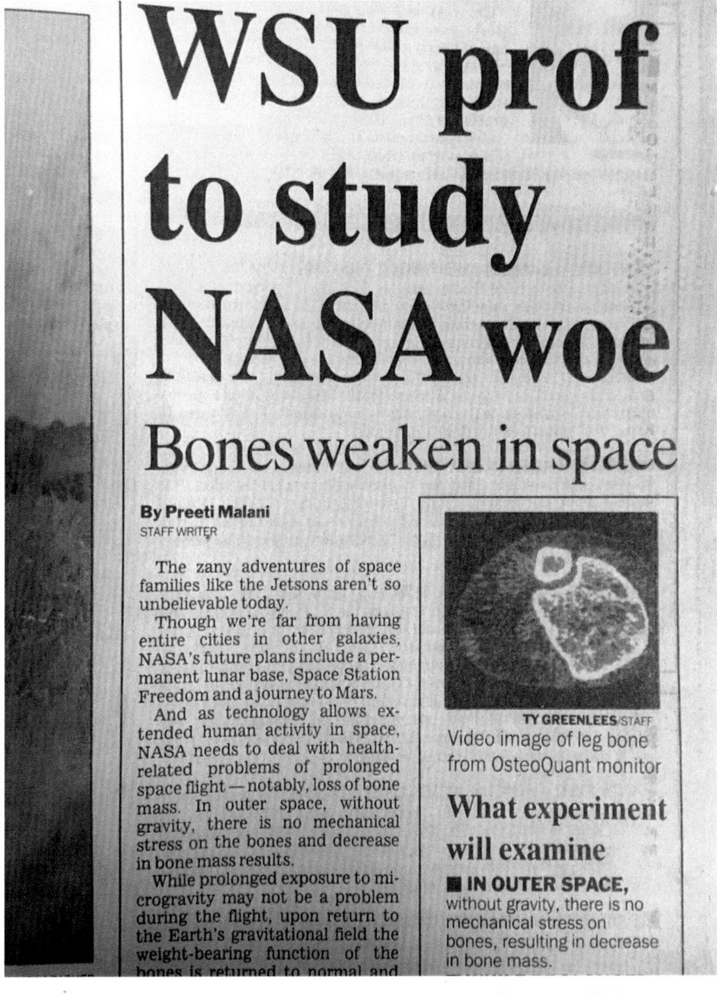

WSU prof to study NASA woe

Bones weaken in space

By Preeti Malani
STAFF WRITER

The zany adventures of space families like the Jetsons aren't so unbelievable today.

Though we're far from having entire cities in other galaxies, NASA's future plans include a permanent lunar base, Space Station Freedom and a journey to Mars.

And as technology allows extended human activity in space, NASA needs to deal with health-related problems of prolonged space flight — notably, loss of bone mass. In outer space, without gravity, there is no mechanical stress on the bones and decrease in bone mass results.

While prolonged exposure to microgravity may not be a problem during the flight, upon return to the Earth's gravitational field the weight-bearing function of the bones is returned to normal and

TY GREENLEES/STAFF

Video image of leg bone from OsteoQuant monitor

What experiment will examine

■ **IN OUTER SPACE,** without gravity, there is no mechanical stress on bones, resulting in decrease in bone mass.

FIGURE 5.1 News article from the Dayton Daily News, 1990.

Wrigleyville every day. It has been 22 years since that time and most of the people I knew have retired or moved on but some of the people I worked with as a student are still there. I had this great experience as a student and it prompted me to wonder what the next step would be for my writing.

I finished medical school and began my residency at the U-M. Despite every intention and desire that I was going to go away and never come back, I ended back up at U-M. So, one of my messages is to be nice to everybody that you come across. The way you behave and the way you act comes back to help you or to bite you later on. You do not know where your life will take you, but people will come back into your life. You are going to be in a position of incredible power at some point whether it is local or national or something else. So, be kind always.

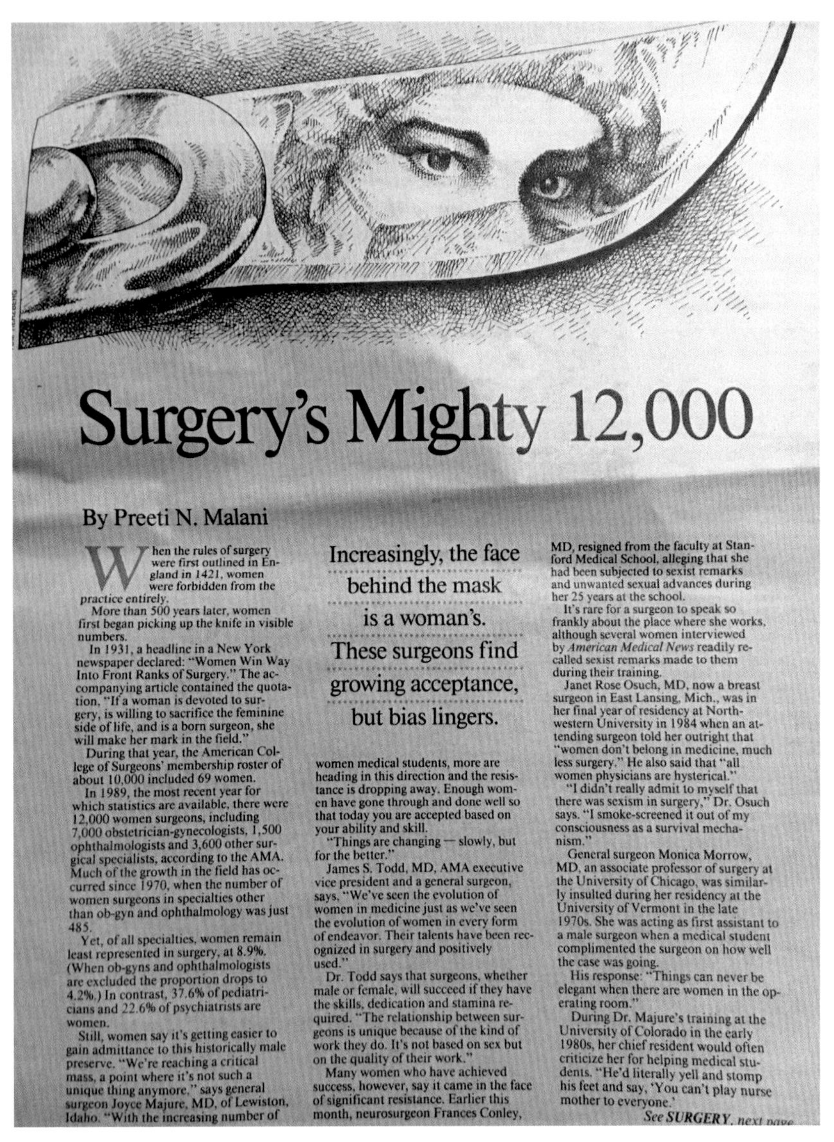

Surgery's Mighty 12,000

By Preeti N. Malani

When the rules of surgery were first outlined in England in 1421, women were forbidden from the practice entirely.

More than 500 years later, women first began picking up the knife in visible numbers.

In 1931, a headline in a New York newspaper declared: "Women Win Way Into Front Ranks of Surgery." The accompanying article contained the quotation, "If a woman is devoted to surgery, is willing to sacrifice the feminine side of life, and is a born surgeon, she will make her mark in the field."

During that year, the American College of Surgeons' membership roster of about 10,000 included 69 women.

In 1989, the most recent year for which statistics are available, there were 12,000 women surgeons, including 7,000 obstetrician-gynecologists, 1,500 ophthalmologists and 3,600 other surgical specialists, according to the AMA. Much of the growth in the field has occurred since 1970, when the number of women surgeons in specialties other than ob-gyn and ophthalmology was just 485.

Yet, of all specialties, women remain least represented in surgery, at 8.9%. (When ob-gyns and ophthalmologists are excluded the proportion drops to 4.2%.) In contrast, 37.6% of pediatricians and 22.6% of psychiatrists are women.

Still, women say it's getting easier to gain admittance to this historically male preserve. "We're reaching a critical mass, a point where it's not such a unique thing anymore," says general surgeon Joyce Majure, MD, of Lewiston, Idaho. "With the increasing number of

Increasingly, the face behind the mask is a woman's. These surgeons find growing acceptance, but bias lingers.

women medical students, more are heading in this direction and the resistance is dropping away. Enough women have gone through and done well so that today you are accepted based on your ability and skill.

"Things are changing — slowly, but for the better."

James S. Todd, MD, AMA executive vice president and a general surgeon, says, "We've seen the evolution of women in medicine just as we've seen the evolution of women in every form of endeavor. Their talents have been recognized in surgery and positively used."

Dr. Todd says that surgeons, whether male or female, will succeed if they have the skills, dedication and stamina required. "The relationship between surgeons is unique because of the kind of work they do. It's not based on sex but on the quality of their work."

Many women who have achieved success, however, say it came in the face of significant resistance. Earlier this month, neurosurgeon Frances Conley, MD, resigned from the faculty at Stanford Medical School, alleging that she had been subjected to sexist remarks and unwanted sexual advances during her 25 years at the school.

It's rare for a surgeon to speak so frankly about the place where she works, although several women interviewed by *American Medical News* readily recalled sexist remarks made to them during their training.

Janet Rose Osuch, MD, now a breast surgeon in East Lansing, Mich., was in her final year of residency at Northwestern University in 1984 when an attending surgeon told her outright that "women don't belong in medicine, much less surgery." He also said that "all women physicians are hysterical."

"I didn't really admit to myself that there was sexism in surgery," Dr. Osuch says. "I smoke-screened it out of my consciousness as a survival mechanism."

General surgeon Monica Morrow, MD, an associate professor of surgery at the University of Chicago, was similarly insulted during her residency at the University of Vermont in the late 1970s. She was acting as first assistant to a male surgeon when a medical student complimented the surgeon on how well the case was going.

His response: "Things can never be elegant when there are women in the operating room."

During Dr. Majure's training at the University of Colorado in the early 1980s, her chief resident would often criticize her for helping medical students. "He'd literally yell and stomp his feet and say, 'You can't play nurse mother to everyone.'

See ***SURGERY***, *next page*

FIGURE 5.2 Feature article published in the American Medical News, June 1991.

COMBINING JOURNALISM AND MEDICINE

After residency, I became an infectious disease fellow at U-M. While there, I had a mentor who recognized that I was interested in writing and she gave me some things to work on. I did the job well, so she gave me more work. I asked people how one becomes an editor because it was what I wanted to do. There was no actual career path because most editors were senior faculty and it was an honorific position. They did not go to school to become an editor. The usual path was to achieve some fame in their field and to be invited to be an editor, even though they had no formal training in how to write. I was a little bit different because I had the journalism training and I liked to write.

So, I did a couple of things. First, I made sure that I was excellent clinically. That is critical. You are taking care of patients. Be the best doctor you can be. I was also fortunate to be surrounded by good role models and I continue to reach out to those people to this day.

One thing led to another and I kept asking people to let me help them with writing. I ended up becoming part of the publications committee for one of my professional societies. I was junior at the time. In fact, when I went to the first meeting a couple of people asked me for directions in the hotel because they just assumed that I worked there.

At the end of one publication meeting, I was invited to have lunch with the editorial board for the journal *Infection Control and Hospital Epidemiology*. All the people who were my heroes were there! And, here is another life lesson: I went up to the editor-in-chief and introduced myself. I handed him my resume and said, "If you ever need help with anything, please call me." He said, "Sure, I'll be happy to call you."

Ten months later (then a geriatrics fellow in Portland, Oregon), I got a random email from the editor. He said, "Hey, I just found your CV at the bottom of this pile of papers. Are you still interested?" This was an opportunity that came to me and it was just because I asked someone to let me help them. Eventually he said to me, "I want to groom you to be the next editor-in-chief at some point down the line." I thought this was insane because people like me were not editors-in-chief. They were famous academic people. But, I did everything he asked me to do. I rewrote poorly organized, horribly written papers. I did detailed peer reviews for him and he made me an assistant editor.

It is a complicated story, but he ended up having to take a medical retirement and another editor came in and I was left in limbo. Then, the journal ended up coming to U-M, but I was clearly not senior enough to be editor. But one of my mentors became editor-in-chief and several of the faculty became of her editorial team for *Infection Control Hospital Epidemiology*. I got a chance to learn a lot, to ask questions and make suggestions. I drew on my journalism background.

In the following years, I edited a couple of books. Any opportunity that came my way, I would just grab it. I think people got to know that, "that person over at Michigan is interested in editing – let's ask her to do something."

In the meantime, I wrote a lot of papers. I understood both the writing/editing and the clinical aspects of medical publishing, but I needed to learn more about research methods and statistics. So, I did a master's in clinical research design and statistical analysis at the University of Michigan School of Public Health. This specialized knowledge is essential to being an editor at a high level. If you are going to be an editor, you need to understand research methods.

You cannot just be a good writer and you cannot just be a nice person. You must understand methodology and clinical trials, especially at the most elite journals.

So, that was a critical piece for me. I was fortunate, but it was the hardest 2 years of my life. It was a really challenging program for me, my kids were small, I was busy clinically, and I felt as though I did not sleep at all for a couple of years. I certainly did not exercise. I ate big bags of Doritos. It was one of those times of my life when I was wondering, "How did this happen? How did I get through that?" But, it was critical to what I do now.

BECOMING AN EDITOR

The next step in my editing journey was *Clinical Infectious Disease*, which is a premier, specialty journal. A few years earlier, I was appointed to that publications committee. The first time I went to one of these meetings I was pretty intimidated. The editor-in-chief was really old school and formal and it took a while for me to get to know him. But after the first year working with him I was able to say more to him. And the following year I was able to say even more, to the point where I was giving him ideas and suggestions. And, he picked up on that.

They posted a job for an art editor, because they publish cover art—and I applied for it. I did not really know that much about art, which was a problem. But, I think I wrote a very nice letter. It was the best personal statement that I could write. And, I had a CV that showed that I knew how to write and edit. I think the editor-in-chief remembered that, even though I did not get that job. And when he came to visit Ann Arbor, I made a point to meet with him and drive him to dinner.

A year later, he called and told me he had an opportunity for me. I was thinking maybe he would ask me to be the book review editor, or a similar low-level thing. But, he told me he wanted to make me deputy editor because he needed some help. I was 41 years old at the time that this happened, and it was at a journal that is a premier journal in my field!

The editor-in-chief was so respectful. He told me that he was going to treat me the same way he treated everyone else. I was not there to just work; I was there to help him. I handled many of the papers that were submitted and, more importantly, every revision that came back from the authors had to get past me before final acceptance. So, I learned a lot about my field and I learned who knows what and who does what. I could say things like, "This language is not right, this is causal and it should be changed to associated with 'X'." I really made people jump through hoops. It was great and I loved it! I really found a great home there. But, I always wanted to get back to JAMA.

RETURNING TO JOURNAL OF THE AMERICAN MEDICAL ASSOCIATION

I wrote several book reviews for JAMA from about 2006 to 2012. I would read a dense book on vacation and write a review on it for JAMA. A position opened for an infectious disease editor and the book review editor recommended me. Initially, the senior editors did not seriously consider because I did not fit the usual mold. But then they needed a geriatrics editor and that is when they decided to take a second look at me. They offered me the job. Once I got there, I started writing for their news section. I also worked to get the best infectious diseases manuscripts to JAMA.

AS AN EDITOR AT JOURNAL OF THE AMERICAN MEDICAL ASSOCIATION

I am an associate editor at JAMA, so I handle manuscripts. But, the other thing I do for the JAMA Network right now is oversee media relations. I have been doing that for a couple of years. I oversee all our videos, press releases, and relationships with various reporters.

People are starting to pay attention to Altmetric scores. They are a relatively new and nontraditional alternative to citation-based impact metrics. Citations can take months and years to accrue. Altmetric scores measure immediate impact by counting the number of times an article is referred to by news outlets that are measured per an algorithm. High profile news outlets, such as the *New York Times*, *The Washington Post*, *USA Today*, and *The Wall Street Journal*, are the primary drivers of Altmetric scores, but they are also increased by social media such as Twitter, Wikipedia, and blogs. Articles that receive a count of 1000 or more are usually in the top 100 for the year. Across the JAMA Network this year we have 20 that are over the 1000 mark.

A recent example would be a paper in JAMA *Pediatrics* about Colorado Recreational Marijuana [1]. It received a very high Altmetric score—941—meaning it is immediately news worthy. This paper discusses the increase in the number of toddlers showing up in the emergency room since recreational use of marijuana was legalized. Papers that score in the 1000 range often end up in the top 100 papers for the year. President Obama published a paper in JAMA about the Affordable Care Act [2]. This paper has the high Altmetric score ever recorded. For more information about Altmetric score, please see Chapter 22 on Measuring Impact.

To understand what social media means from the viewpoint of an editor, it's helpful to take a recent example. We had an amazing paper on gun control from Australia. The paper required a lot of work to make it acceptable for JAMA. My editor-in-chief did the editing on that paper and we published it in the wake of the Orlando nightclub shooting. It was planned around then, but we could get it out a little sooner. The idea was a comparison to what has happened in Australia since they banned assault weapons. They have not had any more mass shootings. And, from closer to my own world of infectious disease, we just published the AIDS theme issue, which I edited. It is incredible to be able to bring these issues to the forefront.

LOOKING BACK—MY ADVICE

This has been a long story, 30 years in the making. A friend recently commented that the dots are lining up now, but that was not apparent while I was going through it all. If pursuing a career like mine appeals to you, my experience suggests the following:

- Above and beyond everything, make sure you are a good doctor.
- Be certain that you know how to write, whether through education, experience, or a combination of the two. Listen to feedback, even when it's hard to hear.
- Do a lot of writing. Write about the things that you know and the things that interest you. Focus on being good at what you do, getting involved with professional societies, volunteering, and maybe get on the publications committee.
- Becoming an editor has traditionally been more of an honorific thing. People did not train in becoming good writers or even necessarily train in formal research methods. Today, to become an editor and not have those two qualifications would be pretty unusual. Get yourself a solid background in research methods and statistics.
- One thing that newer editors sometimes struggle with is the difficulty of reading papers about fields in which they have no experience and do not understand. Please remember that a well-written paper should be understandable. Perhaps some fields are more difficult than others, but you should be able to understand the Introduction and Discussion.
- If you read something that is beautifully written, set it aside and then use it as a model for your own work. I am not suggesting that you plagiarize, but follow the example of a tightly written Introduction. At JAMA, all Introductions are two to three paragraphs and they are never more than 250 words.
- Take advantage of every opportunity that comes your way. Attend conferences; talk to colleagues with similar interests; build a network. If no opportunities are coming your way, make them happen for you.
- If you know or work with someone who is an editor, you can gain experience by offering to do reviews.
- Look for fellowships and internships that provide valuable, hands-on experience for people interested in editing/writing.
- Be willing to learn and to work hard. Build a reputation for your skill set and for your positive personal qualities.
- It is important to understand what nurses and respiratory therapists and all the other members of our clinical teams do. It is just the same in the medical editing world.
- Early on, editorial positions may be voluntary or there might be a small stipend to cover your time. But if this is something you want to do, then this is what you need to do. You cannot expect to be paid well for the privilege of being a scientific editor, especially when you are just starting out.

- Social media editors and some of the lower editing positions are mostly volunteer jobs. Lots of journals need someone to help with social media and the senior people do not know how to do it. So, there are places where you can be more junior and still get your foot in the door.

I recognized that I could bring something to the table, so I did it. Once I had one opportunity, I could get the next one and the next one, and then suddenly you have street credibility. It is getting that first foot in the door that is hard. As an editor, you can shape the conversation, which is an opportunity to change your field.

REFERENCES

[1] Cerdá M, Wall M, Feng T, Keyes KM, Sarvet A, Schulenberg J, O'Malley PM, Pacula RL, Galea S, Hasin DS. Association of state recreational marijuana laws with adolescent marijuana use. JAMA Pediatr 2016. http://dx.doi.org/10.1001/jamapediatrics.2016.3624.

[2] Obama B. United States health care reform: progress to date and next steps [Published online July 11, 2016]. JAMA; doi:10.1001/jama2016.9797.

Chapter 6

The Journal Team: Editors and Publishers Working Together

John P. Bilezikian, MD[1], Jasna Markovac, PhD[2]
[1]Division of Endocrinology, Columbia University, New York, NY, United States; [2]Health Information Technology and Services, University of Michigan, Ann Arbor, MI, United States

INTRODUCTION

As scientists and academic physicians move forward on their career paths, publish their research, and become known in their fields, they are often asked to assume editorial roles on journals that publish their articles. Active engagement in scholarly publications is an important part of an academic career and includes several different roles and responsibilities. Initially, one may be asked to review papers that are being considered for publication. Exceptional reviewers are often invited to join journal editorial boards. Proactive editorial board members, those who are engaged in the journal process and show a real interest in growing and developing the publication, may be offered an editorial role, such as Section Editor or Associate Editor, and ultimately the position of main Editor or Editor-in-Chief (EiC). The size and structure of the journal editorial team varies from journal to journal. Generally, one main Editor, or EiC, will lead a team of editorial board members and, for larger journals or those with a broad scope, a group of Associate or Section Editors. Fig. 6.1 shows a typical masthead page for a scientific journal. This chapter presents information about the role of the journal EiC and the roles and responsibilities of the various members of the editorial team. In addition, the role of the journal publisher is discussed as it pertains to the editorial team.

EDITOR-IN-CHIEF

The main function of the EiC is to maintain and, when appropriate, enhance the reputation, profile, and editorial integrity of the journal. The EiC has final responsibility for all editorial content and the editorial decisions resulting in the publication of that content. The EiC needs to ensure that the journal has enough high quality manuscripts to maintain a regular and sustained publishing schedule. In addition, the EiC needs to work with the publisher to ensure that the aims and scope of the journal reflect the discipline and the

Medical and Scientific Publishing. https://doi.org/10.1016/B978-0-12-809969-8.00006-1

Developmental Biology

Editor-in-Chief **Marianne Bronner**

Editors
Blanche Capel (Durham, NC)
Claude Desplan (New York, NY)†*
Denis Duboule (Geneva, Switzerland)†*
Matthew Gibson (Kansas City, MO)
Richard Harland (Berkeley, CA)
Vivian Irish (New Haven, CT)
Philippe Soriano (New York, NY)
Yoshiko Takahashi (Kyoto, Japan)
Deborah Yelon (La Jolla, CA)

Editorial Board
Markus Affolter (Basel, Switzerland)
Detlev D. Arendt (Heidelberg, Germany)
Kristin Bruk Artinger (Aurora, CO)
Hans R. Bode (Irvine, CA)
Bruce Bowerman (Eugene, OR)
Sarah Bray (Cambridge, UK)
James Briscoe (London, UK)
S.J. Butler (Los Angeles, CA)
Yang Chai (Los Angeles, CA)
Ajay Chitnis (Bethesda, MD)
Ken Cho (Irvine, CA)
Cheng-Ming Chuong (Los Angeles, CA)
David Clouthier (Aurora, Co)
Frank L. Conlon (Chapel Hill, NC)
S.J. Conway (Indianapolis, IN)
Frank Costantini (New York, NY)
E.A. Cowan (Farmington, CT)
Karen Crawford (St Marys City, Maryland)
Peter David Currie (Clayton, VIC, Australia)
Jacqueline Deschamps (Utrecht, Netherlands)
Susanne Dietrich (London, UK)
Andrew Dingwall (Maywood, IL)
Wolfgang Driever (Freiburg, Germany)
Denis Duboule (Geneva, Switzerland)
Gregg Duester (La Jolla, CA)
Jon Epstein (Philadelphia, PA)
Donna M. Fekete (West Lafayette, IN)
Richard A. Firtel (La Jolla, CA)
Scott Fraser (Pasadena, CA)
Laura Gammill (Minneapolis, MN)
Andrew K. Groves (Los Angeles, CA)
Kat Hadjantonakis (New York, NY)
Christine Hartmann (Vienna, Austria)
Richard Harvey (Darlinghurst, Australia)
Carl-Philipp Heisenberg (Klosterneuburg, Austria)
Ali Hemmati-Brivanlou (New York, NY)
Lothar Hennighausen (Bethesda, MD)
Oliver Hobert (New York, NY)
Simon Hughes (London, UK)
Laurinda A. Jaffe (Farmington, CT)
Rulang Jiang (Cincinnati, OH)
Jane E. Johnson (Dallas, TX)
Randy L. Johnson (Houston, TX)
Keith T. Jones (Callaghan, Australia)
Alexandra Joyner (New York, NY)
Vesa Kaartinen (Ann Arbor, MI)
Ryoichiro Kageyama (Kyoto, Japan)
David Kimelman (Seattle, WA)
Hisato Kondoh (Osaka, Japan)
Paul Anthony Krieg (Tucson, AZ)
Robb Krumlauf (Kansas city, MO)
Paul M. Kulesa (Kansas City, MO)
Justin P. Kumar (Bloomington, IN)
Kenro Kusumi (Tempe, AZ)
Carole LaBonne (Evanston, IL)
Patrick Lemaire (Marseille, France)
Brian A. Link (Milwaukee, WI)
Melissa Helen Little (St Lucia, QLD, Australia)
Malcolm Logan (London, UK)
Suzanne L. Mansour (Salt Lake City, UT)
Christophe Marcelle (Marseille, France)
Ralph Marcucio (San Francisco, CA)
Benjamin Martin (Stony Brook, NY)
James Martin (Houston, TX)
Mark Q. Martindale (Honolulu, HI)
Robert E. Maxson (Los Angeles, CA)
Roberto Mayor (London, UK)
David R. McClay (Durham, NC)
Daniel Meulemans Medeiros (Boulder, CO)
Cathy Mendelsohn (New York, NY)
Alan Michelson (Bethesda, MD)
Anne Helene Monsoro-Burq (Orsay, France)
Randall Moon (Seattle, WA)
José Xavier Neto (Sao Paulo, Brazil)
Jenny Nichols (Cambridge, UK)
Lee Niswander (Aurora, CO)
Elke Ober (Copenhagen, Denmark)
Michael B. O'Connor (Minneapolis, MN)
Guillermo Oliver (Memphis, TN)
David Ornitz (St. Louis, MO)
Nipam H. Patel (Berkeley, CA)
Tatjana Piotrowski (Salt Lake City, UT)
S. Steven Potter (Cincinnati, OH)
Olivier Pourquié (Illkirch, France)
Victoria Prince (Chicago, IL)
David W. Raible (Seattle, WA)
Yi Rao (Beijing, China)
Elizabeth Robertson (Oxford, UK)
Nadia Rosenthal (Monterotondo, Italy)
Michael A. Rudnicki (Ottawa, Canada)
Gary Ruvkun (Boston, MA)
Alejandro Sánchez Alvarado (Salt Lake City, UT)
Makoto Sato (Ishikawa, Japan)
Noriyuki Satoh (Kyoto, Japan)
Andreas Schedl (Nice, France)
Alexander Schier (New York, NY)
Thomas F. Schilling (Irvine, CA)
Gerhard Schlosser (Galway, Ireland)
Robert A. Schulz (Notre Dame, IN)
Lilianna Solnica-Krezel (St Louis, MO)
Beatriz Sosa-Pineda (Memphis, TN)
Michelle Southard-Smith (Nashville, TN)
Derek Stemple (London, UK)
Claudio D. Stern (London, UK)
Andrea Streit (London, UK)
Xin Sun (Modison, MO)
Lori Sussel (New York, NY)
Clifford J. Tabin (Boston, MA)
Yoshiko Takahashi (Tokyo, Japan)
Patrick Tam (Westworthville, Australia)
Miguel Torres (Madrid, Spain)
Paul Trainor (Kansas City, MO)
Mark Van Doren (Baltimore, MD)
Jean-Paul Vincent (London, UK)
Talila Volk (Rehovot, Israel)
John Wallingford (Austin, TX)
Deneen Wellik (Ann Arbor, MI)
Carol Anne Wicking (St Lucia, QLD, Australia)
David Wilkinson (London, UK)
Christopher V. E. Wright (Nashville, TN)
H. Joseph Yost (Salt Lake City, UT)
Katherine E. Yutzey (Cincinnati, OH)

***Genomes & Developmental Control Editors**
†Evolution of Developmental Control Mechanisms Editors

Published under the Auspices of the Society for Developmental Biology

FIGURE 6.1 Masthead page of Developmental Biology. *(Published by Elsevier.)*

changes in the community it serves. Journals often need to evolve to meet the changing needs of the authors and readers in a particular field to be able to attract emerging research. The EiC sometimes invites authors to submit manuscripts for the journal. This is often the case when professional conferences highlight new and interesting research that is relevant to the journal.

Special issues or a series of topical collections may also be commissioned by the EiC to reflect newly emerging work.

The EiC is also responsible for the selection of any associate or section editors as well as the editorial board for the journal. Publisher involvement at this stage is variable and depends on the journal as well as on any professional society affiliation it may have. In general, for society journals (those owned by a learned society), the editor and the society will have full editorial control of the content as well as the composition of the editorial team. For publisher-owned journals, the publishing staff may have more input into the selection of the editorial team even though the final decision very often will rest with the EiC.

Once the editorial team is in place, the EiC engages with the editors and the editorial board on the progress of the journal and updates them on ideas for content development. Most journals will hold an annual meeting of the editorial team and the EiC, along with the publishing staff, will present the current status report for the journal as well as encourage discussion about the overall well-being of the journal. The EiC provides strategic input into the journal's development. The publishing contact will be in touch regularly to report on the journal's performance and suggest possible strategies for development. If commercial advertising, supplements, and reprint sales represent important sources of income for the journal, as is the case for many medical journals, the EiC will be asked to be on the lookout for these types of opportunities.

At best, the EiC is not only the operational leader of the team but also the inspiration for new ideas for journal content. The most successful EiCs are remembered for their innovation as well as for their ability to oversee journal operations smoothly and expeditiously.

ADDITIONAL EDITORS

In general, a journal will have multiple editors if it is very large, and the number of submissions is too great for one editor to handle and/or the scope of the journal is so broad that it is not possible for one editor to make informed decisions about submissions in all subject areas.

Multiple editors may sit between the editor(s) and the editorial board and can also be referred to as follows:

- Co-Editors;
- Associate Editors;
- Section Editors;
- Editorial Advisors;
- Editorial Committee Members.

If the EiC is working with additional editors, then papers may be divided between the editors and the EiC on the basis of geographical origin, field of specialization, type of contribution (e.g., reviews), and as a means to equalize workloads.

Multiple editors may have different roles, depending on the journal. The publisher will be able to provide advice on the various options.

THE EDITORIAL BOARD

The editorial board, sometimes known as the (editorial) advisory board, is a team of individuals who are known experts in the journal's field. Some individuals may also belong to the editorial boards of other journals.

The board's role consists of the following:

- Expertise in subject matter;
- Reviewing submitted manuscripts;
- Advising on journal policy and scope;
- Identifying subjects and conferences for special issues which they might also help to organize and/or guest edit;
- Attracting new authors and submissions; and
- Ideally submitting some of their own work for consideration.

The editorial board is selected by the EiC (and/or the Editors), with advice from associate editor(s) where appropriate, and sometimes with input from the publisher. When the journal is sponsored by a scientific society, the publisher has very little input into the selection of the editorial board. However, the publishing staff can be very helpful in that they are able to provide reports about prolific authors, geographical distributions, important institutions, and names of individuals who may have expressed an interest in the journal. The editorial board will generally undergo a rotation every 2 or 3 years by removing some members, inviting others, and renewing some existing members for another term. The EiC can also make changes to the board between rotation cycles if, for example, a board member resigns and there is an immediate need to replace him or her. There are unusual situations in which an editorial board member does not fulfill the expected duties, such as reviewing a certain number of papers per year. There is no ideal size for an editorial board—it will vary by journal as well as by discipline, and by editorial preference. Some editors prefer working with very large boards, while others like smaller groups.

The quality of a journal is judged to some degree by the composition of its editorial board. Along with this point, the EiC should consider the following when forming the board:

- The location of the board members should represent the full geographical appeal of the journal;
- Board members' expertise should represent the complete range of subject areas covered by the journal's scope;
- Representatives should be appointed from key academic or research institutes;
- Former guest editors of special issues and authors of key reviews;
- Nonboard member reviewers whose reviews are of a high standard and/or who have shown an interest in and commitment to the journal;

- Nonboard member reviewers whose reviews are of a high standard and/or who have shown an interest in the direction of the journal;
- Prestigious figures in the field who might not always be very active but whose names might attract submissions;
- Existing board members may wish to suggest peers whom they consider would be a benefit. Retiring board members or those who decide to step down for other reasons are generally happy to make suggestions for their replacements.

REVIEWERS

Editorial board members are not usually responsible for reviewing all submissions to a journal. Most journals, however, expect that a certain percentage of manuscripts that are reviewed in a given year are provided by the editorial board. Even more specific, each editorial board member of most journals is expected to review a certain number of manuscripts per year. The total number of reviewers always exceeds the expected capacity of the editorial board member. Thus, journals commonly call on additional reviewers that constitute repository of reliable reviewers. These reviewers are considered to be members of a wider editorial team. They may be younger scientists who are interested in extending their editorial experience through peer-reviewing activities. Please see Chapter 7 on Peer Review for more information.

THE PUBLISHING TEAM

Successful EiCs generally work smoothly with the publisher on all aspects of the journal. While the publisher has responsibility for production, marketing, and distribution of the journal, editorial input is important and can help guide the publisher to better accommodate the needs of the community served by the journal. The partnership between the publisher and the editorial team is vital to the ultimate success of the journal. As such, it is important for the editor to understand what the publisher does and how its input can be beneficial.

Please note that while the functional responsibilities of the various publishing company staff involved with the journal are similar across companies, the job titles may differ from company to company. For the purposes of this chapter, Elsevier kindly provided detailed information about their journal publishing processes and how the author uses Elsevier's job titles to represent the various functional roles [1]. Fig. 6.2 shows Elsevier's Journal Editors' Hub website.

The main contacts at the publisher's organization are as follows:

- Publishing (in some companies also known as "Editorial")
- Production
- Marketing/distribution

PUBLISHING

For a journal editor, the main line of support at the publishing organization will be the internal publishing staff (also known as the publisher or the publishing

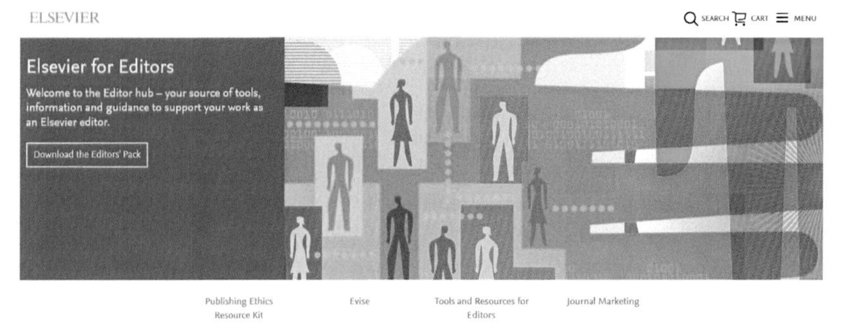

FIGURE 6.2 Elsevier journal Editors hub (https://www.elsevier.com/editors).

editor (PE) and can include publishing assistants). The publisher responsible for the journal will manage the entire publication process and will work with Production and Marketing and Distribution teams to ensure that the end-to-end workflow, from manuscript submission to final publication, is smooth and seamless.

The publishing staff can facilitate the following:

- Work with the editor to set the future journal strategy and ensure that it is acted on;
- Do desk and field research on market trends and competition;
- Communicate publishing policies and procedures, and new developments;
- Arrange the journal's finances, including any editorial payments;
- Officially invite or retire editors and editorial board members;
- Provide key performance information, such as editorial statistics, downloads, citation information, author satisfaction research results, and publication times. This information is then used to prepare reports for editorial board meetings;
- Organize editorial board meetings;
- Develop plans for special issues and/or supplements with the editorial team;
- Guide and support the editor and the editorial team in the event that plagiarism, falsification, or multiple publication is detected;
- Offer expert advice and support about scholarly communications and journal publishing policies.

As experienced journal editors know, the main point of contact at the publishing company is instrumental in the overall editorial experience for the editor and the editorial team. Most publishing organizations with large journal programs are well equipped technically and operationally to publish journals. The quality of the overall editorial experience rests with the people involved (on both "sides"). If the publishing contact is fully engaged and is able to make things happen in the publishing organization, the experience is very positive. Likewise, the EiC needs to be fully engaged and committed to the journal, striving to attract the best possible authors and papers, and be willing to work with the PE, to make the overall process a success. The relationship between the editor and the PE needs to be a partnership for the journal to be successful.

PRODUCTION

Production staff is responsible for all production aspects related to your journal. The journal manager (JM) supports the publisher by facilitating the editorial process and the production of academic journals. The JM manages the production and publication process for both online and printed journals. As process managers, JMs interact with authors, editors, and reviewers and are the single point of contact for their respective journal portfolios during the entire end-to-end process. The JM is supported in this role by the journal administrator. While the publisher (or PE) is responsible for the journal strategy and objectives, the JM manages the editorial and production process of journal articles and issues. The JM is an important resource of information in areas such as editorial and production workflow, editor and reviewer performance, article copy flow, production, and publication planning. If the editor would like to make changes to any of these areas, the JM can facilitate the appropriate adjustments to the production workflow.

MARKETING

This publisher function is the one that tends to generate the most frustration with the editor and the editorial team. The EiC and the team very often feel that the publisher is not doing "anything" or not doing "enough" to market the journal. If editors do not see readily marketing efforts (e.g., brochures, email campaigns, stacks of journals at professional conferences), they assume the publisher is not doing anything at all. This is usually not the case. The marketing staff at the publishing company develops a detailed marketing communications strategy for the journal based on their experience with what is most effective and efficient. This strategy is designed to promote the journal to the disciplines it serves as well as to raise awareness for the journal among the publisher's author communities. To implement the marketing strategy, publishers will use a combination of online, physical, and print channels. Marketing will work with editors and editorial board members to help attract high quality content to the journal, and the editorial team is encouraged to provide input and constructive feedback. Journal editors will receive regular overviews (usually annually) of all marketing activities done for the journal. This marketing report can also be shared with the editorial team during the yearly editorial board meetings. While both the EiC and the Marketing department may not always see eye to eye on this element of journal publishing, it is important to emphasize that both must work in good faith to achieve maximal impact of the journal.

As journal content is primarily accessed online, increased use of digital marketing techniques enables thc publisher to reach the author and reader communities much more effectively and efficiently. The strength of the journal can be demonstrated by using websites, digital advertisements, RSS feeds, social media, search engine optimization, and email communications. Response times are improved as is customer interaction and the ability to offer greater flexibility and quantifiable results.

Journal home pages: Each journal will have its own customized home page on the publisher's site that can be accessed from different subject pages. The home page will also have direct links to full-text articles on the publisher's online platform and will lists the most popular articles as well as news and resources for authors, editors, and readers. It will provide information on the aims and scope, editorial board, impact factor, abstracting and indexing services, guide for authors, and open access information (if applicable). It should also provide a clear and intuitive navigation path for authors during their orientation, submission, and publication process. The journal page can show previews of recent articles, social media post, most-read articles, most-cited articles, and special issues or special collections. It may also show journal-specific metrics such as impact factors. The editor and the editorial team can have significant input into the design and content of the journal home page. Editors should engage with the publisher's marketing staff to ensure that the home page accurately reflects not just the journal's look and feel but also the content being published and the field it represents.

Social Media: Publishers are increasingly using social media channels to communicate with new and existing audiences about their products and services. These channels are among the most effective ways to promote content. Well-established publishing companies can have hundreds of social media channels across a vast range of subject areas and target markets.

Social media journal marketing benefits include the following:

- Reaching new audiences with the journal's best content;
- Engaging directly with individual researchers;
- Amplifying the reach of the research published in the journal;
- Sharing journal news with a wide community;
- Increasing traffic to the journal articles and featured content;
- Increasing awareness for new research;
- Encouraging new submissions from target authors.

The editor and the editorial team can be very proactive in the marketing of the journal by following the publisher's channels and using their own social media channels to repost and distribute the publisher's messages as well as by posting their own journal news. Some publishers have also developed a series of social media guides for editors, with information on key channels, advice on how to pick the right channels, guidance on how to set up profiles, and tips on how each channel can be used by editors and researchers. Elsevier provides these guides at elsevier.com/editors/journal-marketing/social-media.

Email marketing campaigns: Publishers will initiate journal-specific email campaigns periodically, but these are most often geared for a specific purpose or an important announcement related to the journal. Examples include calls for papers, yearly impact factor announcements, and most downloaded or cited article campaigns. Publishers, and editors, need to be very careful with email blasts and the email lists they use, however, so their messages are not be mistaken for spam or unsolicited (and usually unwanted) marketing. Editors can be very effective with email messaging to small distribution lists of their own contacts who may appreciate hearing news about the journal or an important research finding that has just been published.

Exhibits and conferences: Exhibitions and professional and industry conferences offer an excellent opportunity for publishers to meet face to face with editors, authors, reviewers, and readers. The publisher's presence can range from a physical stand in the exhibit hall, flyer inserts in the delegate bags, displaying flyers at relevant sessions, poster campaigns throughout the venue, to an advert in the meeting program. Publishers will also on occasion sponsor a lecture or a session in a specific discipline under the name of the journal to build awareness and encourage submissions.

ALTERNATIVE EDITORIAL MODELS

For many journals, the EiC and the other editors are generally full-time scientists or academic physicians at research universities and/or medical centers. The editorial work they do is considered in part to be a responsibility to the scholarly community. However, several prominent publishing companies have journal programs that include prestigious journals that have academically trained inhouse staff who serve as EiCs and Editors. Examples include Cell Press, Nature, The Lancet, and the PLOS journals (Fig. 6.3), just to name a few. These journals employ scientists and physicians who have decided to leave academia to pursue

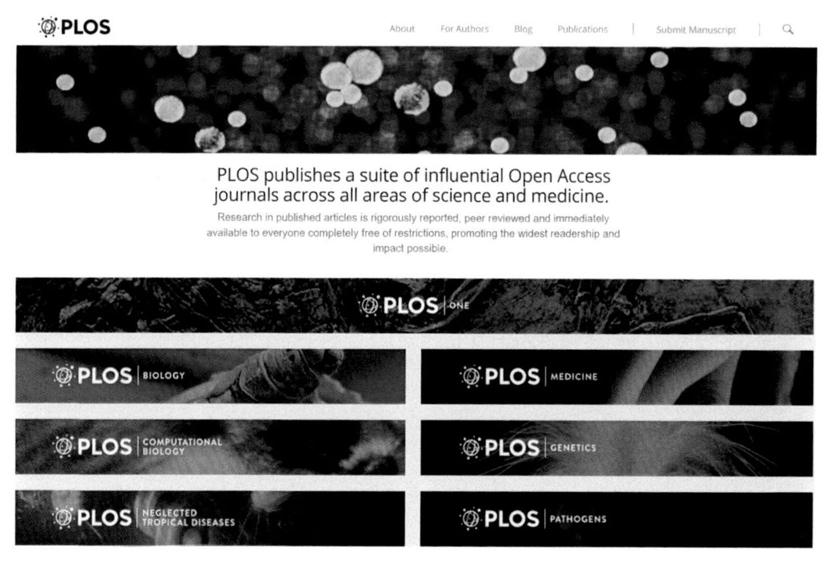

FIGURE 6.3 PLOS Journals website (https://www.plos.org/publications).

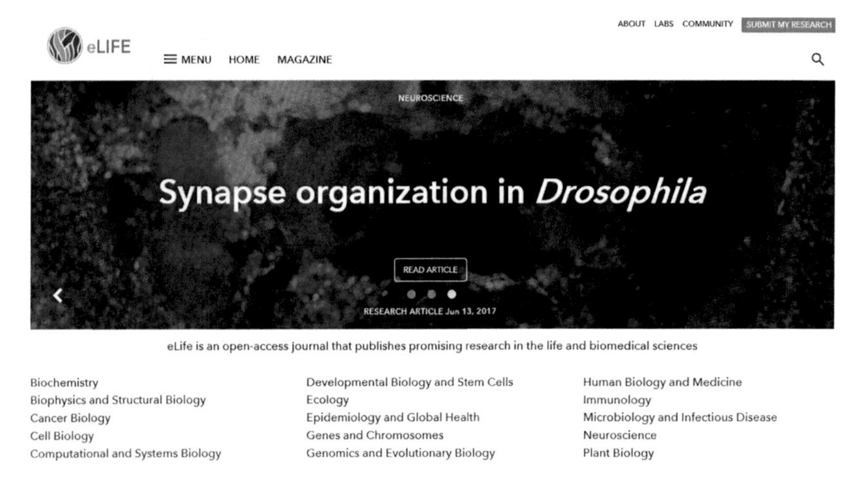

FIGURE 6.4 eLife home page (https://elifesciences.org/).

publishing careers. The in-house editors will then work closely with external editorial advisors, editorial board members, and reviewers (all academic professionals) to make editorial publishing decisions and develop journal strategies. Other journal-related publisher functions, such as production and marketing, are guided and managed by the in-house editorial staff to ensure a smooth end-to-end process.

A relatively new and rapidly growing open access journal, eLife (Fig. 6.4) [2], founded in 2011 by the Howard Hughes Medical Institute, the Max Planck Society, and the Wellcome Trust, uses an interesting editorial model in that its editors are working scientists at very prestigious institutions, but the portion of their time spent on the journal is compensated financially by the journal. This model brings the editorial publishing process back to academics and also allows the editors the luxury to spend significant time working on the editorial quality of the journal.

There are some interesting full- and part-time publishing job opportunities for academic physicians and scientists both as journal editors and as members of the journal's publishing staff. Chapter 5 describes one physician's journey to becoming a journal editor.

REFERENCES

[1] Portions of this chapter were adapted from Elsevier's Editors Hub, with permission. https://www.elsevier.com/editors.

[2] https://elifesciences.org/.

Chapter 7

Peer Review—Past, Present, and Future

Meng H. Tan, MD
Division of Metabolism, Endocrinology and Diabetes, Department of Internal Medicine, University of Michigan, Ann Arbor, MI, United States

INTRODUCTION

In broad terms, peer review is "evaluation of scientific, academic or professional work by others working in the same field" [1]. It is a system/process that assesses the quality of work under consideration by experts in the same field/subject matter to determine whether it meets predefined standards. Peer review is widely used by many to referee quality—journals to evaluate submitted manuscripts for publication (scholarly peer review), professional groups to evaluate performance of their members, grant agencies to evaluate requests for funding, universities to evaluate faculty performance for promotion, and external survey groups to evaluate accomplishments of departments (Fig. 7.1).

This chapter focuses on peer review of original research manuscripts submitted for consideration for publication to scholarly journals in general, and scientific journals in particular. Journal editors do not know everything—they have some knowledge of many subjects but in-depth knowledge of only a few. With the wide spectrum of topics submitted manuscripts can cover, they need evaluation of the manuscript from experts (in that topic) before they decide to accept or reject the manuscript. Here, peer review is defined as "the process of someone reading, checking, and giving his or her opinion about something that has been written by another scientist or expert working in the same subject, area, or a piece of work in which this is done" [2]. The past, present, and future of peer review will be covered.

The Past: How Did We Get Here?

Peer review of professionals dated back to the 9th century—it was described in Ishaq bin Ali Al Rahwi's book *Practical Ethics of the Physician* [3].

Scholarly peer review dated back to 1731 when the Royal Society of Edinburgh published *Medical Essays and Observations* in which "memoirs sent by correspondence are distributed according to the subject matter to those members

Medical and Scientific Publishing. https://doi.org/10.1016/B978-0-12-809969-8.00007-3

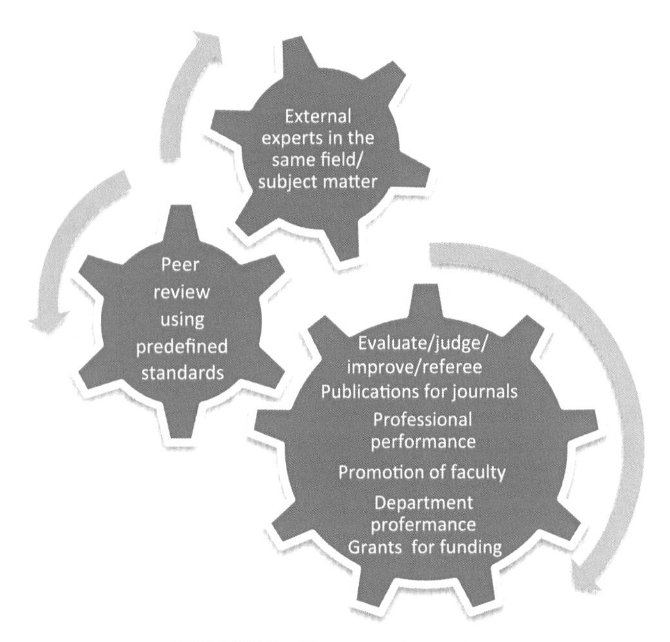

FIGURE 7.1 The peer review system.

who are most versed in these matters. The report of their identity is not known to the author" [4]. Table 7.1 shows selected milestones of scholarly peer review process since then.

The Present: Where Are We Today?

Today, scientific journals use peer review to scrutinize submitted manuscripts to ensure they meet certain standards before they can be accepted for publication. Submission of a new paper initiates a chain of events that eventually leads to the editor deciding whether to publish the manuscript (Fig. 7.2).

Finding Reviewers

Finding the right reviewers who are available to review a manuscript can be very challenging for the editor. The right reviewers should have most, if not all, of these attributes:

1. Be an expert in the manuscript's topic;
2. Published recently on matters related to that topic;
3. Be willing to spend time and review in a fair, balanced, objective, and unbiased way;
4. Be willing to return the evaluation report by the deadline specified;
5. Have no conflict of interest with the author.

TABLE 7.1 Selected Milestones of Scholarly Peer Review Process in Journals

Year	Comments
1665	1. *Journal des scavans* (1st academic journal) was published in January 1665 and edited by Denis de Sallo with no mention of peer review [5]. 2. *Philosophical Transactions* (1st scientific journal) was published in March 1665 by the Royal Society of London. Its first editor, Henry Odenberg, shared his scientific correspondence with members with basically no peer review [6].
1731	Royal Society of Edinburgh's *Medical Essays and Observations* is considered by many as the first peer-reviewed journal [4].
1752	Royal Society of London's Committee on Papers established to review the papers for *Philosophical Transactions* [6].
1893	British Medical Journal implemented peer review [7].
1940	Journal of American Medical Association and Science introduced peer review [8].
1942	Journal of Clinical Investigation introduced peer review [9].
1967	Nature introduced peer publication [10].
1976	Lancet introduced peer review [8].
1989	First International Congress of Peer Review Biomedical Publications [11].

The editor or his/her delegate (usually an associate editor) can identify potential reviewers in many ways:

1. Use the journal's editorial board whose members have previously specified their areas of interest and expertise;
2. Use search engines to match reviewers with topics. Three examples of these are Springerlink [12], ScholarOne Review Locator [13], and EVISE [14];
3. Use publication databases like PubMed, Embase, Web of Science, or Google Scholar to identify individuals who recently published on the same topic;
4. Use the author's list of suggested reviewers;
5. Use the manuscript's reference list to identify potential reviewers whose work overlaps that of the author.
6. The journal's own database of reviewers—those who have reviewed manuscripts previously and/or those who have indicated interest to review.

If a manuscript covers several topics, the editor should consider inviting reviewers for the different topics to get a fair and objective review.

Once experts on the manuscript's topic(s) are identified, the editor invites them to review the manuscript. The editor can ask them first before sending

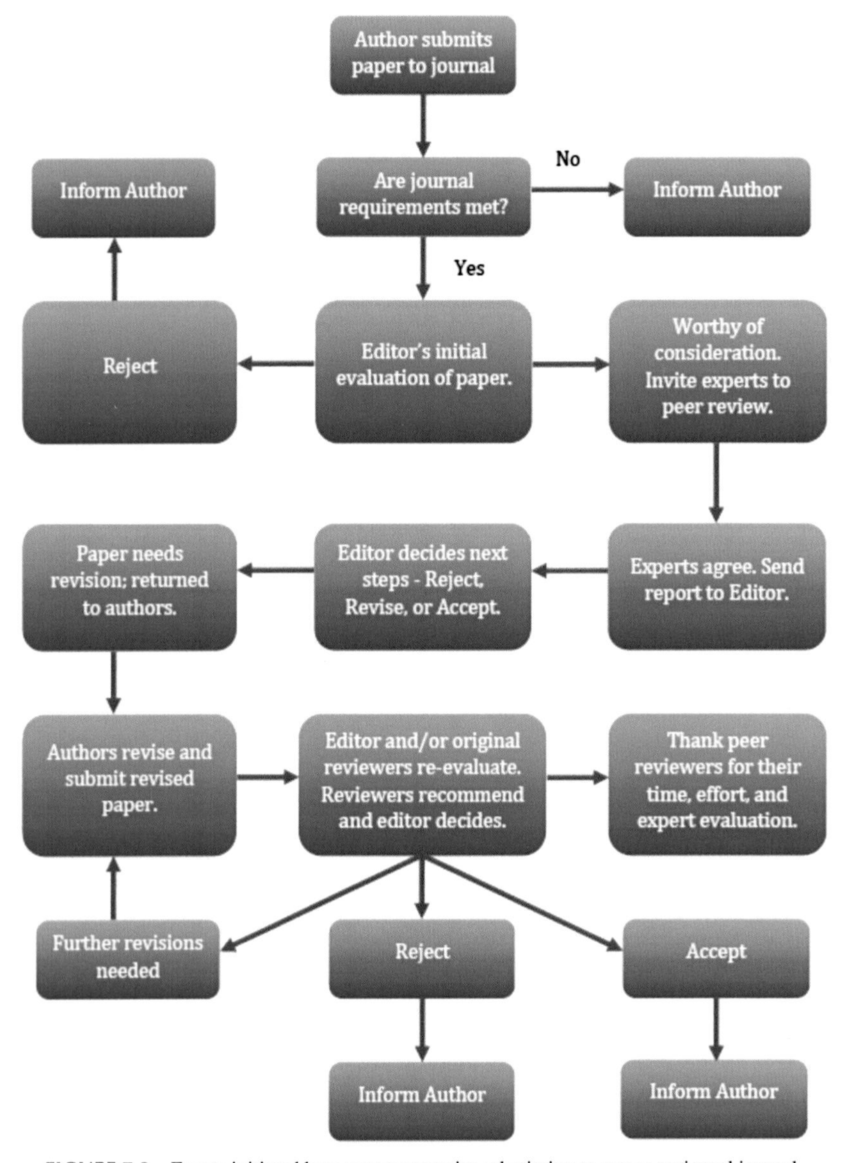

FIGURE 7.2 Events initiated by a new manuscript submission to a peer-reviewed journal.

them the manuscript or just send them the manuscript hoping they will review. The ask-first approach had a higher decline rate than the just-send approach [15]. However, they returned their reviews faster than the other group. The quality of the reports did not differ between the two groups.

When invited, the reviewer should respond promptly whether to accept or decline the invitation. To decide, the reviewers should (1) be knowledgeable in the manuscript topic to give an expert opinion; (2) have no conflicts of interest; (3) have the time to review within the timeline given; (4) know the journal's scientific integrity; and (5) be willing to review for the journal despite its reputation.

Many experts decline to review manuscripts. One survey reported lack of time due to reviewers' workload (they have a full-time job with many other deadlines) to be the main reason for doing so [16]. Other reasons given included, but not limited to, conflict of interest, deadline for review completion too tight, topic not reviewers' area of expertise, reviewing too many manuscripts, and scholarly peer review not recognized and rewarded by institution. Another survey reported manuscript topic outside the reviewer's expertise to be the main reason followed by lack of time [17].

Often two reviewers differ in their recommendations to the editor. Perfect congruence between two reviewers' recommendations for 496 manuscripts submitted to New England Journal of Medicine was 41.8% compared with 30%, as determined by chance [18]. In view of this and that reviewers often decline an invitation to review, it is prudent and economical (in time and money) to initially invite at least four reviewers to referee a manuscript.

HOW TO REVIEW PAPERS

Publishers [19,20], journals [21,22], and reviews [23–26] have shared guidelines on how to review manuscripts. Below is a synopsis of some of these suggested guidelines.

Guidelines for beginning the review include the following: (1) Read the entire manuscript first before evaluating it. Some then evaluate it the same day. Others evaluate it the next day. (2) Initial decision whether the manuscript's findings will advance/update current knowledge.

Guidelines for continuing the review include evaluating each section of the manuscript:

1. **Abstract**: (1) Does it accurately, clearly, and concisely summarize the aim, method, results, and conclusion? (2) Does its content reflect that in the main text? (3) Can it be understood without reading the entire paper?
2. **Introduction and study aim**: (1) Is the Introduction clear and concise? (2) Is the rationale for the study based on what knowledge is known and what gaps there are in current knowledge? (3) Is the question to be answered/hypothesis to be tested important?
3. **Methods and analysis**: (1) Are the study design and methods appropriate? (2) Can they answer the question/test the hypothesis in a scientific manner? (3) Is the subject population with appropriate controls defined? Is power calculation for sample size done to show a difference, if one is there? (4) In laboratory techniques, are data on precision and accuracy provided? Are the measurements valid? (5) Can another researcher reproduce the results using the same methods? (6) Are the analyses correctly done? (7) Do the results need to be reviewed by a statistician? and (8) Are statistical differences clinically/biologically important?

4. **Results**: *text, figures, and tables*: (1) Are the data clearly, concisely reported in a well-organized manner? (2) Are the figures properly labeled, presented, and supported by data? (3) Do they clearly show the important results? Would a different figure be clearer? Are there results not mentioned earlier in the Methods section? (4) Do the tables appropriately describe the results? Do the tables duplicate the text? (5) Are supplementary data available elsewhere?
5. **Discussion**: (1) Is it concise and does it make sense? (2) Is it balanced and do the data support the claims? Are alternative explanations given for the findings? (3) Are limitations and flaws of the study addressed? (4) Is the hypothesis verified? Is the question answered? If not, why not? (5) For unexpected results, do the authors explain them?
6. **Conclusion**: (1) Are the conclusions justified by the results reported? (2) Are more experiments needed to extend reported findings?
7. **References**: (1) Are relevant references cited to support salient points? (2) Are there missing relevant references that should be cited? (3) Are there errors in the references? (4) Are the references cited in the journal's format?
8. **Journal requirements**: (1) Is the study approved by Institutional Ethics Committee? (2) Are conflicts of interest declared? (3) Are informed consents obtained for human studies? (4) Are funding sources mentioned?

Prepublication peer review of a new paper has two main purposes—help the editor decide whether the paper meets predefined standards of quality and help the authors improve their revised manuscript for resubmission, if invited, to the same journal or, if rejected, to another journal.

With this mind, a helpful and useful critique has information for both the editor and authors. There are different ways to write such a report but the journal's format should be used. The reviewer can begin by writing a summary of the paper, indicating whether the findings will advance/update current knowledge. Or, is the paper repackaging published literature analogous to bottling old wine in a new bottle. This is followed by a "Major Comments" section which, in turn, is followed by a "Minor Comments" section. In both sections, each comment should state the problem with suggestions on how they can be possibly addressed. They should also be accompanied by identification of the page, paragraph, and line. All these will help the editor and author understand and follow the critique.

It is not the responsibility of the reviewer to detect plagiarism/recycling of text/duplication in new manuscripts. However, if the reviewer suspects these concerns, he/she should alert the editor.

The reviewer should make a recommendation on next steps for the paper after weighing the strengths and weaknesses of the paper—accept, reconsider after major revisions, reconsider after minor revisions, or reject. This can be based on (1) whether the study question is important/novel; (2) is the experimental approach valid; (3) are the results believable and properly reported;

(4) are the findings novel and important? and (5) is the conclusion based on data? This rating should not be in the "Comments for the authors," but in the "Comments to the editor," who will decide the next steps.

Almost all reviewers will also submit their own work to journals for consideration for publication. When they do, they expect to receive reviews, which are balanced, fair, and objective, unbiased together with constructive criticisms with suggestions to address them. They hope not to read abusive, derogatory, and sarcastic comments. They should treat their colleagues, other authors whose work they are reviewing, the same way.

Original research manuscripts are not the only type of manuscript submitted to scientific and medical journals for consideration for publication. There are many other article types including, but are not limited to, hypothesis, methods, reviews, systematic reviews and metaanalysis, case reports, clinical trials, perspectives, and others. Obviously, different article types will use different guidelines to review them as each will have different requirements. This chapter will not cover them.

OTHER ASPECTS OF PEER REVIEW

Types of Peer Review System

There are four major types (Fig. 7.3) of peer review systems [27]. Journals use different systems, depending on their policy and procedures. The most frequently used system is the single blind followed by the double blind, with increasing use of the open and postpublication. A 2008 survey reported 84% of reviewers had experience with single blind, 45% with double blind, and 22% with open peer review system [28]. Double-blind peer review is growing as reported in a 2013

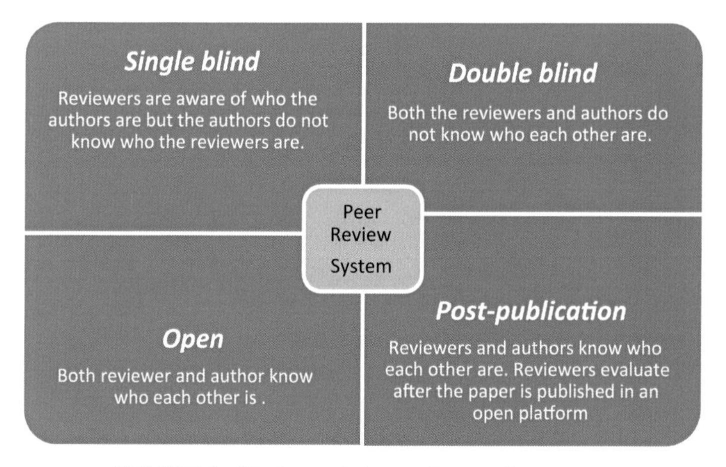

FIGURE 7.3 The four major types of peer review systems.

survey [17], and Nature announced availability of double-blind peer review in its journals in 2015 [28]. The peer review landscape is shifting.

Table 7.2 lists each system's strengths and weaknesses [29]. Except for the postpublication system, peer review functions to screen out bad science from being published and to help authors improve the quality of their paper before publication. Except for the postpublication system, peer review systems delay publication of eventually accepted papers as finding the appropriate and available peer reviewers and the back and forth evaluation processes take time.

TABLE 7.2 Strengths and Weaknesses of Different Peer Review Systems

Peer Review System	Strengths	Weaknesses
Single-blind	1. Reviewers may be more willing to be critical of the manuscript. 2. Protects reviewers.	1. Reviewers may not be as critical on "big name" authors' manuscripts and this may affect the quality of their reviews. 2. Reviewers may unethically delay review completion to publish their own (similar) work first.
Double-blind	1. Prevents bias/opinions. 2. Eliminates "big name" effect, gender, and racial bias. 3. Protects young reviewers from criticism by "Big names."	1. Getting paper ready for review is more time-consuming and costly. 2. In specialized research areas, it is not difficult to identify reviewer or author. 3. Reviewers may be more irresponsible.
Open	1. Fully transparent. 2. Less likely to be biased and more likely to be higher in quality. 3. Reviewer gets credit for the review as it is published with paper. 4. Eliminates "hiding behind anonymity" perception of some authors.	1. Reviewer may be reluctant to be too critical of paper.
Postpublication	1. Flaws of peer review reports easily identified and criticized. 2. Good science rapidly shared.	1. Bad science can be rapidly published and journal's reputation may be tarnished.

Time Spent in Peer Review

The median time for evaluating a manuscript is 5h, with a mean of 9h [29]. As almost all manuscripts are evaluated by two reviewers and by the editor, greater than 10h are spent in reviewing a new manuscript for publication. This can be costly.

Conflict of Interest of Peer Reviewers

Reviewers must observe the code of ethics [30] including, but not limited to, conflicts of interest of reviewers. They must declare their duality of interest when considering/accepting to review new manuscript(s) for a journal. A reviewer must not be from the institution as the author, be recent collaborators/coauthors of the author(s), be able to profit financially from the publication of the manuscript, and must not be biased against the author(s) work.

Comparison of Different Systems

There are advantages and disadvantages of maintaining anonymity (single- and double-blind systems) of peer reviewers or making them known (open and post-publication systems) to the authors. Does the chosen system affect the quality of the reviewer's report?

Single- Versus Double-Blind Systems

Reviewers in the single-blind system may be biased against certain authors. Is a double-blind system better? A randomized-controlled trial revealed no difference in the quality of review between single- and double-blind peer review of 92 manuscripts from 5 biomedical journals [31]. A recent metaanalysis of six randomized control trials comparing blinding with no-blinding showed no difference in quality of review and rejection rates [32].

In contrast, in a larger study involving 531 manuscripts and 1182 masked reviews, no difference in the quality of reviews between the two groups was found, but those who reviewed double blind had lower score and fewer were published [33].

Open Versus Anonymous Reviews

A recent metaanalysis reported that the open system improved the quality of the review report (of 1252 manuscripts) and decreased the rejection rate (of 1182 manuscripts) significantly when compared with those reviewed anonymously [32].

What Happened to Rejected Manuscripts?

Manuscripts rejected by one journal are usually submitted (after revision based the reviewers' comments) to other journals. About 40%–60% are eventually published. Studies on what happened to rejected manuscripts include the following:

1. American Journal of Ophthalmology: Of the 1444 manuscripts rejected in 2002–03, about 50% were eventually published in 94 different PubMed listed journals, most with lower impact factor with exceptions [34].
2. American Journal of Neuroradiology: Of the 554 manuscripts rejected, about 56% were eventually published about 16 months later in 115 journals, most have a lower impact factor but with exceptions [35].
3. Journal of the American Academy of Dermatology: Of the 489 manuscripts rejected, about 41% were eventually published 28 months later in 55 journals, most with lower impact factor with exceptions [36].
4. American Journal of Roentgenology: Of the 254 manuscripts rejected, about 64% were eventually published about 15 months later in 57 different journals, most with lower impact factor [37].

Rejection of a manuscript by a journal does not preclude its publication in a different journal.

Quality of Reports From Author-Suggested Versus Editor-Chosen Reviewers

A retrospective rater-blinded study reported the quality of 100 evaluations of manuscripts submitted to BMC journals did not differ between author-suggested and editor-chosen reviewers [38]. However, the former group recommended acceptance of the manuscript more often.

Similar findings were reported in a larger observational study involving 329 manuscripts in 10 biomedical journals. Each manuscript was reviewed by an author-suggested reviewer and an editor-chosen reviewer. There was no difference in the quality of the review between the two groups; however, the author-suggested reviewer group made more favorable recommendations for publishing the manuscript [39].

Fake Reviewers Manipulate Peer Review

An increasing number of journals, prestigious and otherwise, have retracted published papers because they discovered retroactively the peer review evaluations of these papers were manipulated by fake reviewers [40,41]. At the time of submission, the authors suggest the names (usually real) and email addresses (usually fabricated) of their preferred reviewers. It turned out the recipients of invitations to review a manuscript were either the authors themselves, their colleagues, or outside agencies (who had provided paid manuscript editing for the authors). All three would then provide a favorable evaluation of the manuscript to the editor, increasing the chances of acceptance of the manuscript. Regretfully, this "scam" has tarnished the peer review system for journals. To contain this scam, some publishers have discontinued their author-suggested peer reviewer's option [42].

Do Peer Review and Editing Improve Research Reporting?

In a before–after study of 111 consecutive original research papers published in the Annals of Internal Medicine, peer review and editing improved the quality of these papers in many ways, especially in areas deemed important by readers for generalizability of findings reported. Manuscripts scored in the bottom 50% improved more than those scored in the top half [43].

Can We Shorten the Peer Review Process?

The journey between submission of a manuscript to its publication follows a long, winding path in which peer review is a segment. Analysis of PubMed listed publications (with submission and acceptance dates) up to 2015 reported a median of 100 days, essentially unchanged between 1980 and 2015, the study period [44]. For certain journals, the median time actually increased. In the peer review segment of the journal (Fig. 7.2), the many steps take time to complete. No "magic bullet" solution is currently available.

Can the Current Peer Review System Be Sustained?

Can peer review be sustained as STEM literature continues to grow by leaps and bounds? In 2015, Scopus reported over 2.45 million citable items published in journals, books, and abstracts. In the same year, Web of Science and Scopus reported 1.8 and 1.7 million citable journal articles, respectively [45]! Assuming 2 unique reviewers per journal article, 3.6–4.9 million reviewers are needed. Assuming 50% of reviewers evaluate 2 papers in 2015, 1.8–2.45 million reviewers were needed. The increasing number of scientists and authors together with the proliferation of journals will make the burden of peer review heavier.

THE FUTURE: WHERE ARE WE GOING?

Peer review has been changing since it was first introduced in 1731. Today, we have prepublication and postpublication as well as anonymous and open peer reviews. Where it will be tomorrow is difficult to predict.

In the digital age there is a steady stream of new technologies, some of which can be used in enhancing peer review. Artificial intelligence, with its capacity and capability to rapidly complete tasks, can be used in peer review in many ways [46]. These include identifying potential peer reviewers, detecting plagiarism/text recycling, checking required formatting, discovering missing essential data, identifying wrong statistical analysis, and uncovering data fabrication. When implemented, all these can improve the quality and speed of peer review.

Retraction of published papers is rising partly due to manipulation of peer review by fake reviewers [40,41]. A few ways to minimize this unethical practice include (1) stop inviting authors to suggest reviewers; (2) search the Internet to verify these reviewers' information (names, publication records, and email addresses, especially noninstitutional addresses); (3) use search engines like ScholarOne to link to PubMed, Google, and Web of Science to find out whether these reviewers have published in the manuscript's topic; (4) use the web to check whether the reviewers are colleagues or collaborators of the authors; (5) check whether a reviewer's email address is identical to that of a coauthor; and (6) consider requesting the reviewer's ORCiD [47,48].

In the era of evidence-based medicine, a Cochrane Database Systematic Review in 2007 concluded there was little evidence to support peer review as a way to ensure quality of biomedical research [49]. A 2016 metaanalysis and systematic review of editorial peer review reported more randomized-controlled trials were available to assess peer review interventions, but pointed out the need to have more [32]. There is now more peer review research being conducted and more peer review journals to publish in. New knowledge uncovered by these will lead to new ways to improve peer review.

REFERENCES

[1] https://www.en.oxforddictionaries.com/definition/peer_review.
[2] http://www.dictionary.cambridge.org/ dictionary/english/peer-review.
[3] Al Kawi MZ. History of medical records and peer review. Ann Saudi Med 1997;17:277–8.
[4] Kronick DA. Peer review in 18th-century scientific journalism. JAMA 1990;263:1321–2.
[5] Banks D. Approaching the Journal des Scavans, 1665–1695: a manual analysis of thematic structure. J World Lang 2015;2:1–17.
[6] Spier R. The history of the peer-review process. Trends Biotechnol 2002:357–8.
[7] Burnham JC. The evolution of editorial peer review. JAMA 1990;263:1323–9.
[8] Benos DJ, et al. The ups and downs of peer review. Adv Physiol Educ 2007;31:145–52.
[9] Brainard ER. History of the journal of clinical investigation. 1924–1959. I. Personnel and policies. J Clin Investig 1959;38:1865–72.
[10] http://www.nature.com/nature/history/timeline_1960s.html.
[11] http://www.peerreviewcongress.org/pdf/1989/program_1989.pdf.
[12] https://www.springer.com/gp/products/springerlink.
[13] http://mchelp.manuscriptcentral.com/gethelpnow/.
[14] https://www.elsevier.com/editors/evise.
[15] Pitkin RM, Burmeister LF. Identifying reviewers: randomized comparison of asking first or just sending. JAMA 2002;287:2795–6.
[16] Tite L, Schroter S. Why do peer reviewers decline to review? A survey. J Epidemiol Community Health 2007;61:9–12.
[17] Mulligan A, Hall L, Raphael E. Peer review in a changing world: an international study measuring the attitudes of researchers. J Am Soc Inf Sci Technol 2007;64:132–61.
[18] Ingelfinger FJ. Peer review in biomedical publications. Am J Med 1974;56:686–92.
[19] https://www.elsevier.com/reviewers/how-to-conduct-a-review.

[20] https://authorservices.wiley.com/Reviewers/journal-reviewers/what-is-peer-review/the-peer-review-process.html.
[21] www.neurology.org/site/misc/info_review.xhtml.
[22] journals.plos.org/plosone/s/reviewer-guidelines.
[23] Benos DJ, Kirk KJL, Hall JF. How to review a paper. Adv Physiol Educ 2002;27:47–52.
[24] Provenzale JM, Stanley RJ. A systematic guide to reviewing a manuscript. Am J Roentgenol 2005;185:848–54.
[25] Seals DR, Tanaka H. Manuscript peer review: a helpful checklist for students and novice referees. Adv Physiol Educ 2000;22:52–8.
[26] Black N, van Rooyen S, Godlee F, Smith R, Evans S. What makes a good review and how to write a good review for a general medical journal? JAMA 1998;280:231–3.
[27] http://www.editage.com/insights/what-are-the-types-of-peer-review.
[28] Announcement nature journals offer double-blind review. Nature 2015;518:278.
[29] Ware M, Mabe M. The STM report. An overview of scientific and scholarly journal publishing. 4th ed. International Association of Scientific, Technical and Medical Publisher; March 2015.
[30] http://www.councilscienceeditors.org/wp-content/uploads/ entire_whitepaper.pdf.
[31] Justice AC, Cho MK, Winkler MA, et al. Does masking author identity improve peer review quality? A randomized controlled trial. JAMA 1998;280:240–2.
[32] Bruce R, Chauvin A, Trinquart L, Ravaud P, Boutron I. Impact of interventions to improve the quality of peer review of biomedical journals: a systematic review and meta-analysis. BMC Med 2016;14:85.
[33] Isenberg SJ, Sanchez E, Zafran KE. The effect of masked manuscripts for peer review process of an ophthalmic journal. Br J Ophthalmol 2009;93:881–4.
[34] Liesegang TJ, Shaikh M, Crook JE. The outcome of manuscripts submitted to the American Journal of Ophthalmology in 2002–2003. Am J Ophthalmol 2007;143:551–60.
[35] McDonald RJ, Cloft HJ, Kallmes DF. Fate of submitted manuscripts rejected from the American Journal of Neuroradiology: outcomes and commentary. Am J Neuroradiol 2007;28:1430–4.
[36] Armstrong AW, Idriss SZ, Kimball AB. Fate of manuscripts declined by the journal of the American Academy of Dermatology. 2008;58:632–5.
[37] Chew FS. Fate of manuscripts rejected for publication in the AJR. Am J Roentgenol 1991;156:627–32.
[38] Wagner E, Parkin EC, Tamber PS. Are reviewers suggested by authors as good as those chosen by editors? Results of a rater-blinded, retrospective study. BMC Med 2006;4:13.
[39] Schroter S, Tite L, Hutchings A, Black N. Differences in review quality and recommendation for publication between peer reviewers suggested by authors or by editors. JAMA 2006;295:314–7.
[40] Ferguson C, Marcus A, Orlansky I. Publishing: The peer-review scam. Nature 2014;515:480–2.
[41] www.retractionwatch.com.
[42] Callaway E. Fake reviews prompt 64 retractions. Nature 2015:18202.
[43] Goodman SN, Berlin J, Fletcher SW, Fletcher RH. Manuscript quality before and after peer review and editing at annals of internal medicine. Ann Intern Med 1994;121:11–21.
[44] Powell K. Does it take too long to publish research? Nature 2016;580:148–51.
[45] Newton A. The sustainability of peer review. SpotOn report. What might peer review look like in 2030? May 2017. p. P14–6.
[46] DeVoss CV. Artificial intelligence applications in scientific publishing. SpotOn report. What might peer review look like in 2030? May 2017. p. P4–6.

[47] https://hub.wiley.com/community/exchanges/discover/blog/2017/05/01/5-tips-to-help-prevent-reviewer-fraud?referrer=exchanges.

[48] http://editorresources.taylorandfrancisgroup.com/peer-review-match-fixing-author-suggested-reviewers/.

[49] Jefferson T, Rudkin M, Folse SB, Davidoff F. Editorial peer review for improving the quality of reports of biomedical studies. Cochrane Database Syst Rev 2007;(2):MR000016.

Chapter 8

Pursuing Scholarship: Educating Faculty and Students on How to Publish Their Academic Work

Paula T. Ross, PhD
Office of Medical Student Education, University of Michigan, Ann Arbor, MI, United States

SCHOLARSHIP

In 1990, Ernest Boyer challenged the traditional view of scholarship beyond research and publication and fully recognized the scope of academic work [1]. These four dimensions of scholarship include the following: (1) discovery—the generation of new knowledge; (2) integration—the infusion of knowledge across disciplines; (3) application—the use of knowledge to solve current problems; and (4) teaching and learning—the public sharing and peer review of reproducible teaching and learning methods [1,2]. These four dimensions broaden our conceptualization of scholarship and provide a clear framework with which to use in thinking about our work.

Medical education scholarship is the cornerstone of academic medicine [3]. It provides us with new, peer-reviewed resources that advance the field of understanding how physicians learn and develop professionally along the continuum of medical education [4]. Medical education scholarship covers teaching and learning, curriculum development, learner assessment, mentoring, advising, education leadership, and administration [4]. In medical education, scholarship in this area consists of original research, teaching materials and methods, assessment methods, and academic performance outcomes. Scholarship can focus as early as precollege through practicing physician and everything in between. Because of the broadness of medical education, the potential for quality scholarship is vast.

GET STARTED

Review What You Already Have

Medial educators often have materials that can be used to launch a scholarship agenda (courses you have taught, lectures you have given, content you have

Medical and Scientific Publishing. https://doi.org/10.1016/B978-0-12-809969-8.00008-5

been asked to provide for a larger session, panels you have served on as a content expert). Review these materials and give some thought of potential places this information could be shared. Some questions to consider are as follows: Who was the audience? Who will benefit most from the results of your work or your perspective?

Identify Topics of Interest

Topics can emerge from everyday encounters with students, patients, or peers as well as a scholarly curiosity about the way something is done. Topics of interest can also emerge from new topics that are critical to innovations in medical education and education scholarship. These range from assessment methods, pedagogy, curriculum content, policy or regulations that guide accreditation, and licensing. While hot topics may be more likely to get published, chasing these topics may minimize your ability to focus your thinking and explorations in a single domain and becoming someone viewed as an expert by your peers. Nonetheless pursue topics that are important to a larger audience, relevant and novel.

Conduct a Literature Review

The purpose of a literature review is to describe and evaluate the work done in a particular area [5]. The literature review is the foundation of the problem exploration and conceptualization; therefore, it is important that it is adequate and complete. In addition to identifying what is known and unknown, the literature review contributes to scholarship by demonstrating evidence of preparation, selecting appropriate methods, avoiding future pitfalls, and duplicating existing, published works [6]. The goal is to collect, read, and organize the meaning of the current literature in a meaningful way and identify gaps in this area.

Consult Key Resources

Informationists and statisticians are the resources that will improve the quality of your scholarship. Informationists are knowledgeable about resources in specific education and clinical areas and possess expertise in search engines and search strategies that can help you identify relevant literature. This is an important step in identifying leading scholars in a particular field, seminal articles, what has already been done, and how your work can contribute to this literature.

Statisticians are also paramount to quality scholarship products. Not only can they help with the analysis but they can also ensure your approach is methodologically sound. In their article, "You Can't Fix with Analysis What You've Spoiled by Design" Rickards et al. [7] outline key steps to developing and using a survey as a data collection method and emphasize how appropriate preparation can influence research outcomes. If you lack expertise in this area, identify individuals who possess skills in designing survey instruments, developing research questions, or conducting statistical analysis or qualitative analysis to ensure you produce a high quality project.

Identify Collaborators

Identifying collaborators, either within your own institution or outside, will make the process less overwhelming. Moreover, identifying others in your area of interest provides the opportunity to discuss and brainstorm your ideas and gain different perspectives on the topic.

When thinking about potential collaborators, consider reaching out to other health professions (e.g., nursing, dentistry, pharmacy, kinesiology, social work, etc.) and other disciplines (e.g., education, statistics, sociology, public health, information, law, engineering, business, etc.). Also consider learners as potential collaborators. Due to their proximity to the content being studied, learners provide a unique perspective, lend credibility to curriculum reform projects, and help communicate innovations to their peers [8].

Determine If Institutional Review and Informed Consent Are Needed

Institutional review boards (IRB) exist to protect human subjects/research participants. Become familiar with the IRB staff at your institution and feel free to contact them for advice or with any questions you may have about your project. Each institution has its own review requirements. Your project may be exempt; however, this categorization should be determined by institutional review [9,10].

Projects that meet the following criteria may fall within the exemption categorization: conducted in commonly accepted educational settings, involve research on regular and special education instructional strategies; explore the effectiveness of or the comparison among instructional techniques, curricula, or classroom management methods; or involve the use of educational tests (cognitive, diagnostic, aptitude, achievement), survey procedures, interview procedures, or observation of public behavior [11].

Because education research is often conducted using existing data (students' evaluation and/or performance data), obtaining informed consent may seem unnecessary, especially when individual students are not identified or when data are reported in aggregate. However, government regulations, 45 CFR 46.11b(d), indicate that four criteria must be met before an IRB can waive the requirement for informed consent [11]:

1. The research involves no more than minimal risk to the subject
2. The waiver will not adversely affect the rights and welfare of the subjects
3. The research could not practically be carried out without the waiver or alteration
4. Whenever appropriate, the subjects will be provided with additional pertinent information after participation.

It is strongly recommended that students be informed and provided consent whenever possible when their data will be used for research purposes [12].

Develop Your Expertise

Joining a writing group or journal club is an expedient way to build your expertise and expand your knowledge and scholarly activity. Other options for more specific training include the following:

- Workshops on topics relevant to scholarship (e.g., research design, literature review, data analysis, survey design, etc.) at your institution
- Coursera courses (online learning)
- Faculty development and education fellowships [13].

Secure Funding

Some education scholarship requires funding to execute. For example, if your project would benefit from the expertise of a statistician, participant incentives, or expert consultation, funding for medical education scholarship is available [14]. Table 8.1 provides some potential funding sources for investigations in medical education. National organizations often remain consistent in their topic areas, foundations generally have specific areas that they fund and often require a letter of interest as a first step; your institution may have ad hoc funding

TABLE 8.1 Funding Sources for Medical Education Scholarship[a]

National Organizations

- Association of American Medical Colleges (AAMC)—Group on Educational Affairs (GEA) and Regional GEAs
- National Board of Medical Examiners (NBME) Stemmler Grant
- Patient-Centered Outcomes Research Institute (PCORI)
- American Educational Research Association (AERA)

Foundations

- Josiah Macy Jr. Foundation
- Arthur P. Gold Foundation
- Alfred P. Sloan Foundation
- W.K. Kellogg Foundation
- Robert Wood Johnson (RWJ) Foundation
- Spencer Foundation

Your Institution

- Explore funding for Teaching and Learning

[a]*Not an exhaustive list [14].*

available for smaller funding projects as well as calls for larger funding. Small grant programs may also help launch larger projects and allow you to collect preliminary data [15]. Securing funding also lends validation to your program of research.

PUBLISH YOUR WORK

Clarify Authorship

Because individuals often possess differing views on authorship criteria and expectations for author order/placement, the ideal time to discuss author order is before the writing begins and should be agreed on, along with expectations by each authors' level of contribution. The International Committee of Medical Journal Editors lists the following four criteria for authorship [16]:

- Make substantial contributions to the conception or design of the work; or the acquisition, analysis, or interpretation of data for the work; AND
- Draft the work or revising it critically for important intellectual content; AND
- Have final approval of the version to be published; AND
- Agree to be accountable for all aspects of the work in ensuring that questions related to the accuracy or integrity of any part of the work are appropriately investigated and resolved.

Contributors who fail to meet all four of the criteria should not be listed as authors, but they should be acknowledged [16].

Author positioning can be challenging. Authors should be listed in order of their contribution to the paper [16]. Author order is based on the degree to which authors engage in the process [17]. When learners (students, interns, residents, etc.) are members of your team, maintain their up-to-date contact information to ensure they remain included as authors reflective of their contribution, even after they leave your institution or program [18].

Establish Clear Rules of Engagement

Determine the rules of engagement with your coauthors that clearly establish parameters for the use of the project data, including conference presentations and publications. This may be of particular importance when your collaborators are from other institutions and/or disciplines who do not share the same professional networks and opportunities arise to present the results to multiple audiences.

Ask for Feedback on Your Work

Ask someone (experienced colleague, writing coach, or copy editor) to read your work for clarity, grammatical or spelling errors, and content gaps. Clarity in writing is an essential element of getting your work published. Submitting abstracts to national, regional, and local conferences are good testing grounds and great opportunities to receive feedback on your work before submitting for publication.

Give careful thought to your title, abstract, and keywords. Keep your abstracts informative, use the most important words in the title that relate to your topic. Use appropriate keywords to ensure those searching on your topic can find your work [19,20]. The frequency of certain keywords and phrases can impact how easily someone is able to locate your article. For more information about how to prepare a journal article, please see Chapter 9 (Ibrahim, Dimick chapter).

What to Publish

Various formats exist to publish your work: commentaries, perspectives, brief reports, innovations, letters to the editor, and teaching materials. Table 8.2 contains descriptions of these format types.

Where to Publish

Consider your intended audience. Who will benefit most from your work? Will you focus on undergraduate, graduate, or continuing medical education?

TABLE 8.2 Publication Formats

Format	Description
Original research	A presentation of primary research generally using the traditional (Background, Methods, Results, Conclusion) format
Brief reports	A shorter article covering matters of topical interest or work in progress
Systematic review	A literature review conducted for the purpose of critically analyzing and summarizing multiple exiting research studies or papers.
Commentaries	A brief discussion on a timely issue that reflects the author's views as well as covering what is known on the topic. Generally fewer than 1000 words, with few references.
Perspectives	Similar to a commentary, in that it is primarily the author's views, but this format presents an argument that is not essentially based on practical research. The author will provide a new hypothesis or theorize the implications of a not yet implemented program or innovation.
Curriculum innovations	Introduces a solution to a large-scale problem, a new teaching method, or guidance on structuring courses and assessing performance
Letters to the editor	Either a response to articles in the journal, a reply to other letters, or a very brief commentary on an issue of importance in academic medicine

Narrow your focus so you can become more familiar with the journals and literature in a particular area.

In addition to informationists, Journal/Author name estimator (JANE) (http://jane.biosemantics.org/) is a helpful resource for identifying potential journals for submission of your work. JANE is resource to help with where to submit your article to. Do not hesitate to submit your work to international journals, but keep in mind your wording may need to be modified for international audiences. Investigate the credibility of potential journals. The following are questions that will help you determine journal credibility and avoid predatory journals:

- Where is the journal indexed (e.g., PubMed, MEDLINE, Scopus, etc.)?
- What is the journal's impact factor? This number reflects the yearly average number of citations of articles published in this journal.
- What is the journal's format (e.g., print online, open source, online only, etc.)?
- Does the journal require a publication fee if your manuscript is accepted? Fees can range from $100 to over $1000.

Identify 2–3 possible journals before your first submission to avoid becoming discouraged if it is rejected.

Keep up with journals in your scholarship area. Pay special attention to the types of articles they public (original research, innovation reports, case reports, etc.). This will help you identity your target journals. Note that some journals are general, while others are specialty (anesthesiology, surgery, etc.) or learner level (UME, GME) specific. Before you submit your article to specific journal ask yourself:

- Does my article fit the audience for this journal?
- Has this journal published similar work?
- Is my article formatted to the style required by the journal?
- Have I followed the journal's author instructions/guidelines, including references?

Think of your work beyond academic medicine, as other disciplines may be interested on your topic. For example, teaching methods may be of interest to other education audiences and clinical settings. Journals in other disciplines (e.g., education, teaching, research methods, or interprofessional journals) may also be interested in your work.

Email listservs are also good sources of information about requests for proposals or special journal issues and other unique opportunities to publish your work.

Embrace Less Traditional Avenues for Publication

Other avenues exist for publishing your teaching materials. Online repositories house educational materials that others can download for their own use, such as MedEdPortal, icollaborative, MedEdWorld [21], and your institution may also have its own. Writing for well-known blogs such as FutureDocs https://futuredocsblog.com/, The Conversation https://theconversation.com/us, or Reflective MedEd https://reflectivemeded.org/ may not contribute toward promotion requirements; however, this can help you gain credibility in the area, get your topic to a larger audience, and connect you to potential collaborators or speaking opportunities. Social media outlets such as Twitter are becoming increasingly popular [22] for gaining interest and raising awareness, generating buzz around you and your scholarship ability.

Revise and Resubmit

Receiving a "revise and resubmission" can be bittersweet. Try to approach the revisions from a positive perspective, as this request illustrates some interest in your work. Meyer, Carline, and Durning [23] provide key steps to approach your resubmission:

- Ask for clarification
- Address every comment
- Do your best to respond
- Follow the advice of resources in the literature
- Be professional and tactful in your reply
- Be timely in responding
- Make changes easy to follow

Rebound From a Rejection

Do not get discouraged and stop after rejection. It may take a few attempts to locate the right home for your paper. The rejection may not reflect the quality or relevance of your work, but the volume of material and priorities of the journal. Nonetheless, seriously consider any how you may address feedback you received prior to resubmitting to another journal.

Engaging in education scholarship is a worthy endeavor and allows others to benefit from your experiences. These fundamentals will help you along your journey. With time, commitment, and persistence, you can share your innovative activities with others in your community.

REFERENCES

[1] Boyer EL. Scholarship reconsidered: priorities of the professoriate. 1990. Princeton (NJ).
[2] Shulman L. The scholarship of teaching. Change 1993;31:11.
[3] Grigsby RK, Thorndyke L. Recognizing and rewarding clinical scholarship. Acad Med 2011;86:127–31.
[4] Simpson D, Fincher RE, Hafler JP, et al. Advancing educators and education by defining the components and evidence associated with educational scholarship. Med Educ 2007;41:1002–9.

[5] Sewell JL, Maggio LA, Artino AR. Planning the literature review. Acad Med.
[6] Maggio LA, Swell JL, Artino AR. The literature review: a foundation for high-quality medical education reserach. JGME 2016:297–303.
[7] Rickards G, Magee C, Artino Jr AR. You can't fix by analysis what you've spoiled by design: developing survey instruments and collecting validity evidence. JGME 2012;4:407–10.
[8] Burk-Rafel J, Jones RL, Farlow JW. Engaging learners to advance medical education. Acad Med 2017;92:437–40.
[9] Tomkowiak JM, Gunderson AJ. To IRB or not to IRB? Acad Med 2004;79:628–32.
[10] Johansson AC, Durning SJ, Gruppen LD, Olson ME, Schwartzstein RM, Higgins PA. Medical education research and the institutional review board: reexaming the process. Acad Med 2011;86:809–17.
[11] Department of Health and Human Services (DHHS) Human Subject Protections Regulations, 45 CFR 46. Available from: https://www.hhs.gov/ohrp/regulations-and-policy/regulations/45-cfr-46/.
[12] Henry RC, Wright DE. When do medical students become human subjects of research? The case of program evalution. Acad Med 2001;76:871–5.
[13] Searle NS, Hatem CJ, Perkowski L, Wilkerson L. Why invest in an educational fellowship program? Acad Med 2006;81:936–40.
[14] Gruppen LD, Durning SJ. Needles and haystacks: finding funding for medical education research. Acad Med 2016;91:480–4.
[15] El-Sawi NI, Sharp GF, Gruppen LD. A small grants program improves medical education research productivity. Acad Med 2009;84:S105–8.
[16] Uniform requirements for manuscripts submitted to biomedical journals: ethical considerations in the conduct and reporting of research: authorship and contributorship. ICMJE. Available from: http://www.icmje.org/ethical_1author.html.
[17] Roberts LW. Addressing authorship issues prospectively: a heuristic approach. Acad Med 2017;92:143–6.
[18] Lypson ML, Philibert I. Residents and authorship: rights, obligations, and avoiding the pitfalls. JGME 2012:138–9.
[19] Cook DA, Boardage G. Twelve tips on writing abstracts and titles: how to get people to use and cite your work. Med Teach 2016;38:1100–4.
[20] Wiley Author Services. Optimzing your article for search engines help readers find you. Available from: https://authorservices.wiley.com/author-resources/Journal-Authors/Prepare/index.html.
[21] Cohen LG, Sherif YA. A comparison of three health care education collaboration and publication portals. Acad Med 2014;89:1425.
[22] Twitter is trending in academic medicine. 2017. Available from: https://news.aamc.org/medical-education/article/twitter-trending-academic-medicine/.
[23] Meyer HS, Carline H, Durning SJ. Ten tips to move from "revisions needed" to resubmission. Acad Med 2016;91:e15.

Part II

Writing

Chapter 9

Writing for Impact: How to Prepare a Journal Article

Andrew M. Ibrahim, MD, MSc[1], Justin B. Dimick, MD, MPH[2]
[1]*Department of Surgery, University of Michigan, Ann Arbor, MI, United States;* [2]*Department of Surgery, Center for Healthcare Outcomes and Policy, University of Michigan, Ann Arbor, MI, United States*

I would not give a fig for the simplicity this side of complexity, but I would give my life for the simplicity on the other side of complexity.
Oliver Wendell Holmes, Jr., United States Supreme Court Justice, 1902–1932.

WHY YOU SHOULD GET SERIOUS ABOUT YOUR WRITING

Consider a few of our highest impact scientific journals, such as *New England Journal of Medicine* and the *Journal of the American Medical Association*. Publishing in these journals, or other high-impact factor journals, can lead to significant changes in clinical practice and policy. What do these high-impact publications have in common? Besides having a great idea and a well-executed study, they also have a clear and compelling narrative that makes the research accessible to their audience. There are a countless number of important scientific discoveries that never realize their potential impact because they are buried within poorly written manuscripts.

The importance of writing a clear and compelling manuscript applies beyond the top tier publications. Even if you are a seasoned writer and researcher, most of your work will not be in these journals. But you should still write with the same clarity and focus as this will increase the impact of your work no matter where it is published. The impact of your research is limited by your ability to effectively communicate the findings and implications of the work.

The impact of your research is limited by your ability to effectively communicate the findings and implications of the work.

Perhaps the most valuable reason to get serious about writing goes beyond manuscripts. Becoming a more effective writer will teach you how to communicate complex ideas into a logical and clear narrative. Such a skill is necessary to other responsibilities often encountered by academic researchers:

Medical and Scientific Publishing. https://doi.org/10.1016/B978-0-12-809969-8.00009-7

public speaking, grant writing, or institutional leadership positions. We point out the transferable nature of writing skills to overall professional development to help you justify putting in the time necessary to become an effective communicator.

The remainder of this chapter is organized into two sections. First, we outline how to structure the key content that should be included in a scientific manuscript. It draws on seminal work from Gil Welch—"Preparing Manuscripts for Submission to Medical Journals: The Paper Trail" [1]—that we have adopted and tailored on over time. Second, we offer some practical advice on how to improve your writing process. These lessons come directly from our own learning curve as authors, our observations as peer reviewers, and experience working with mentees.

THE CONTENT OF A SCIENTIFIC MANUSCRIPT

Scientific manuscripts submitted to academic journals are generally organized in the following order:

- Abstract
- Introduction
- Methods
- Results
- Discussion
- Tables and Figures

There is some variation from journal to journal on the details that should be included within each section. On the website of each journal you will find "Instructions for the Authors" that will detail any deviation from this format.

We discuss below each section separately.

Abstract

What is in an Abstract?

The *abstract* section of a manuscript is a summary (often 300 words or less) of the research article. It typically follows the same format as the article (i.e., introduction, methods, results, and conclusion) but in an abbreviated form.

Although your main manuscript may include multiple findings, the abstract only has space to focus on one or two key findings. As such, you should spend time thinking about which is the most important. Take time to ensure that your introduction, methods, results, and conclusion are consistent *within* your abstract. For example, your paper may examine multiple outcomes (e.g., complications, mortality, costs) but you only plan to focus on mortality and costs in the abstract. Your introduction, methods, results, and conclusion should all be tailored to those two outcomes. Readers will be very confused if state in your

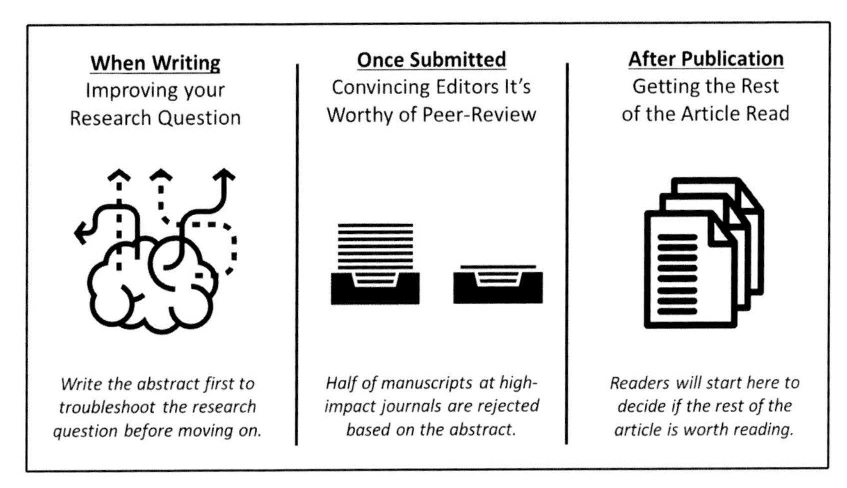

FIGURE 9.1 The three roles of an abstract across the manuscript timeline.

abstract that studied three outcomes, but then only report on two in the results of the abstract.

The Three Roles of an Abstract Across the Manuscript Timeline

An abstract takes on three different roles from the time you start writing, once it's submitted and after it is published (Fig. 9.1).

1. When Writing: Improve Your Research Question.
   We recommend that you write the abstract first because it helps you refine the narrative of the project. We even encourage doing so before you even have data with placeholder results (e.g., "XX%," "YY%") assuming a number of possibilities. This exercise will help you focus the research question clarify which outcomes you want to evaluate and assess if your study design and data are appropriate. If you cannot troubleshoot these issues and write a compelling abstract with placeholder results, you should stop. This is a sign you need to refine or change your research question before wasting time executing the work plan.
2. Once Submitted: Convince Editors It's Worth of Peer-Reviewed.
   The abstract is where journal editors will look first to decide if the manuscript should be sent out for peer review. At high-impact journals, more than half of the submissions will be rejected based primarily on the abstract. A common mistake here is to overstate the importance of your findings with a "conclusion" that is not supported by the results. Editors have a sharp radar for this type of "overreach" and it gives them an easy reason to quickly reject your work. Remember, this last section of the abstract is labeled "conclusions" not "editorial overreaching."

3. After Publication: Getting the Rest of the Article Read.
   The abstract is the first section that readers encounter to decide whether or not they want to read the entire article. Many readers may never read past the abstract, so it is important to make sure you have communicated your key message. A poorly written abstract will not entice readers to spend time on more poorly written prose.

Introduction

The purpose of the introduction is to give context to the question, create a knowledge gap, and preview your study plan. We feel this is done more effectively with three distinct paragraphs (Fig. 9.2).

Paragraph 1: Give Context to the Problem

The first paragraph of the introduction should get the reader to care about the topic. It needs to bring the reader up to speed on the why the topic is important. For example, if your paper is evaluating a federal payment policy, you will need to help the reader quickly understand why the policy was created and what is important about it now.

Common mistakes here are to give context that is too broad or too narrow for your audience. Most people start too broad and tell their audience things they already know. For example, let us consider a manuscript about colorectal cancer. Starting off with, "Colorectal cancer is the biggest killer in America" is not good. Almost all papers start that way, but you lose a huge opportunity because you are telling people things they already know. The only time it is okay to start a manuscript with a sentence like, "Every year in the United States there are 100,000 cases of XXX" is when you are writing about epidemiology and you are going to say that number is wrong—it is actually 200,000.

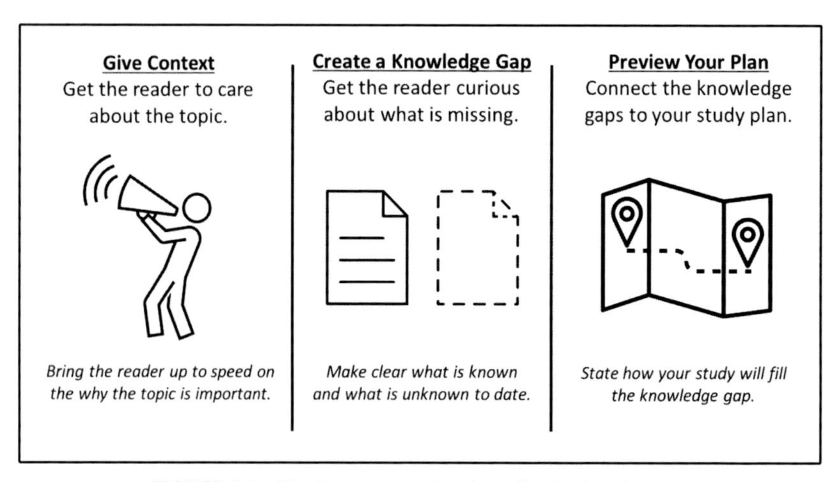

FIGURE 9.2 The three paragraphs of an effective introduction.

You have to establish the right entry point for your topic. If you start too broad you (A) put everyone to sleep and (B) will take up too much writing space getting people all the way up to your knowledge gaps.

Paragraph 2: Create a Knowledge Gap

The second paragraph needs to get the reader curious by creating a knowledge gap between what is known and unknown. You should not summarize all the literature on the topic here, but highlight the areas that have tension or uncertainty related to your study question. The knowledge gaps you introduce in this paragraph should directly correlate with the outcomes that your study will address.

This is the hardest paragraph of the introduction to write for a few reasons. First, you actually have to know exactly what is known and unknown. Second, that knowledge gap needs to be exactly what your study is designed to do. Third, you need to put those both together in a compelling narrative that convinces the reader it is an important gap in the literature that needs to be addressed. For example, if your paper is about the long-term outcomes of colorectal cancer patients after surgery, you need to set up related knowledge gaps. Did previous studies not follow patients long enough? Are most of the studies focused on narrow subpopulations? Whatever gaps you choose to highlight here should play right into the strengths of your study (e.g., longer follow-up, more representative study participants, etc.). Ideally, by the end of this paragraph, the reader should be thinking, "If only there was a study with longer follow up and a more representative sample, we would understand this topic so much better." Bingo—then you tell them (Paragraph 3) that is exactly what your study will do!

The knowledge gaps you introduce in this paragraph should directly correlate with the outcomes that your study will address.

Paragraph 3: Preview Your Work Plan

The third paragraph of the introduction should preview your work plan, i.e., briefly explain how you will close the knowledge gap discussed in the prior paragraph. Save the details for the methods section, but simply state the database and the outcomes you are going to use. Again, the outcomes should directly line up with the knowledge gaps you just created. If you wrote the first two paragraphs correctly—motivated why the topic is important, highlighted areas where there are knowledge gaps—then this should be an easy paragraph to write.

If you are having trouble with paragraph 3, go back and look at paragraph 2 again. A common mistake is to highlight too many knowledge gaps. You get the reader curious about so many controversial areas in the topic, then provide a huge let down in paragraph 3 when they realize your study is only going to fill one of them.

Methods

The methods sections should explain how the study was conducted. There are different conventions on what needs to be reported here for different study designs (e.g., randomized control trials, survey data, qualitative interviews, etc.). We recommend looking at previous articles from the journal your targeting and/or your mentor to see how this section was organized.

For many papers, the methods sections will include these subsections:

- Data Source—what data did you use? (e.g., Medicare Claims)
- Patient Population—who did you study? (e.g., all patients undergoing surgery for colorectal cancer)
- Outcomes—what did you measure? (e.g., 30-day complications, readmissions)
- Statistical Analysis—what methods did you use? (e.g., multivariable logistic regression)

Each of those headings, on average, will be two paragraphs. Again, we recommend following precedent from previous papers with similar methodological approach to guide you here. Chances are your mentor would have used most of these same methods before. A detailed reading of your mentor's prior work will likely yield most of the methods that you will need. However, you do not want to simply plagiarize prior work. Rewrite them in your own voice, with an eye toward creating a clear linear narrative, emphasizing those methods most relevant to your current study. There may be a few areas that are entirely new, and those are the ones you should spend the most time crafting with your mentor.

> A detailed reading of your mentor's prior work will likely yield most of the methods that you will need...Rewrite them in your own voice, with an eye toward creating a clear linear narrative.

Results

The results section details the findings from the analysis. This should be reported in multiple paragraphs starting with one that describes the patient characteristics. If applicable, the next paragraph should describe the hospital characteristics of the groups being compared.

The next 2–3 paragraphs should describe the outcomes of the study. These should be stated objectively. Avoid phrases such as "Surprisingly, we found…" or "Contrary to what we expected…" This section should simply present the information without any editorializing or interpretation.

You should present the outcomes in the same order that you introduced them in the methods. Start with your primary outcome, then your secondary outcomes as appropriate.

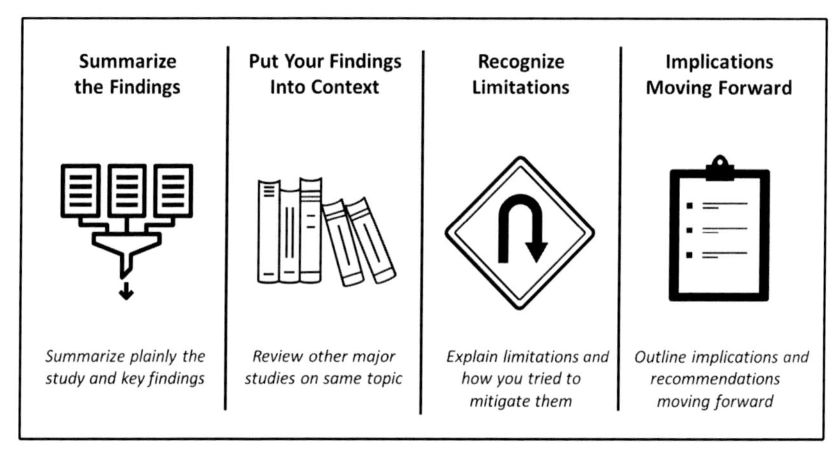

FIGURE 9.3 Four components of a compelling discussion.

Each of the tables and references in the article should be references in the results section. In fact, this is a good strategy to avoid repeating lines and lines of results that are already clearly presented in the tables and figures.

Discussion

The discussion section is your opportunity to connect the previous three sections—introductions, methods, and results—together and put them into the broader context of the topic. We typically use a five-paragraph approach for this section that includes four components (Fig. 9.3).

Paragraph 1: Summarize the Findings

The first paragraph of the discussion should be used to summarize the one or two key findings from the study. You have taken the reader on a long journey so far, so this is a good time to "refresh" in plain language what this study was about and what the key findings were.

Paragraphs 2–3: Put Your Findings Into Context

The next two paragraphs are used to place your main findings into context. You may have referenced some of this information in the introduction, but this is your chance to take a deeper dive. In addition to summarizing previous similar studies, end each paragraph with an additional sentence about how your research builds or adds to this prior work. It may challenge previous findings or extend a deeper understanding of them. If you cannot write that sentence because your research demonstrates the exact same findings as eight prior studies on this topic, do not write the paper.

An important style point here: authors who have written on the same topic will likely be a reviewer of your paper. So make sure you reference them appropriately and describe their study accurately.

Paragraph 4: Recognize Limitations

No study is perfect, including yours. The easiest way to annoy an editor or a reviewer is to ignore the limitations of your study.

Limitations are design features of your study that threaten the validity of the findings. You want to discuss 3–5 main limitations, which fall into the three main categories of threats to validity: Chance, bias, and confounding.

- Chance is random error. Addressing random error means making sure the statistical comparisons are adequately powered and analyzed with appropriate tests.
- Bias is systematic error. Addressing systematic error means discussing which strategies you used to ensure that these biases did not make your study results invalid, e.g., making sure you have a strategy for addressing selection bias.
- Confounding is when there are variables that are associated with the exposure and outcome that are actually driving your results, rather than a true relationship between exposure and outcome. Addressing confounding includes a thorough discussion of how you were able to address confounding with study design and/or methodologically.

To really take advantage of this section, you will want to provide a counterpoint about how you tried to mitigate that limitation or why it may not threaten your entire study. You can think of it as prophylactically addressing concerns you think will come up from reviewers. It will demonstrate you were thoughtful about the study design and are not overreaching your conclusion.

Paragraphs 5: Implications Moving Forward

The last paragraph of the manuscript should discuss the implications of your findings. An extremely common mistake here is to simply conclude "more research is needed." Do not do that. It makes everyone mad and cheapens your value as an author. Take a more sophisticated and detailed perspective with your recommendations. Demonstrate you have really thought about the subject matter and genuinely want to see your field advance based on the findings.

The four P's of the Discussion: How Will this Study Impact-Patients? Providers? Payers? and Policymakers?

To help brainstorm the implications of your study, we often think about "the four P's": patients, providers, payers, and policymakers. How will your study affect each of them? Does this change how patients choose treatments? How providers practice? How payers should reimburse? How policymakers regulate? This is your chance (within reason) to make a call for action based on your work. Having a coauthor with deep experience in the topic area can be particularly valuable for help in writing this section.

Tables and Figures

Although tables and figures in final print are included within the manuscript, during submission they should all be placed at the end after the references. All data and figures should be referenced within the results section of the manuscript.

A clear table or figure takes a long time to create. It is worth looking at other manuscripts who have done this effectively and learn what made them useful to the reader. The most effective tables have clear headings, identical spacing, and logical organization of information.

IMPROVING YOUR WRITING PROCESS

This next section offers some practical tips and advice we have acquired over time to improve your writing process.

Learn What Is Tried and True

You are unlikely to discover something new without a lot of practice on old stuff.

Richard Feynman Ph.D., Winner of the Nobel Prize in Physics in 1965.

If you are new to writing scientific manuscripts, you will want to start with a lot of reading. Ask your mentor for a handful of important articles in your field and read them closely, sentence for sentence. Learn the style, tone, and conventions that are used within your field. After a few articles, then reread the articles alongside the guide above and identify key paragraphs within each section. Soon the template above will become second nature, and you will quickly hone in exactly what each paragraph of a manuscript is designed to achieve.

Write in 20-Minute Bursts and 2-Hour Blocks

For many people, sitting in front of a blank page can be intimidating. It still is for us. That is why a template like the one we have given you above is helpful. When you have writer's block, what can you do? You can write *one* paragraph. Chip away at the paragraphs where you know what the content is supposed to be like the methods paragraph about the data source. If you sit down and write a paragraph every morning for 20 min for 2 weeks, guess what you have? You have an entire paper that took you 20 min a day to write. It is not very good because you have not edited it, but you do have a paper.

The 20-minute bursts can be effective for some writing, but not sufficient. We also recommend setting aside significant blocks of time in your schedule (e.g., 2 h) for writing. This should be uninterrupted time to deeply focus on a single paper. The abstract, introduction, and discussion sections particularly benefit from these longer writing periods.

Stick to a Parallel Writing Structure

As you read more and more scientific articles, you will observe that they follow a very clear style and pattern. You will want to develop that same habit in your own writing. The easiest place to start here is making sure that you introduce content within each section of your manuscript in the same order. For example, if the title of your paper is the "Complications and Costs of Rural Surgery," then your introduction should first introduce complications, then costs. Similarly, your methods should first define complications then costs. The results should then be reported in that order too, complications then costs. And finally, the discussion should first discuss the findings about complications, then about costs. Being diligent about keeping the same order throughout every section will make your manuscript easier to read and follow.

Be Consistent With Terminology

Use the same terminology throughout the manuscript. Scientific manuscripts are different than other forms of writing where you want to use variety to keep it interesting. This is the opposite. If you are calling something, for example, "Decline in Applicants," do not call it "Diminution in Med Students Interested in Applying to Urology" later on. Call it the same thing everywhere. It is too hard to read a paper when you are inconsistent. Switching the terminology or topic around is something that editors dislike because it makes the paper seem unfocused or confusing. The goal here is for the reader to understand the content of your research—not dazzle the editors with rhetorical flourishes.

Getting the Most From Feedback

A good mentor wants to see your writing early on and help you iterate. It is your mentor's job to help improve the way you think, and to do that, they need to see what you are doing. Frequent short meetings are best. And, record everything! If you sit down with your mentor for 20 min to look over your paper, bring your audio recorder (or your smartphone) because they will say twenty things in that meeting and you will walk out remembering only two. If you try and write it down, you may walk out with five of the twenty. If you record the conversation (with their permission, of course), you walk out with all twenty. Your mentor might even say things such as "Why don't you try something like this?" and it will be the perfect sentence that pulls it all together. Put it into your own words if you like, but that is the best use of your time with your mentor. We can recall countless times we have gone back to audio and rediscovered optimal phrasing or ideas to put into a revision. We also can recall asking mentees who did not record the conversation, "Didn't we talk about changing this when we met last time? Why is it still the same?"

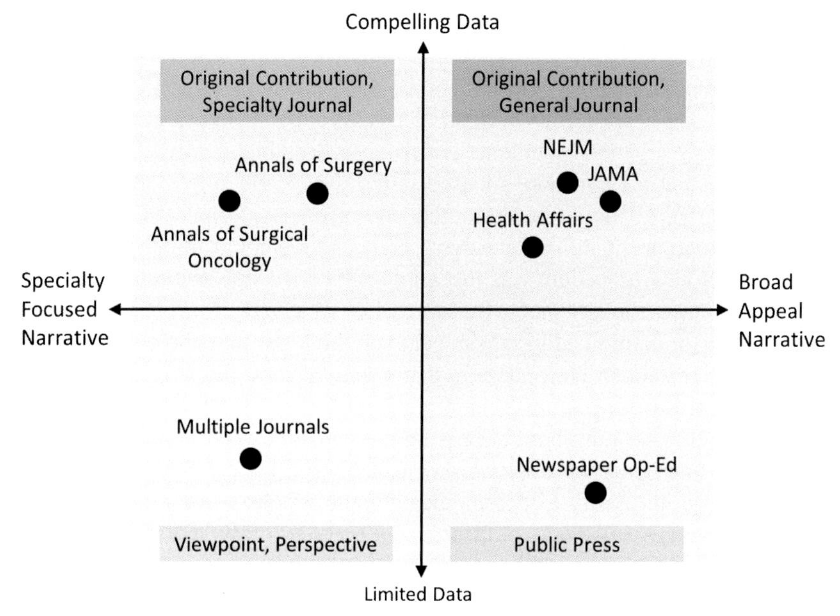

FIGURE 9.4 Where should you submit your manuscript? *NEJM*, New England Journal of Medicine; *JAMA*, Journal of the American Medical Association.

Write As You Go

When can you start writing these various sections? You can write an abstract without data. You can write an introduction anytime because it helps you understand if your research question is good. Particularly, the second paragraph of the introduction where you identify knowledge gaps. If you are thinking about a research question, try to write your introduction. If it is not compelling, then you may want to shift your research question.

Write the methods as you are doing the project so you do not forget details, especially if you do a lot of complicated analysis or make a lot of assumptions.

You have to wait for your results to finish your tables and figures, but you can mock up tables and figures. In this way you can think about the table, free of any data. You can develop a good structure for presenting your data.

Diversity Your Writing Portfolio

As you write a manuscript, you may find it not going as planned. The data you had were more limited than you initially thought to make a compelling argument. Or, after thinking through your knowledge gaps, you realized your research question is more appropriate for a specialty audience. Rather than hit these road blocks and scrap the manuscript, we suggest using that as an opportunity to refocus the manuscript to a different submission target.

We use the quality of the data (compelling vs. limited) and scope of the narrative (broad vs. specialty) to help determine where the manuscript should be submitted (Fig. 9.4). When there is compelling data and a narrative with broad appeal (e.g., Medicare claims evaluating a national payment policy), we recommend targeting a general medicine journal as an original contribution. Similar quality data, but with a more focused topic (e.g., evaluating readmissions in oncologic surgery) should be submitted to a specialty journal. When there are not great data to evaluate the question—perhaps they not available yet because the policy or new procedure was just announced—we still encourage developing the idea into a thought piece. Building a narrative in the form of a viewpoint, commentary, or opinion editorial can help you think through knowledge gaps in that area and establish your name in that space. Several journals, newspapers, and online forums support this type of publication.

Every author serious about improving their writing should intentionally target to hit all four quadrants of articles regularly. Writing for difference audiences will sharpen your ability to communicate complex ideas in a clear narrative. Moreover, it takes advantage of effort spent on early drafts that do not all end up where initially planned.

Eat Some Humble Pie

What most determines whether or not you will be a good writer? It has to do with how often you seek feedback and how you respond to that feedback. The people who are the best writers and who produce the best manuscripts are those who are the most open to feedback. If you think your writing is great and that your mentor's criticisms of it are unfounded, then you will probably not become a great writer. Put your best foot forward in listening to what they say.

We try to write iteratively, and let a paper unfold over a long time horizon. If you write over a long period of time, you can see your flaws more clearly because you can set them aside and come back to them. We also try to bring in someone who has a different perspective than us, just to get their feedback. Often, the best ideas emerge from these conversations, instead of written feedback. Try not to be defensive, just focus on understanding the problem. And, you do not need to take every suggestion. You want to make it your own. So, eliminate your defensiveness, hear the problem, and come up with a solution. The more iteration you invest in your paper, the better it will be.

REFERENCE

[1] Welch HG. Preparing manuscripts for submission to medical journals: the paper trail. Eff Clin Pract May–Jun 1999;2(3):131–7.

Chapter 10

Writing for Success

Hunter Heath III, MD
Department of Medicine, Indiana University School of Medicine, Indianapolis, IN, United States

> *"...scientific understanding is inseparable from the written and spoken word. There are no boundaries, no walls, between the doing of science and the communication of it; communicating is the doing of science."*
>
> Scott L. Montgomery, in The Chicago Guide to Communicating Science, 2003.

INTRODUCTION

Whatever careers physicians and scientists undertake, writing is an important—often critical— part of their professional work. In this chapter, I shall define *good writing,* point out some failings in bad writing, and outline approaches that an individual can take to improve the ability to write well. This topic is highly important for biomedical professionals because poor writing skills lead to weak documents. This is especially important for *persuasive writing*, that is, of documents intended to sell ideas, projects, or even oneself. Most professional documents fall into this category. Bad writing can lead to professional failure, for example, causing rejection or delay of applications for training or grant support and academic or business promotions (Table 10.1).

Poorly written abstracts and research papers can blunt the impact of great data. The ability to write well—clearly even if not elegantly—provides a competitive edge for an individual, a work group, an institution, or a company. To be clear at the outset: no one is born a great writer, but writing well is a learnable skill. The process I went through to reach a tolerable level of writing ability involved mentoring, criticism by colleagues, and personal study. It was worth the effort.

Many medical and scientific articles and grant applications are poorly written, but this is not a new problem. In 1900, nearly 120 years ago, the editor of the *Journal of the American Medical Association* complained:

> *The majority of articles submitted for publication could be cut down by one-half, and not a thought be eliminated... The repetition of well-known facts, padding..., and words, words, words, too often constitute the papers... [appearing] in*

Medical and Scientific Publishing. https://doi.org/10.1016/B978-0-12-809969-8.00010-3

TABLE 10.1 The Costs of Poor Medical and Scientific Writing

Bad writing leads to the following:

- Rejection or delayed acceptance of research papers and abstracts
- Unfavorable reviews of grant and project applications
- Possible failure of a drug's or device's development, registration, or commercialization
- Failure to be accepted for training or professional position
- Confusion about messages
- Time wasted on rewriting

All of the above are tied to loss or wastage of money

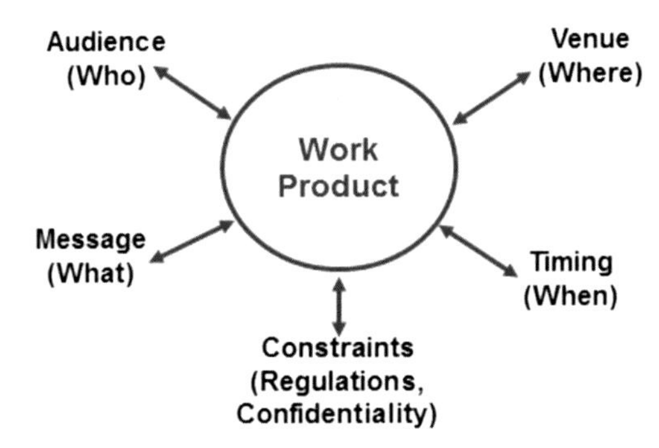

FIGURE 10.1 Major factors affecting the writing of a document.

medical journals. And if the editor presumes to... [reduce the text], the majority of authors consider it an insult [2].

Excessive length—too many words—continues to bedevil medical and scientific writing. But there are many other important failings, including poor textual organization, lack of focus, overuse and underdefinition of jargon, overly complex sentences, pretentiousness, lack of clarity, misuse of words, vast overuse of the passive voice, overuse of "that" phrases, unclear references, and unintentionally dumb-sounding or funny phrases. Fig. 10.1 illustrates the factors that need to be considered in writing a manuscript.

What follows are illustrative samples of such problems in writing and approaches to avoiding them. As you read, consider my four key rules for preparing a persuasive document:

1. Tell me a story.
2. Be brief.
3. Be clear.
4. Make me care.

Everything from here on is intended to help the writer achieve those four goals.

EXAMPLES OF GOOD AND BAD WRITING

Here is an example of dismal scientific writing:

> *The results of the present study suggest that in addition to the manifestation of aberrant homeostatic patterns of neuro-humoral activity following the cessation of noxious stimulation, the neurotic may be further characterized by atypical autonomic responses to an increase in the level of appetitional drive.*

Why is this bad writing? First, it uses too many words and too much obscure jargon; the sentence is too long for easy comprehension; and it uses "fancy" or pretentious words needlessly. Good writing is concise, crisp, clear, and flows logically.

Here is an example of very good writing in a less formal context:

> *The laboratory director who seeks a program of quality control sometimes feels like Goldilocks at the bears' home: Some suggested programs are Too Big, others are clearly Too Small. This handbook outlines procedures and suggests combinations that can be Just Right for each individual laboratory.*

The writer breaks sentences cleanly and with purpose; uses analogy and humor to convey meaning; and is clear, direct, and forceful. While humor is often not a part of biomedical persuasive documents, the other elements make any writing stronger.

Sometimes logical inconsistencies creep into sentences:

> *All data, except for the 6 hour blood draws, will be collected from patients at the study endpoint.*

Obviously, we do not "draw" or "collect" data from patients. "Data" are not "blood draws." And the phrase "blood draws" is awkward hospital jargon.

Better:

> *All blood samples, except for the 6 hr specimens, will be collected at study endpoint.*

Inadvertent humor does not help the reader stay focused on a document's messages:

> *The pneumonia was treated with penicillin, then changed to erythromycin.*

One has to chuckle as the writer suggests that "pneumonia" can be "changed to erythromycin." Less funny, but communicating more accurately,

> *The pneumonia was treated first with penicillin, then erythromycin.*

A common failing in biomedical writing is abuse of the horrific "that" phrase, frequently preceded by "It:"

> *It has long been known among international experts, through an extensive literature and vast experience <u>that</u> hyperglycemia in diabetes mellitus leads to numerous complications involving many organs.*

In this made-up example, 15 words stating the obvious precede the intended communication, separated by "that." The key ideas of the sentence are better communicated as follows:

Hyperglycemia in diabetes leads to many complications.

"That" phrases generally convey little additional meaning, but always add words and complexity, sometimes adding pomposity. One strategy for controlling one's use of "that" phrases is to use the computer to search one's document for the word, "that," then to look at each instance to determine if there is a better and more concise option. A search for "it" can be similarly rewarding.
The following example illustrates the wooden dullness of the passive voice:

Numerous data were derived from these studies and X, Y, and Z were shown.

The active voice has more clarity and force and often uses fewer words:

Our data showed x, y, and z.

The passive voice is often ambiguous:

My first patient interview will always be remembered.

By whom? Me? The patient? The ward nurses? The hospital legal department? Passive voice can be clear but cumbersome:

My first patient interview will always be remembered by me.

Active voice is clearer, cleaner, and more forceful:

I shall always remember my first patient interview.

For reasons unknown to me, most scientists and physicians think that scientific documents must be written primarily in the passive voice, perhaps to avoid the personal "I" or "we." However, all experts on persuasive writing agree that the benefits of using the active voice far outweigh any disadvantages. There is nothing wrong and much right in such writing as "After reviewing the literature, we designed an experiment as follows... We recruited X patients and Y control individuals... At the conclusion of the study, we analyzed the data in accord with the statistical plan.... We interpret our findings in light of other reports...." I am not against all use of the passive voice; sometimes it is the best way to go. However, use it judiciously and mixed with active voice.

Ambiguity is a common problem in medical and scientific writing. Here is a rather silly example, but which makes the point: "Labs were checked weekly." Does "labs" refer to laboratory tests, a place where experiments are done, or Labrador retrievers? Did the dogs have checkered coats? Putting one's written product aside for a while, then reading it as if you had never seen it before, can reveal ambiguous statements; colleagues reading the work will also point out ambiguities.

Yet another common problem in medical writing is *monotony.* It can result from repetition of words, for example, beginning many sentences with "it," or "these," or "this;" overuse of clichés such as "stands to reason," "few and far between," or "in the long run;" or a lengthy run of very short sentences. There are other causes of monotony in writing. Fresh rereadings of one's own work can reveal some.

APPROACHES TO IMPROVING WRITING SKILLS

The reader may be thinking, "I was not an English major; how can I deal with all this?" Here are some working strategies.

To write well, it helps to *read* good writing; it need not be only scientific writing, either. Wide reading in thoughtful journals such as *The Atlantic* or *The New York Times Review of Books* will give ample exposure to skillful explanatory or persuasive writing. When reading in the medical or scientific literature, look for writing that is unusually clear, direct, and forceful. Analyze those good examples and emulate them. Because writing is so critical to professional success, become a student of good writing. There are numerous books on scientific writing, which vary in quality and specifics, but any of them will provide similar, useful approaches to self-improvement in writing.

Another crucial approach is to seek out good writers in one's own environment and talk to them about writing techniques. If one of them is willing to be a mentor for your writing, take advantage of this generosity. You may find that good writers love to talk about and teach writing. As a corollary to this, it is extremely valuable to have colleagues who will provide critical reviews of one's documents. When asking for such reviews, one should be specific in requesting frank criticism of not just the data, but the quality of the communication. (it is better to hear the bad news from a friend than an uninterested and potentially hostile reviewer!) A reviewer who gives a document back marked, "Looks good," has not helped the writer. In my personal experience, having multiple critical reviewers is best because one reader will see a problem others have missed.

Finally, to become a good writer, it helps to practice. I urge the beginning medical or scientific writer to practice writing well in every communication, be it a paper letter, email, or other electronic missive. Writing skill is analogous to muscle: use it to gain strength.

Let us turn to the first steps in preparation of a document; these ideas should have been conveyed during junior high and high school courses but, apparently, often are not.

PLANNED STRUCTURE AND LENGTH

It is absolutely necessary to do some planning before setting pen to paper or fingers to keys. First, who are you writing for? What is your message? What is the desired timing of delivering the information? What constraints are there on sharing this information (e.g., regulations or confidentiality agreements)?

Where is the information to be shared (e.g., before an audience or in a journal)? If it is to be published, what journal—print or online—is the best fit? Each of these factors will affect the structure and length of the document. It is essential to read and understand any guidelines and requirements of a granting agency, scientific organization, or journal to be sure your document will fit their criteria.

Starting to write without an outline, at least in your mind, leads to meandering, confusing prose, and lost messages. Remember: you will be telling a story, with a beginning, a middle, and an end. Thinking and planning the approach helps the writer achieve logical flow and internal consistency, as well as clarity. A written outline is a helpful starting point.

Tables 10.2 and 10.3 provide length guidelines' manuscripts for various purposes.

FINGERS TO KEYS

I have found it helpful to have all the supporting material at hand before putting down the first word. This might include reprints of articles to be

TABLE 10.2 Suggested Length Guidelines for Small- to Medium-Sized Clinical Trials or Moderate Research Reports

A report on a small- to medium-sized clinical trial or moderate experiment should usually not be longer in each section than:

- Introduction—1–2 printed pages
- Methods—2–5 printed pages, depending on complexity of the study; this is after all the crux of the report
- Results—1–5 printed pages, depending on complexity
- Discussion—1–4 printed pages
- Minimize the use of figures and tables; usually, 2–3 figures and 2–3 tables will do the trick
- Seldom use more than 25–30 reference citations, except in review articles.

TABLE 10.3 Tips for Reports From Large Teams

Reports on "Big Team" Science

- The 20-author study will always be long relative to smaller studies with fewer authors
- But word count and length still matter, especially to editors who must control journal page count!
- Length matters to readers who have limited time to examine new articles
- The same principles apply to all parts of all reports: brevity, clarity, flow, and correctness of grammar are essential

quoted and cited and charts, tables, and graphs of the data to be presented. A cup of fresh coffee is not a bad idea, either! Different writers do things differently, so you will need to find your own best approach. In my case, I set aside a block of time and write entire papers from start to finish, leaving blanks where I am unable to provide information or analyses, to be filled in later. Others will write the methods section first, then the data section, then the introduction, and so forth. For me, writing straight through helps assure that my paper tells a coherent story. Having an outline prepared will help keep the writer on course. Generally, it is advisable to write straight through a section rather than stopping and backtracking to correct typos or refine sentences. Get the whole section or the whole paper set down, then come back through to touch up the draft. When the skeleton of the document is complete, that is the time to work on filling in any missing information, meanwhile looking for "that" phrases, passive voice, awkward or overlong sentences, and any unintentional humor or offense. Make sure that "this thing" and "that thing" references are clear.

GET SOME DISTANCE

Once a clean draft paper has been completed, and blanks filled in, it is good practice to set the document aside. I recommend printing the document and literally putting the draft in a drawer or on a shelf—one way or another, put it out of sight and out of mind (as much as possible!). Depending on time sensitivity, it may not be possible to set the document aside for very long, but even a lunch break or overnight will help. Even better is to let a few days or a week go by before rereading the document. The beginning writer is often stunned to see mistakes, inconsistencies, typos, overlong sentences, and other structural and linguistic problems almost pop off the screen or page on such a delayed reading. This is also a good time to ask whether each section has been pruned down to the minimum length needed for communication. Once all the newly found issues are fixed, it is time to obtain independent, frank critiques from colleagues, local or distant.

WHAT TO DO WITH RETURNED CRITICAL COMMENTS?

Having several reviews in hand can be confusing because they might not all identify the same errors or problems, may interpret the information differently, or may misunderstand things. In any case, it is wise to read carefully and take each review seriously because you do not want readers of the finished product to react similarly. My practice has been to put the document up on screen, with the reviews at hand, and go through the document from top to bottom, making changes in response to all the reviews simultaneously in each section. Once that is done, you probably have a much-improved document that is ready for submission.

GENERAL THOUGHTS

Recall the 1900 complaint about "words, words, words," and strive to make communications concise. When helping colleagues edit their writing, sometimes I have been able to reduce the length of a document by two-thirds, without losing a single idea from the original. How could this miracle happen? Simple: the writer made each point three times in the original. Try to find such redundancies in your own work. Critical reviewers will often point out duplications between and within sections. Keeping documents short makes them easier to read and often much clearer and more persuasive. Some documents must be long, but nonetheless they should be only as long as needed to communicate the key points.

Because most professional documents are intended to sell something—a hypothesis, a planned project, experimental results, or oneself—and because the readers of one's documents may be under time pressure, the first sentence of an abstract or grant proposal or research paper should be a "grabber." That is, the first sentence should make the reader want to keep reading. Furthermore, the first paragraph should draw the reader into the second, and so on. Dull, poorly constructed, ambiguous, or incomprehensible first sentences or paragraphs can sink the goals of the document. Here is a made-up example of a "grabber" first sentence: "Most patients having lung cancer are dead within 5 years of diagnosis, regardless of treatment. Here, we present clinical trial findings with a new approach that has dramatically extended median disease-free survival of patients having small-cell carcinoma of the lung." Similarly attractive first sentences can be crafted even for less important material than in the example.

Do not forget the key points in writing a strong persuasive document:

1. Tell me a story.
2. Be brief.
3. Be clear.
4. Make me care.

In conclusion, individuals and groups in medicine and science cannot afford to tolerate bad writing. Good writing provides a competitive edge for the individual, the work group, and the organization. The well-written project proposal, grant application, abstract, or research report will stand out from the vast mediocrity around it and have a greater chance of success. Take heart: writing well is a learnable skill. Study writing as carefully as you would study a new research method. Almost certainly, there are good writers around you: find them and learn from them. Throughout your career, commit to helping each other to improve the quality of your writing.

For additional information on how to write specific types of manuscripts, please Chapter 9 (for journal articles) and Chapter 11 (for medical book chapters).

RESOURCES

Here I list some recent books on medical and scientific writing, plus some stalwarts from earlier times that have served me well in learning to write. Any of them would be valuable if read and referred to over years. My all-time favorites are items 4 and 5.

1. *The Complete Guide to Medical Writing*. Mark C. Stuart. Pharmaceutical Press, 2007 (491 pp). This book has an excellent chapter on writing that reflects and expands on many ideas presented in the present chapter. However, much of the book is about larger issues in preparing, submitting, and publishing various kinds of medical or scientific documents.
2. *Medical Writing: A Guide for Clinicians*, 2nd ed. Robert B. Taylor, Springer (New York), 2011. The author is a widely published academic family practitioner. I find the text very accessible, almost chatty, but he knows whereof he speaks. Even an experienced specialist writer would find useful information here.
3. *AMA Manual of Style: A Guide for Authors and Editors, 10th ed.* Multiple authors. This classic has been updated regularly over the years and is a valuable resource for writers seeking publication in peer-reviewed journals. It should be on every medical writer's bookshelf (or hard drive). Available at Amazon or through the American Medical Association.
4. *The Elements of Style, 4th ed.* William Strunk and E.B. White. Pearson Publishers, 1999 (105 pp). This slender book, updated for the current century, has been for many the first and best writing guide of their lives. Some dislike the later version—I had the original and found great value in it, but admittedly, it was dated in places—but there is so much good in its 105 pages that the current edition remains the best bargain ever in learning how to write clear, concise, communicative English. BUY IT!
5. Unfortunately, the most useful, practical book on medical writing I have is now out of print, but if you can track down a copy, I highly recommend it. It was published in a cheesy paperback form that fell apart quickly. Because the book was so valuable to me, I took its remains to a bookbinder for binding into hardback form. The book is *Scientific Writing*. Lester S. King and Charles G. Roland. American Medical Association, 1968 (133 pp). The chapters are short, the examples compelling, and the solutions straightforward.
6. Another out-of-print gem is *Why Not Say It Clearly?* Lester S. King. Little, Brown, and Co., Boston, 1978 (186 pp). Read in conjunction with citation 5, above, for a broader understanding of good and bad scientific writing. Track one down and treasure it.

REFERENCES

[1] Montgomery SL. In: The Chicago guide to communicating science. 2003.
[2] Editorial (anonymous). JAMA 1900;35:626.

Chapter 11

How to Write a Book Chapter: Skip the History, the Histrionics, and the Howevers

Michael W. Mulholland, MD
Department of Surgery, University of Michigan, Ann Arbor, MI, United States

As a practicing surgeon, frequent author, and editor of surgical textbooks, I assure you that the ability to write clearly is incredibly important. Many famous academic physicians are surprisingly challenged by the prospects of stringing sentences together, and most doctors do not count writing among their strengths. The ability to write lucidly is a huge competitive advantage. Fig. 11.1 shows the author with textbooks edited by University of Michigan Department of Surgery faculty members.

Over the years, I have established some rules for my own writing that you may find useful. If you are an accomplished author—articulate and stylish—if ideas and their expression flow effortlessly, you may stop reading now. Still, you might want to take a look…

START WRITING RIGHT AWAY

For most physicians, the most difficult sentence is the first sentence. The problem lies not in framing the topic; the trouble comes with the commitment that writing requires. Most chapters are assigned with a due date of 6, 9, or 12 months in the future. The author agrees, placated by the knowledge that this is a generous timeframe and he/she can easily finish on time. No further thought is given to the topic until a pleading note arrives from the editor after 9 months (or more) inquiring about the delinquent manuscript. Every author believes that the chapter will begin to write itself tomorrow. We all lie to ourselves in this way. There are no exceptions. My advice: Doctor, take your dose of castor oil. Write the first paragraph the first day.

KEEP WRITING, EVEN A LITTLE BIT, EVERY DAY

Writing is a skill sharpened with practice, and perfected with continuous practice. For me, operative surgery reinforces this dictum. The physical skills, sense

Medical and Scientific Publishing. https://doi.org/10.1016/B978-0-12-809969-8.00011-5

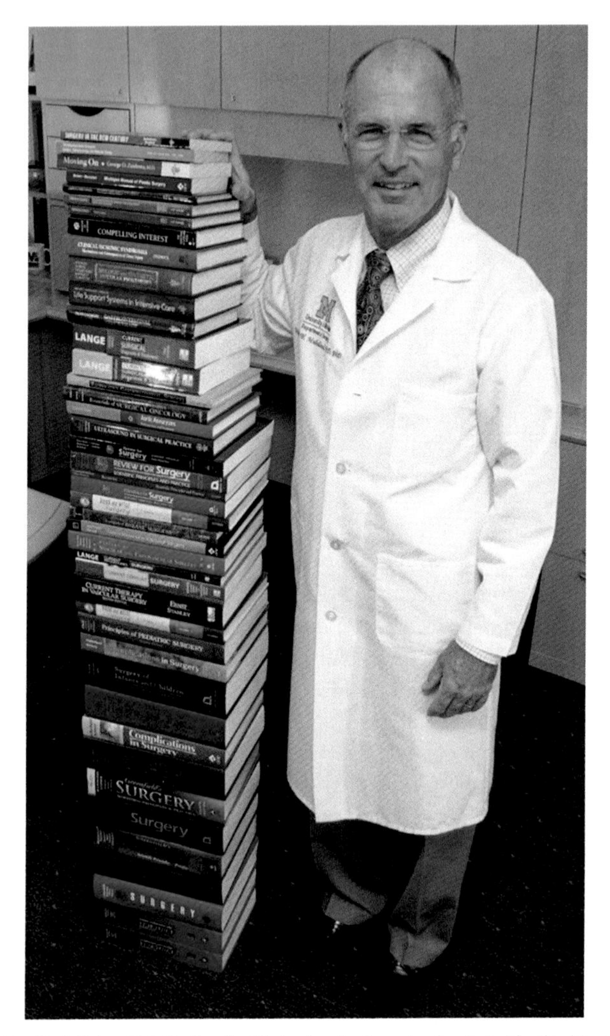

FIGURE 11.1 Dr. Michael Mulholland with textbooks from his bookshelf edited by University of Michigan Department of Surgery faculty members.

of prioritization, organization, personal confidence, and intuition of the accomplished surgeon result from attention to the craft. That is the reason it is called the practice of surgery. The same holds true for most physicians, regardless of their specialty.

A medical career is not a sprint, it is a marathon. The way to prepare for a marathon is to run every day. It is not helpful to think that you have to run a marathon in 8 months so you will start running the month before. That will not work. It is not okay to take a week off. Start writing, and keep writing. The pen (really the keyboard) becomes friendlier with use.

BAN THE WORD "HOWEVER"

The word "however" is used as an apology. An idea is expressed and then must be modified or constrained in some way. "However" is often inserted at this point to warn the reader that, "I've just told you something, and now I must change the idea slightly, so prepare yourself psychologically." Most doctors are used to being surprised, and some even like surprises.

Consider the following: *Embolization of bleeding vessels in the colon has been reported to be safe and effective in the majority of patients; however, there is a definite risk of ischemic complications.*

Why not: *While embolization of bleeding vessels in the colon has been reported to be safe and effective in the majority of patients, there is a definite risk of ischemic complications.*

Or this: *Peritoneal adhesions account for more than half of small bowel obstruction cases. Lower abdominal procedures such as appendectomy, hysterectomy, colectomy, and abdominoperineal resection are common precursor operations to adhesive obstruction. Adhesions form after any abdominal procedure, however, including cholecystectomy, gastrectomy, and abdominal vascular procedures.*

Try this: *Peritoneal adhesions account for more than half of small bowel obstruction cases. Lower abdominal procedures such as appendectomy, hysterectomy, colectomy, and abdominoperineal resection are common to precursor operations to adhesive obstruction, but adhesions may form after any abdominal procedure, including cholecystectomy, gastrectomy, and abdominal vascular procedures.*

NEVER MAINTAIN A HIGH INDEX OF SUSPICION

Every profession has its clichés. Football coaches play one game at a time. Lawyers declare that everyone deserves their day in court. Funeral directors say that the deceased looks so peaceful. And doctors maintain a high index of suspicion. Physician authors who write this intend to say that biological variation exists, clinical signs may be subtle, and that surgical practice involves fallibility. For most of the reading public, maintenance of a high index of suspicion is an indication for psychiatric therapy.

SKIP THE HISTORY

Modern medicine is forward looking, seeking to improve the care of current patients and to prevent disease in future patients. Given the pace of modern biomedical research, no one individual can be expected to find, read, synthesize, and apply all new knowledge relevant to any clinical problem. In the midst of this information overload, the recognition by Reginald Fitz, in 1886, of appendicitis as a surgical disease is difficult to retain [1]. In addition, young Americans, including young doctors, are not interested in history. If that assertion seems harsh, consider this. Polls have found that many Americans consider the more recent presidents "the greatest"—who tops the list, of course, will depend on the poll and the demographic but in nearly all surveys, earlier presidents fall lower on the list than those of recent years. Skip the history lesson.

DO NOT READ WILLIAM FAULKNER

Stephen King is an amazingly prolific and successful American author with book credits such as "Bag of Bones," "Carrie," and "The Green Mile." King has written that "If you want to be a writer, you must do two things above all others: read a lot and write a lot." If you want to be a prolific and successful American academic physician, you must also read a lot. But be careful not to be seduced. A medical chapter is not literature.

In William Faulkner's novel, "Go Down Moses," one of the main characters is Ike McCaslin [2]. The author introduces Ike in the book's first three sentences. For purposes of illustration, each of the three sentences is offered here separately.

Sentence 1: Isaac McCaslin, 'Uncle Ike,' past seventy and nearer eighty than he ever corroborated any more, a widower now and uncle to half a county and father to no one this was not something participated in or even seen by himself, but by his elder cousin, McCaslin Edmonds, grandson of Isaac's father's sister and so descended by the distaff, yet notwithstanding the inheritor, and in his time the bequestor, of that which some had thought then and some still thought should have been Isaac's, since his was the name in which the title to the land had first been granted from the Indian patent and which some of the descendants of his father's slaves still bore in the land.

Sentence 2: But Isaac was not one of these: a widower these twenty years, who in all his life had owned but one object more than he could wear and carry in his pockets and his hands at one time, and this was the narrow iron cot and the stained lean mattress which he used camping in the woods for deer and bear or for fishing or simply because he loved the woods; who owned no property and never desired to since the earth was no man's but all men's, as light and air and weather were; who lived in the cheap frame bungalow in Jefferson which his wife's father gave them on their marriage and which his wife had willed to him at her death and which he had pretended to accept, acquiesce to, to humor her, ease her going but which was not his, will or not, chancery dying wishes mortmain possession or whatever, himself merely holding it for his wife's sister and her children who had had lived in it with him since his wife's death, holding himself welcome to live in one room of it as he had during his wife's time or she during her time or the sister-in-law and her children during the rest of his and after not something he had participated in or even remembered except from the hearing, the listening, come to him through and from his cousin McCaslin both in 1850 and sixteen years his senior and hence, his own father being near seventy when Isaac, an only child was born, rather his brother than cousin and rather his father than either, out of the old time, the old days when he and Uncle Buck ran back to the house from discovering Tomey's Turl had run again, they heard Uncle Buddy cursing and bellowing in the kitchen, then the fox and the dogs came out of the kitchen and crossed the hall into the dogs' room and they heard them run though into the kitchen again and this time it sounded like the whole kitchen chimney had come down and Uncle Buddy bellowing like a steamboat blowing and this time the fox and the dogs and the five or six sticks of firewood all came out of the kitchen with Uncle Buddy of them hitting at everything in sight with another stick.

Sentence 3: It was a good race.

William Faulkner won the Nobel Prize for literature. Since I will not, I have got to admire that third sentence. In preparing a chapter for a medical book, be brief and to the point.

IT IS OKAY TO READ ERNEST HEMINGWAY

Nick sat against the wall of the church where they had dragged him to be clear of the machine gun fire in the street. Both legs stuck out awkwardly. He had been hit in the spine. His face was sweaty and dirty. The sun shone on his face. The day was very hot. Rinaldi, big-backed, his equipment sprawling, lay face downward against the wall. Nick looked straight ahead brilliantly. The pink wall of the house opposite had fallen out from the roof, and an iron bedstead hung twisted toward the street. Two Austrian dead lay in the rubble in the shade of the house. Up the street were the other dead. Things were getting forward in the town. It was going well. Stretcher bearers would be along any time now. Nick turned his head and looked down at Rinaldi. "Senta, Rinaldi, senta. You and me, we've made a separate peace." Rinaldi lay still in the sun, breathing with difficulty. "We're not patriots." Nick turned his head away, smiling sweatily. Rinaldi was a disappointing audience [3].

Hemingway's prose was lean and compelling. Like me, you will not win the Nobel Prize for literature, but Hemingway's style is a good one to emulate. Grab your audience and do not let go.

PUT YOURSELF IN THE OTHER PERSON'S SHOES

Many clinicians and nearly all researchers apply for grants during their careers. I have been in a position of evaluating or editing grant applications, and I often find myself struggling to understand what the investigator proposes. "What are they saying? What is the hypothesis? What is the applicant trying to do?" Reading turgid scientific prose is like making your way through a jungle with a machete. And then comes the application that is crystal clear, and I say to myself, "I could have thought of that. I could do that." The writer gets bonus points when the application is written with such clarity that it is easy to understand exactly what the investigator wants to accomplish.

Another important aspect of academic writing is understanding the composition, structure, and purpose of a chapter versus a manuscript. Manuscripts are narrowly focused. Manuscripts provide methods and discrete data. These data become the building blocks of general ideas. Chapters are broadly focused. A chapter is meant to be an expert compilation that provides a balanced view of a field of study or the solution to a general problem. Manuscripts typically offer new research or contribute new ideas to an existing body of research. Book chapters make a contribution to the education of students and colleagues. To write a book chapter you first need to become expert on a given subject—think deeply about it, study it, and provide your readers with a balanced view. There is real value in having the aforementioned expertise, and close-to-equal value in perfecting your ability to write effectively.

STAY IN TOUCH

If you are invited to develop and edit a book with multiple chapters, part of your job as editor will be to encourage your chapter authors to write clearly and effectively. And to deliver their work on time. As an editor of a book, stay in touch with your authors. Do not simply invite them and then assume they will be vigilant in their writing and deliver their manuscripts on schedule. Authors need attention and updates so that they know their timely contributions are important.

KEEP THE REFERENCE LIST SHORT [4]

REFERENCES

[1] Fitz RH. Perforating inflammation of the vermiform appendix with special reference to its early diagnosis and treatment. Am J Med Sci 1886;92:321–46.
[2] Faulkner W. Go down moses. Random House (US); 1942.
[3] Hemingway E. In our time. New York: Boni & Liveright; 1925.
[4] Leonard E. Easy on the adverbs, exclamation points and especially hooptedoodle. Writers on writing, volume II. In: Smiley J, editor. New York (NY): Times Books; 2003. p. 143–6.

Chapter 12

A Case for Case Studies

Robert M. Cermak, MM[1], Hanna Saltzman[1], David Fessell, MD[2]
[1]*Medical Student Leadership Program, University of Michigan, Ann Arbor, MI, United States;*
[2]*Department of Radiology, Medical Student Leadership Program, University of Michigan, Ann Arbor, MI, United States*

CASE STUDIES IN MEDICINE

Introduction

Case studies, also commonly referred to as case reports, have a long tradition of use in both medical education and medical scholarship. A case study can broadly be defined as an "in-depth study undertaken of one particular 'case', which could be a site, individual or policy" [1]. The first known case study reported the diagnosis and treatment of jaw dislocation and was written circa 1600 BC on an Egyptian papyrus. Since that first report, physicians and scholars including Hippocrates, Galen, Osler, and Freud have used case studies to communicate their findings [2]."

In medical education, "case-based teaching strategies—through which instruction and learning occur through discourse around specific, contextualized cases"—have customarily utilized *clinical* case reports, dealing primarily with diseases, symptoms, and treatments [3]. These contextualized studies have long contributed to the recognition of new maladies, diagnoses, and treatments, and to the ongoing discussion of complex symptoms and challenging patient cases in the training of generations of medical practitioners [3].

Likewise, in medical research and scholarship, "case study research—a qualitative research strategy that investigators within health professions education may apply—represents an effective methodology for examining a phenomenon within its real-life context [3]." Case studies common to other fields of study, such as business, law, administration, leadership, policy, and ethics, are also useful tools in both discussions of the changing health-care landscape and in the training of well-rounded physicians. Such case studies will hereafter be referred to as *nonclinical* case studies.

Medical and Scientific Publishing. https://doi.org/10.1016/B978-0-12-809969-8.00012-7

Criticisms and Limitations

Despite the long-standing convention of using case reports for medical teaching and research, "medical case reports have fallen out of favor in the era of the impact factor [2]." Indeed, anyone working in medical education has likely heard someone challenge the utility of case studies, asserting "Don't make too much of that article. It's not a real study, just a case study [4]." Such denigrations, now all too common, are perhaps to be expected, given the emphasis placed on statistical power. This shift is often viewed as incompatible with the use of case studies, whose main differentiator is a conscious focus on singular real-life contexts where "the number of variables of interest far outstrips the number of data points [5]."

Additional criticisms include a perceived lack of representativeness and rigor, largely stemming from their small sample size [3]. This limited focus has led to the interpretation in some quarters that clinical case reports engage in overinterpretation and an overt emphasis on rare diseases and cases [2].

Defense and Uses

In medical scholarship, case study research, when pursued with rigorous design, data collection, analysis, and quality control procedures, can enable insights into innovations and interventions and their impact on patient care. In unique or emerging contexts, where sufficient data for a larger study may not exist, and in situations where the interpretive power of the report is highly contextualized, case studies remain a distinctively useful means by which new knowledge can be disseminated. Case study research can thus yield unique information that would not be attainable using other methods [3].

Throughout their long history, case studies have contributed to the recognition of new diseases and medical breakthroughs. The report by Gottlieb et al. [6] on Kaposi's sarcoma in young homosexual men proved pivotal in the recognition of acquired immune deficiency syndrome, or AIDS. More than a century and a half earlier in 1817, James Parkinson reported on a case of "shaking palsy" that first documented the disease that came to bear his name [7]. Likewise, case studies first noted a link between certain anorexic agents (fenfluramine and dexfenfluramine) and primary pulmonary hypertension, stimulating further studies that led to their removal from the market [8,9]. The American Medical Association chose 51 papers from Journal of the American Medical Association that had "significantly changed the science and practice of medicine during the 150 years of the organization's existence. Five of these papers were case reports" [10]. These examples illustrate that the findings and principles laid out in case studies can, in some instances, have a powerful and lasting impact. In such instances case studies play a unique role, laying the groundwork for larger investigations and clarifying which questions to ask.

Similarly, in medical education, Kanter notes that "case studies that are analytic and penetrating, that illuminate fundamental precepts and concepts, and that reveal new avenues for research or theory development have the potential to broaden and deepen knowledge and understanding [among medical students] in a way that might not be available otherwise [4]." This utility runs the gamut from clinical case studies to studies about medical schools and teaching hospitals and the full range of nonclinical case studies [4]. Additionally, research has shown that case-based teaching can effectively enable students to fulfill learning objectives across various disciplines. A study by Bonney of New York University working with undergraduate biology students suggests that case studies, compared to class discussions and textbook readings, are more effective at increasing performance on examination questions and promoted oral and written communication skills [11].

The question then becomes not whether case studies have their limitations. Clearly, they do. Does this eliminate their usefulness in all instances? No, it does not. Rather, the question is, in which situations can case studies abet research or teaching? For those interested in authoring medical case studies and engaging in case study research in academic medicine, key considerations should include the number of data points present, the feasibility of research methods available to researchers, and the goals of the researchers. Crowe et al. advise that when selecting the most appropriate study design, researchers should consider, "whether it is desirable or indeed possible to undertake a formal experimental investigation in which individuals and/or organizations are allocated to an intervention or control arm" [1]. If so, a controlled experimental design is likely most appropriate for the study. If, on the other hand, the researchers are interested in obtaining a more "*naturalistic* understanding of an issue," in its observed context or in documenting qualitative or narrative information, then a case study design is more appropriate [1].

Not only is case study research a useful, and at times the most appropriate, form of medical scholarship, but the very act of conducting such research and authoring case reports can be of immense value to both medical students and residents. Packer et al. ran a "case report program" for third-year clerkship students at a Veterans Affairs hospital. They noted several educational benefits for students which included "improving writing and critical thinking skills, 'collaborative writing' with faculty mentors, gaining experience with the peer review process, and developing the skills necessary to produce scholarly publications [2]." Additionally, they identified "Five Observed Educational Benefits of Case Reports:

- Observation and pattern recognition skills
- Hypothesis-generating skills
- Understanding of patient-centered care: The case report as a 'hybrid narrative'
- Writing skills and rhetorical versatility
- The case report as a 'mini-thesis' [2]"

The "subject" of a case study could be a fellow student, a resident, faculty member, department, division, school, or hospital [4]. The numerous demonstrable benefits of using and writing case studies, combined with the wide range of applicable subjects for inquiry, render case studies a valuable avenue by which medical students and practitioners can grow their skills and enlarge their scholarly experiences.

LEADERSHIP DEVELOPMENT IN MEDICAL EDUCATION

Across today's rapidly evolving health-care landscape, physicians are expected to provide leadership at every level and stage of patient care, including to themselves, to the members of their multidisciplinary teams, to patients, and to patients' families. Changes in modern health care's delivery, scope, and workforce present a persuasive case for an increased emphasis on leadership development in medical education. Such changes are increasingly transforming the skills required of physicians. The tradition of physician autonomy is gradually being superseded by multidisciplinary teams working amid complex social change and technological innovation [12]. The scope and pace of revisions to the physician's role, and truly the physician's *identity*, in patient care demand an increased focus on and reorientation toward leadership at all levels of health care, from the self on up to the broader health-care system.

Unfortunately, however, despite the evident need, most medical schools continue to offer little in the way of explicit leadership training. Medical education has traditionally focused on developing a clinical acumen, with little time dedicated to addressing systemic issues or team collaboration in modern health care [12]. Other disciplines, predominantly in the fields of business and law, already have a proven track record of dedicating resources and instructional time to leadership development for their students. This is often realized through specialized institutes and centers, such as the Sanger Leadership Center at the University of Michigan's Stephen M. Ross School of Business, and specialized leadership courses and programs, such as Harvard Law School's executive education program on Leadership in Law Firms [13,14]. This commitment to leadership development extends to the scholarly output at such institutions, including case studies. For example, the Harvard Business Review, Harvard Business School's preeminent management publication, features numerous case studies including those on leadership in health-care contexts, such as the case of Dr. Laura Esserman, a surgeon who leveraged her leadership style and professional network to drive change within her medical organization [15].

Knowles et al., of the University of Queensland, observe that, "professional medical bodies and specialty colleges, nationally and internationally, increasingly emphasize leadership experience at all levels within the profession as part of their criteria for career progression. In this context, leadership, teamwork, and communication skills are assuming a progressively recognized status as essential competencies for graduating medical students to demonstrate throughout their clinical training [12]." Since 2015, Michigan Medicine has included the Medical Student Leadership Program in our core MD curriculum. This longitudinal program, with touch points across all 4 years of medical school, emphasizes five "critical competencies" for physicians to positively influence change across health care.

1. Leading yourself
2. Communicating and influencing others

FIGURE 12.1 Team approach to medical education and health care (iStock.com/AndreyPopov).

3. Building teams
4. Executing and problem-solving
5. Impacting systems [16]

The photo in Figure 12.1 illustrates the team approach to medical education and health care.

In each year of medical school, students at Michigan Medicine participate in a series of prescribed activities designed to help them develop these competencies. Frequently employed activities include faculty mentorship, hands-on sessions, guest speaker events, and case-based learning. During the fall term of the 2016–17 academic year, the Case Study Session was the Medical Student Leadership Program's highest-rated session among first-year medical students. More than three-quarters of student respondents either agreed or strongly agreed that the Case Study Session had contributed to their learning.

The positive response rate among our students, combined with both the growing emphasis on leadership development in medical education and the aforementioned benefits of utilizing and authoring case studies, supports the utility of nonclinical case studies and case-based teaching. With that in mind we at Michigan Medicine seek to formalize a process through which our students can play an active role in the writing and publishing of leadership-themed case studies in medicine.

LEADERSHIP CASE STUDIES AT MICHIGAN MEDICINE

Our use of medical leadership–themed case studies as part of our educational mission at the University of Michigan Medical School has been limited by a shortage of relevant studies in the published literature. Such leadership case studies written by and for students would, we feel, resonate with our students and potentially offer lasting insights and wisdom they could draw from during their educational journey and beyond.

To grow the range of case studies for our own use in teaching, as well as to provide our students with opportunities to take part in scholarly research, we are developing a process through which medical students at Michigan Medicine can write and publish case studies dealing with leadership in both medical education and medicine more broadly. Such case studies can be a valuable way to introduce medical students to publishing, sharing their perspectives, and collaborative scholarship. This process would also represent "leadership in action" for our students, directly supporting the Medical Student Leadership Program's core mission.

We anticipate that our case study program will largely be *student-driven*. Medical students, from across all 4 years of study, will be invited to take the lead in authoring, individually or in collaboration with faculty mentors and/or other students, and navigating the publication process. In this way, our students would not only augment their visibility in the academic medical literature but also engage in meaningful practice of core leadership competencies, from effective written communication and time management to collaboration and team building. This process will also provide our instructors with a larger body of nonclinical case studies to draw on when conducting case-based teaching as part of the Medical Student Leadership Program.

We have a network of advisors and mentors at the ready, drawn from the faculty body at Michigan Medicine, including faculty teaching in the Medical Student Leadership Program. Such mentors can assist students in choosing a suitable case, outlining the scope, editing, and navigating the publication process. In addition to traditional journals which publish case studies or perspective articles, such as Academic Medicine and BMJ Case Reports, we are considering pursuing a partnership with the Michigan Journal of Medicine, a completely student-run publication.

CONCLUSION

These are exciting times in medical education. The rise of leadership development as part of medical school curricula makes the development of innovative practices in case-based leadership scholarship both timely and appropriate. Our aim is to share our experience and aspirations in the hope that others may find them valuable and instructive. For those medical practitioners and institutions interested in producing case studies, or in encouraging their students and residents to do so, we have included both a sample case study (below) and a case study template in Figure 12.2. Case studies represent important information for the medical profession and publication of these reports is a very effective way to disseminate that knowledge.

Template

Title Page

Title:

Running Header:

Authors:
a) Name, academic degrees and affiliation
b) Name, academic degrees and affiliation) ...

Name, address and telephone number of corresponding author

Disclaimers (if pertinent)

Statement that patient/subject consent was obtained (if pertinent)

Sources of financial support (if any)

Key words: (limit of five to ten)

Abstract: (maximum of 150 words)

Body

Introduction: Provide context for the case and note any related cases in the literature.

Case Presentation:
a) Introductory sentences
b) Describe the essential nature the case
c) Further development of context of the case
d) Describe challenges and compounding factors relevant to the case
e) Summarize the results of the case

Discussion:
a) Synthesize foregoing sections
b) Summarize the case and any lessons learned

Appendixes

References: (limit of 10 to 15 unless explicitly justified)

Legends: (brief descriptions accompanying tables, figures or images)

Permissions: (letter of permission from publisher for any borrowed tables, figures, or images)

FIGURE 12.2 Case study template. *(Adapted from Budgell B. Guidelines to the writing of case studies. J Can Chiropr Assoc 2008;52(4):199–204. https://www.ncbi.nlm.nih.gov/pmc/articles/PMC2597880/,[17].)*

LEADERSHIP CASE STUDY

It was the fourth breast surgery of the day, and third-year medical student Katie Mehari stood by the operating table, sleepy from staying up late the night before to prepare for the day's cases. It was the beginning of her first surgery rotation, which is notorious for its long hours and demanding requirements. Katie avidly heeded the advice of her peers who had already made it through the surgery gauntlet: do not speak out of turn; do not contaminate anything in the operating room; do prepare endlessly for your cases; do know both the anatomy and your patients inside and out.

Knowing every detail about your cases, however, requires working late into the night and then waking up before dawn, often just a couple hours later, to head back to the hospital. Thus while Katie stood by the table that afternoon observing her fourth safety check of the day, she was slightly zoned out. Sterilization? Check. Anesthesia? Check. Right-sided breast? Check. The surgeon guided the resident in tracing a pen around the patient's breast to make a blueprint for the incision.

Glancing over at the marking, Katie suddenly jolted out of her daze. Something was not right. The patient was scheduled for a lumpectomy, which involves removing just a modest amount of tissue—a procedure done for milder forms of breast cancer that are easily contained within a small radius. But the pen had looped around the breast in a wide swoop, much larger than the markings Katie had seen for the lumpectomies earlier in the day. This looked more like the incision for a mastectomy: the complete removal of the breast, as is required to battle more invasive cancers. Katie's heart started to race. It seemed impossible, but was there any chance that the surgeon was accidentally about to do a mastectomy when the patient only needed a lumpectomy?

She scanned the room anxiously. Nobody else—the surgeon, the surgery resident, the anesthesiologist, the nurse, the surgical technician—was batting an eye. All these people knew far more than Katie about the proceedings that should happen in an operating room, living it day in and day out. These were experts, and Katie herself only was beginning to know the bare essentials. But she knew two things for sure: her patient was scheduled for a lumpectomy, and a lumpectomy did not involve the entire breast.

She wracked her brain, trying to conjure up as many details as possible about this patient's surgery, the images she had looked at in a midnight haze last night while prepping after finishing up with yesterday's cases. She also tried to remember the cases from earlier today—what had those markings looked like again? Was she sure these current markings did not look like a lumpectomy should? There must be multiple ways to do lumpectomies, right?

Or what if she had mixed up her patients. The day's cases had started to blur together. Could she be mistaking this patient with the next one—maybe this woman was the one scheduled for a mastectomy, and the next patient, case 5, was the one due for a lumpectomy? But no, she definitely remembered talking to this patient about a lumpectomy. Her stomach started to churn.

She tried to catch the resident's eye, but the resident was busy making the markings on the breast and did not look up. She wanted to interrupt the physician and double check the surgery, but felt paralyzed. As a medical student, the bottom of the totem pole in a room full of experts, it would be mortifying to speak up and be wrong. The physician would think she had not prepared. It could reflect negatively on her. Perhaps she would even get a reputation as the unprepared one, unfit to be a clinical medical student. Perhaps her grade for this rotation would suffer. Plus, she had been warned by the older students not to speak out of turn. And with nobody else in the room questioning the markings, everything was probably fine.

"Scalpel, please." The surgical technician handed the resident the scalpel. But what if I am right? Katie panicked. What if I am right and this is the wrong surgery, and this patient loses her breast? She felt like there was an angel on one shoulder and a devil on the other: "Say it!" "No, don't say it!" "Yes, say it!" The resident started the incision.

Watching the scalpel cut into the skin, Katie could not deliberate any longer. "W—wait!" She blurted out. She then phrased her concern as a question because then if she were wrong perhaps, it would come across as pure naivety rather than an accusation. "I thought this was a lumpectomy; are you doing a different type of incision for it?" She consciously phrased her concern as a question, hoping that if she were wrong it would come across as pure naivety rather than as an accusation.

The surgeon looked at her. "Stop," he said to the resident. "Stop everything right now." He asked the nurses, who seemed surprised at the delay, to check the consent form. The form confirmed that the surgery should be a lumpectomy. But the incision had, indeed, been for a mastectomy.

The surgeon redrew the markings and started over, leaving no harm done apart from a small incorrect incision at the start. After the surgery, the surgeon spoke to the patient and explained the error behind the extra scar. Yes, it was an error with consequences, and yet those consequences were nothing compared to what would have happened if Katie had not spoken up and the woman's breast had been removed entirely.

One might expect that on realizing that she, a lowly medical student, had saved this woman's breast, Katie would feel a surge of relief, pride, and elation. But she felt none of those positive emotions. Instead, she felt guilt: guilt for not speaking up sooner, guilt for the fact that this woman received an unnecessary incision when it could have been avoided. Since the moment she saw the initial markings for the incision, Katie knew in her head and her gut that it was wrong surgery, and yet it took the actual scalpel cutting skin before she gathered the strength to voice that concern. How could it possibly have crossed her mind, she wondered, that the potential embarrassment over being wrong caused her to pause, risking harm to a patient? Letting her self-doubts prevent her from speaking up now seemed so selfish. Despite ultimately saving this woman from harm, Katie did not feel like a leader. She felt like a coward.

It has been 7 years since then, and Katie the medical student is now Dr. Mehari, the attending physician in obstetrics and gynecology. She is now the one at the head of the operating room table, teaching residents and medical students how to operate. When it comes to responsibility over medical errors, the buck now stops with her.

When she first reflected on her experience through the lens of leadership in medical education, her reaction was that her story did not exemplify leadership in the slightest, given the guilt she had felt for not speaking up sooner. But as she thought about it more, lessons in leadership started to emerge. Leadership is often thought of in a "capital L" framework, defining a leader as someone directing a group, managing a project, and being in charge. But leadership can also be thought of from a "lowercase l" lens, defining leadership in terms influence—of affecting others' lives through everyday actions.

From an "everyday leadership" perspective, perhaps opportunities for leadership emerge precisely in the moments when you feel your farthest from being a leader. Perhaps, Katie has reflected, leadership opportunities also come in those moments when you feel tiny, disempowered, like a scared child in the corner of the room. It's those moments that give you the opportunity to face your fear and rise above it, to consciously decide that you are going to do what is best for your patients, regardless of how scary it feels. The photo in Figure 12.3 illustrates physician discussions as one of the important ways to optimize patient care.

Now, as a physician who still viscerally remembers what that fear of being wrong felt like, Dr. Mehari urges medical students to speak up. Her advice to students: You may be the least experienced person in the room, but you still know a lot, and sometimes you will see things others have missed. You may have less procedural knowledge than those around you, but you'll have more time to spend with your patients and thus more information about them. If something doesn't seem right, ask questions to figure out what's going on. You are not just a learner; you are part of the care team, and you need to take responsibility for that. Taking responsibility means you need to speak up, even when it's hard (and being at the bottom of the medical hierarchy, it certainly will feel hard). Sometimes when you speak up you'll be wrong, and sometimes people

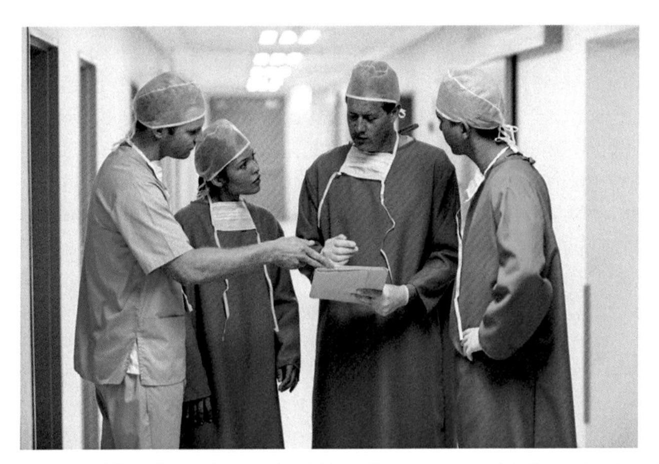

FIGURE 12.3 Physician discussions and collaborations are essential for optimal medical care (iStock.com/Wavebreakmedia).

more powerful than you will judge you for being wrong. That judgment may make it even harder to speak up the next time. This is how the medical world works. And yet: you still must speak up. For sometimes you will be right, and your being right will rescue your patients from harm.

REFERENCES

[1] Crowe S, Cresswell K, Robertson A, Huby G, Avery A, Sheikh A. The case study approach. BMC Med Res Methodol 2011;11.

[2] Packer CD, Katz RB, Lacopetti CL, Krimmel JD, Singh MK. A case suspended in time: the educational value of case reports. Acad Med 2017;92:152–6.

[3] Bunton SA, Sandberg SF. Case study research in health professions education. Acad Med 2016;91(12).

[4] Kanter SL. Case studies in academic medicine. Acad Med 2010;85(4):567.

[5] Yin RK. Enhancing the quality of case studies in health services research. Health Serv Res 1999;34(5 Pt 2):1209.

[6] Gottlieb GJ, Ragaz A, Vogel JV, Friedman-Kien A, Rywlin AM, Weiner EA, Ackerman AB. A preliminary communication on extensively disseminated Kaposi's sarcoma in young homosexual men. Am J Dermatopathol 1981. U.S. National Library of Medicine https://www.ncbi.nlm.nih.gov/pubmed/7270808.

[7] Goetz CG. The history of Parkinson's disease: early clinical descriptions and neurological therapies. Cold Spring Harbor Perspect Med September 2011. Cold Spring Harbor Laboratory Press https://www.ncbi.nlm.nih.gov/pmc/articles/PMC3234454/.

[8] Douglas JG, Munro JF, Kitchin AH, Muir AL, Proudfoot AT. Pulmonary hypertension and fenfluramine. Br Med J Clin Res Ed 3 October, 1981. U.S. National Library of Medicine https://www.ncbi.nlm.nih.gov/pmc/articles/PMC1507127/.

[9] Atanassoff PG, Weiss BM, Schmid ER, Tornic M. Pulmonary hypertension and dexfenfluramine. Lancet (Lond Engl) 15 February, 1992. U.S. National Library of Medicine https://www.ncbi.nlm.nih.gov/pubmed/1346703.

[10] Wáng Y-XJ. Advance modern medicine with clinical case reports. Quant Imaging Med Surg December 2014. AME Publishing Company https://www.ncbi.nlm.nih.gov/pmc/articles/PMC4256250/#r10.

[11] Bonney KM. Case study teaching method improves student performance and perceptions of learning gains. J Microbiol Biol Educ 2015;16(1):21–8.

[12] Knowles L, O'Dowd C, Hewett DG, Schafer J, Wilkinson D. The university of queensland medical leadership program: a case study. Ochsner J 2012;12:344–7.

[13] Sanger leadership center. Michigan Ross. University of Michigan. https://michiganross.umich.edu/sanger.

[14] Leadership in law firms. HLS Executive Education. Harvard Law School; 2017. https://execed.law.harvard.edu/llf/.

[15] Pfeffer J, Chang V. Dr. Laura Esserman (A). Harvard Business Review; 2003. https://hbr.org/product/dr-laura-esserman-a/OB42A-PDF-ENG.

[16] Leadership program. University of Michigan Medical School. Michigan Medicine; 25 January, 2017. https://medicine.umich.edu/medschool/education/md-program/curriculum/longitudinal-learning/leadership-program.

[17] Budgell B. Guidelines to the writing of case studies. J Can Chiropr Assoc 2008;52(4):199–204. https://www.ncbi.nlm.nih.gov/pmc/articles/PMC2597880/.

Chapter 13

Threats to Experimental Hygiene

Sagar Deshpande[1,2], Laura Ostapenko, MD[2,3], Michael Englesbe, MD[1,4]
[1]University of Michigan Medical School, Ann Arbor, MI, United States; [2]John F. Kennedy School of Government, Harvard University, Cambridge, MA, United States; [3]Department of Surgery, Brigham & Women's Hospital, Boston, MA, United States; [4]Section of Transplantation Surgery, Department of Surgery, University of Michigan, Ann Arbor, MI, United States

EXPERIMENTAL HYGIENE

Research is about relationships. Research studies strive to isolate the relationship between "X" and "Y," holding all other things constant. This is easier said than done. Experiments achieve this by assigning "X" to a treatment group, while assigning the absence of "X" to a control group. Observational studies look at a group with "X" compared to a group without "X." Both experiments and observational studies measure "Y." If "X" is the only thing that differs between the two groups, then "X" is responsible for any difference in "Y" that is observed.

In experiments, researchers create groups so that "X" is the only thing that varies between the two groups. In observational studies, study authors try to find groups in which "X" is the only thing that differs between the groups. By creating groups, researchers can control the characteristics of those groups (Fig. 13.1A and B). But when observing groups, researchers must take the characteristics of the groups as given. It is rare that groups differ only in our variable of interest, "X." If endogenously occurring groups differ only in our variable of interest without any additional intervention, that observational study is a **natural experiment**.

Threats to the validity of an experiment or an observational study occur whenever our two groups—those groups with "X" and those groups without "X"—differ in any way other than in "X." Despite the best intentions of researchers, systemic flaws in studies create threats to validity, causing researchers to draw erroneous conclusions from their data. Internal validity is the extent to which researchers can draw meaningful conclusions from their data. Fig. 13.2 depicts discussion among physicians to share knowledge. External validity is the extent to which these conclusions can be generalized to other groups or situations.

Threats to validity can occur during study design, operationalization, or during subject participation. This chapter identifies and discusses these threats and presents two case studies that illustrate problems. By examining and

Medical and Scientific Publishing. https://doi.org/10.1016/B978-0-12-809969-8.00013-9

FIGURE 13.1 (A and B). Photos depicting sample forms (iStock.com/teekid) and medical records (iStock.com/fotofrog) that may be useful in setting up groups for clinical trials.

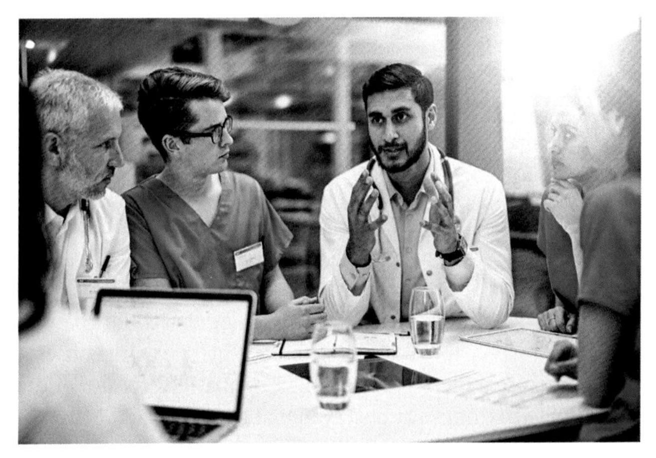

FIGURE 13.2 Photo depicting physicians sharing knowledge, for better outcomes (iStock.com/Cecilie_Arcurs).

understanding these threats, journal editors can protect their publications from inaccurate conclusions as well as scientific misconduct.

THREAT #1: UNBALANCED GROUPS (SETTING UP A STUDY)

Participant randomization is the gold standard for controlling bias within an experimental study. By giving all members of a participant pool the same opportunity to be assigned to a control or treatment group, you can control for the numerous observable and unobservable factors that influence the response to the treatment. This allows the researchers to isolate the impact of a treatment variable, and any difference in outcome metrics can be attributed to that treatment.

Most scientists and study designers are aware of the need for randomization and perform it as a matter of course. However, the intended randomization is not always executed. The risk of nonrandomization increases with experimental conditions that are emotionally difficult. Experiments that address as yet uncured disease or perceived deprivation may induce staff to violate randomization. In some cases, patients may not be properly randomized due to bias, discrimination, or even emotional sympathy from a staff member. Less charitably, scientific fraud is a rare but serious event. It is essential that every clinical study demonstrate the success of their randomization through an *unbalanced group analysis*.

Unbalanced group analysis is a technique to prove the success of randomization. In this technique, a number of variables are identified that the scientist thinks may influence the outcome metrics. For example, in a clinical study examining the effects of a new antihypertensive drug on long-term blood pressure control, good metrics should include age, weight, race, percent with diabetes, and other such factors that may contribute to overall hypertensive control. By ensuring there is no difference between the groups in these metrics, the scientist and editor can verify that across factors that are known to contribute to hypertension, the treatment and control groups are the same.

However, not all unbalanced group analysis metrics should relate to the outcome of interest. Randomized groups should be equal across all features, and so some unbalanced group metrics should be uncorrelated to the treatment or outcome of the experiment. This helps prove that the groups are not designed by the study authors to have a desired conclusion. For example, rate of automobile ownership would be a good metric, as it captures information about economic status but should be uncorrelated to the treatment and study outcomes.

It should be noted that as the number of unbalanced group metrics increases, so, too, does the probability that one is spuriously significant. For example, in a clinical study where 20 demographic variables are compared between a control and treatment group, using a significance threshold of 0.05, the chance that one will be positive by chance is 0.377.[1] Therefore, if there is a significant difference in one of many unbalanced group metrics, this does not indicate that the study is invalid.

THREAT #2: DURATION BIAS (SETTING UP A STUDY)

Research studies are typically performed for a limited period of time—generally 1–5 years. The period of study is defined in advance. It is typically inflexible, due to protocol or budget constraints. It is therefore paramount that data collection time points are considered to ensure that they capture all effects of the treatment. Failure to do so is referred to as duration bias.

Duration bias exists in two forms: early and late. In early duration bias, data collection points are chosen that are before some effects of an intervention. This can confound a study by causing researchers to miss the effect of an intervention and erroneously conclude the intervention is ineffective. Some drugs have early beneficial effects and late detrimental effects. Other drugs may have early detrimental effects and late beneficial effects. Studies with early duration bias may fail to capture the entirety of a treatment response, causing researchers to

---

1. This can calculated by calculating the cumulative probability distribution of a binomial series of 20 trials with a probability of success = .05.

understate or overstate the benefits. On the other hand, in late duration bias, data collection points are chosen that are after some effects of an intervention. Similar confounding effects can manifest with this type of bias.

Early duration bias is generally a larger concern to the scientific community, as drugs and interventions may be approved before their full, long-term consequences are known. Moreover, the early treatment effect of an intervention may not be permanent. Especially with behavioral studies, participant response to temporary programs may differ substantially from participant response to a permanent intervention. An adaptation response can occur even in nonbehavioral studies, which diminishes the treatment over time. Subsequent long-term analysis of any implemented program must be conducted to ensure that the treatment response to the intervention of the study is durable.

THREAT #3: IMPERFECT COMPLIANCE (RUNNING A STUDY)

In most studies, study designers take measures to ensure compliance to a protocol to maintain the balanced groups created by randomization. For example, ideal studies are double-blinded, wherein both researchers and participants do not know which patients are in the treatment and control groups. This uncertainty helps prevent participants from pursuing the treatment outside the experiment or from dropping out of the study entirely. However, scientists cannot force participant compliance with experimental randomization. Ethical standards for clinical research, including the principle of self-determination, prevent researchers from regulating participant behavior to this extent.

The threat of imperfect compliance is ever-present but is substantially increased in studies in which blinding is difficult or impossible. Surgical studies frequently are burdened with this challenge, as it is difficult to blind a patient to whether they had a surgery or to blind a surgeon to the type of surgery he or she will perform. Another common situation in which blinding is not viable is when the intervention is a physical or financial product. For example, in a study examining the effect of health insurance on health outcomes, it is impossible to blind patients to the type of health insurance they have.

Even in the total absence of threats to blinding or randomization, imperfect compliance undermines experimental hygiene. There are two mechanism of imperfect compliance: no-shows and crossovers.

No-shows are patients who are randomized to the treatment group, but do not adhere to the treatment regimen. This includes patients who never start treatment, patients who do not comply with the treatment regimen and timeline, and patients who discontinue treatment early. This can have the effect of mixing some "control" patients into the treatment group, decreasing the apparent difference between groups and increasing the standard errors of the groups, raising the barrier of statistical significance. Authors of a study with only no-shows may erroneously conclude there is no statistical significance between two groups, creating a Type 2 error (false negative).

In contrast, **crossovers** are patients who are randomized to the control group, but acquire the treatment by other means. In a study of a pharmaceutical for FDA approval, patients who suspect they were randomized to the control group may travel to other countries to ensure they receive the study drug.[2] Absent testing serum drug levels in control patients, researchers can only discover this threat by fostering an environment in which patients voluntarily reveal this information. Crossover patients have the effect of mixing some "treatment" patients in the control group, similarly decreasing the apparent difference between the groups and increasing the barrier of statistical significance, driving Type 2 error.

Many studies circumvent this threat by performing an *intent to treat* analysis, which ignores any deviance of patients from their randomized groups. However, these studies do not measure the effect of treatment; rather, they reflect the effect of being randomized into the control or treatment group. Therefore, effect sizes generated from these studies may not accurately reflect the true effect of treatment.

THREAT #4: ATTRITION (RUNNING A STUDY)

Another way in which implementation of a study can introduce bias is through attrition. Attrition occurs when study coordinators are unable to follow participants from the beginning through to the end of a study.

If experiments lose members of either the control or treatment group, the groups may become unbalanced. Groups are balanced through random chance or careful assignment. However, after group balance is achieved, any nonrandom process that affects the membership in one group in a different way than the membership in the other group unbalances the groups.

If possible, it would be best to have no attrition. The second most preferable scenario is to have balanced attrition, randomly distributed across groups. The third best scenario is attrition due to a stressor equally felt by both control and treatment group. The worst threat to validity is when attrition is from a stressor that affects one group differently than another group.

For example, consider that you have created two balanced groups, one group assigned the treatment and one group assigned the control. In this hypothetical scenario, the study office is on the fifth floor of a building and your participants have to walk up four flights of stairs to get to the study office. Both treatment and control participants must visit the study office every week. At the end of 3 weeks, you find that 60% of participants made it to the front door of the building but only 40% of participants in each group made it up to the study office. A less fastidious author may assume their groups to be balanced due to the initial

2. In the case of the Oregon Health Insurance study, patients who did not win the "Medicare lottery" acquired Medicare coverage through other legitimate means [1].

randomization. But, a nonrandom intervention—walking up several flights of stairs—affected both groups. It is possible that this affected both groups in the same way, which would maintain the internal validity of our study. If this is the case, it must be proven. But, it would challenge the external validity of our study as the conclusions are only valid for people who are able to walk up four flights of stairs.

But, this process may affect the two groups in different ways. The control group may visit the first floor, and the treatment group may visit an office on the fifth floor. Alternatively, both offices may be on the fifth floor, but the intervention may make it substantially easier for patients to climb stairs and make it to the office. In both cases, it is inappropriate to assume that this non-random process that differentially affects our groups still results in balanced groups.

What to do with attrition? In the best possible world, all the characteristics of the people that left both groups are known. That bird's eye view can then be used to rebalance our groups. However, as seen in the section on Unbalanced Groups, that bird's eye view is rare. The best tool available is randomization, and attrition cannot be randomized.

Other tools must be used to assess the risk to our study. One way to assess the risk to the study is to see if the participants who left the control group differ on key characteristics from the participants who left the treatment group. This might allow us to understand if the process—or processes—responsible for attrition affected the two groups in different ways.

THREAT #5: SPILLOVER EFFECTS (ENSURING DATA INTEGRITY)

Bias can be introduced in the way a study is set up and the way that a study is implemented. It can also be introduced in participants' response to a study. Spillover effects occur when a treatment's effects extend to the control group. Our treatment group receives the treatment. Our control group receives no treatment. Yet, when spillover effects occur, our control group is affected by the treatment.

Spillover effects are important; they impair our ability to isolate the effect of "X" on "Y." To conduct a proper study, experiments need a group exposed to "X" and a group not exposed to "X." If the group that is not exposed to "X" is somehow exposed to "X" through interactions with the group actually exposed to "X," our study breaks down.

Spillover effects can happen if the benefits of a treatment are transmitted from the treatment group to the control group or due to changes to the environment shared by the treatment and control groups. Spillover effects can also happen as a result of behavioral choices made by members of the treatment group as a result of participation in the study. This specific subset of spillover effects will be discussed in the next section.

THREAT #6: BEHAVIORAL RESPONSES (ENSURING DATA INTEGRITY)

Participating in a study can lead to behavioral change. Participants may change their behavior due to their expectations about the treatment's effect or just due to the fact that they are being observed. This can happen within the treatment and within the control group.

When participants in the treatment group change their behavior due to the existence of observation, this is known as the **Hawthorne effect**. When participants in the control group change their behavior due to the existence of observation, this is the **John Henry effect**.

For example, imagine a study that tested a drug that was to prevent death after a traffic accident. Participants in a study are randomly assigned to either the treatment or control group. During routine data monitoring, you find that the number of participants in the treatment group who have died in car accidents is lower than baseline estimates. However, the number of participants in the control group who have died in car accidents is higher than baseline estimates. And, the overall number of traffic accidents has risen from baseline. What is going on?

The overall number of traffic accidents likely rose because the participants—in both the control group and in the treatment group—assumed they were protected by the drug, started to take more risks while driving, and became involved in more traffic accidents. The risky driving of the participants in the treatment group is an example of the Hawthorne effect. The risky driving of the participants in the control group is an example of the John Henry effect.

CONCLUSION

Journal editors are the final guardians of scientific integrity for their publications (Fig. 13.3). It is their responsibility to be well-apprised of the statistical

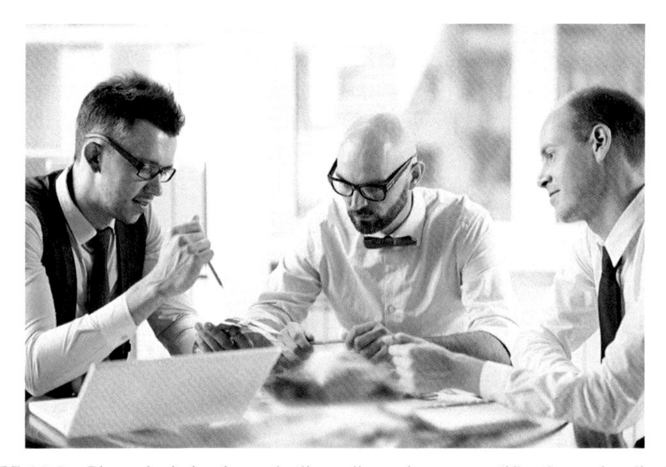

FIGURE 13.3 Photo depicting journal editors discussing papers (iStock.com/mediaphotos).

techniques and tools necessary to verify the integrity of the data in their pages. This chapter provides information on some of the basic fundamental threats that can violate the integrity of studies submitted to their publications. It is not exhaustive, and medical editors are highly encouraged to pursue more formal and rigorous training in statistics to better understand the myriad ways that experimental hygiene can be subverted.

CASE #1: CROSSOVER

Citation: Simeon DT, Grantham–McGregor SM, Callender JE, Wong MS. Treatment of *Trichuris trichiura* infections improves growth, spelling scores, and school attendance in some children. J Nutr 1995;125:1875–83.

Study Summary

This study [2] looked at the effect of giving deworming medication on school attendance in Jamaica. Their population of interest was 407 elementary school children in 14 public schools who were initially infected with worms. The study authors hypothesized that by giving deworming medication, a variety of metrics of student performance would improve, including school attendance and test scores. The study found no significant increase in school attendance ($P = .40$), reading test scores (.65), spelling test scores (.43), or Arithmetic test scores (.92). While some subgroups did see improvements (children with student growth did see a significant improvement in school attendance), overall the study authors did not find their primary end points to be significant.

Where They Went Wrong

In this study, the authors randomized students into placebo and control groups by *individual*. However, students who were in different treatment groups continued to share their environment—their school. As all 14 schools had students who were being treated for *T. trichiura*, the incidence of the worms in that community was decreased, and students in the placebo group benefitted from that change in their environment. This was confirmed by a subsequent 2004 study in Kenya [3], in which students demonstrated substantial benefits from being treated for worms. In the Kenyan study, students were randomized into control and treatment groups by *school*, so that all students in a given school were either in the control or treatment group.

CASE #2: UNBALANCED GROUPS

Citation: MacMahon B, Yen S, Trichopoulos D, Warren K, Nardi G. Coffee and Cancer of the Pancreas. N Engl J Med 1981; 304:630–3.

Study Summary

This study [4] was a case-control study that examined the effect of coffee on pancreatic cancer. The authors surveyed 369 patients with pancreatic cancer (case) and 644 patients without pancreatic cancer (control) regarding their use of a variety of substances, including cigarettes, cigars, pipe tobacco, alcoholic beverages, and tea. While they found a modest association between pancreatic cancer and smoking, they found a much stronger association between coffee consumption and pancreatic cancer in both sexes, even after controlling for cigarette usage. No association was found between pancreatic cancer and cigars, pipe tobacco, alcoholic beverages, or tea. The study authors concluded that coffee consumption may be responsible for a substantial fraction of pancreatic cancer cases in the United States.

Where They Went Wrong

In this study, the means by which the authors selected their control group led to their erroneous conclusion. The experimental group was selected as the group of pancreatic cancer patients who were diagnosed and hospitalized by a group of physicians at hospitals across Boston and Rhode Island. The control group were the *other* patients currently hospitalized by the *same* group of physicians. The physicians that diagnose and treat pancreatic cancer tend to diagnose and treat other gastrointestinal disorders as well, and so gastrointestinal disorders were significantly overrepresented in the control population. For many nonpancreatic cancer gastrointestinal disorders, coffee can worsen symptoms—so these hospitalized patients drink coffee at a much lower rate than the general population. Thus, the coffee usage of the pancreatic cancer group was compared to a control group with an artificially depressed coffee usage, leading to an erroneous conclusion.

ACKNOWLEDGEMENT

This work was based on a course at the Harvard Kennedy School taught by Dr. Amitabh Chandra, whom we gratefully thank and acknowledge for his contributions in the development of this chapter.

REFERENCES

[1] Baicker K, Taubman SL, Allen HL, Bernstein M, Gruber JH, Newhouse JP, Schneider EC, Wright BJ, Zaslavsky AM, Finkelstein AN. The Oregon experiment—effects of medicaid on clinical outcomes. N Engl J Med 2009;368:1713–22.

[2] Simeon DT, Grantham-McGregor SM, Callender JE, Wong MS. Treatment of *Trichuris trichiura* infections improves growth, spelling scores and school attendance in some children. J Nutr 1995;125:1875–83.

[3] Miguel E, Kremer M. Worms: identifying impacts on education and health in the presence of treatment externalities. Econometricia 2004;72:159–217.

[4] MacMahon B, Yen S, Trichopoulos D, Warren K, Nardi G. Coffee and cancer of the pancreas. N Engl J Med 1981;304:630–3.

Chapter 14

A Picture Is Worth a Thousand Words: The Benefit of Medical Illustration in Medical Publishing

Evan Oxner
Freelance Medical Illustrator, New York, NY, United States

Both medical students and practicing professionals are eager to stay informed, to be aware of what is new and on the cutting edge. Everyone wants the latest knowledge from the best and brightest in their fields. On the flip side, doctors/researchers hope to have their work published in highly respected, peer-reviewed journals. Both groups benefit when an author's ability to command language creates a compelling story. But when it comes to the sciences, and human medicine in particular, seeing truly is believing, and observation is decidedly tricky. There are a number of obstacles to keeping the subject of an anatomical study alive or unchanging, along with a host of ethical and technical complications. Enter the medical illustration field. Medical illustration allows for the visualization and study of every part of human biology that is too complicated, messy, rare, or small to be seen in the flesh.

BEFORE OBSERVATION

For much of human history, mythology, speculation, and conjecture ruled our understanding of our bodies, how they function, and how to heal them. Hippocrates attributed the causes of illness to an imbalance of four humors within the body: blood, phlegm, yellow bile, and black bile leading to the practice of bloodletting as a common treatment [1]. Aristotle purported that the heart was the center of thought, motion, emotion, and vitality in the body claiming the heart acted as a sort of furnace, while the other organs only existed to cool it [2]. The discrediting of such claims was stymied since religious institutions forbade dissection of human cadavers; thus religious and philosophical authorities were the ones providing answers to anatomical and medical questions.

This practice changed during the European renaissance when artists, fueled by a desire for greater realism and knowledge in their depiction of the human body, took matters into their own hands and began to dissect cadavers and rob graves.

Medical and Scientific Publishing. https://doi.org/10.1016/B978-0-12-809969-8.00014-0

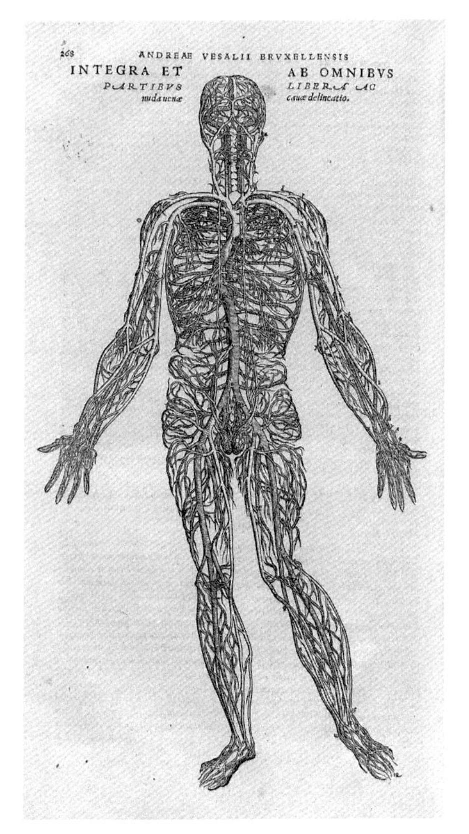

FIGURE 14.1 Page 368 of Andreas Vesalius' *De corporis humani fabrica libri septem*: *Fabrication of the human body in seven books* illustrating the circulatory system in isolation, a visual that would be impossible to bring to reality in the 16th century [3].

During this time a doctor by the name of Andreas Vesalius [3] was pioneering the use of dissection as a teaching tool. His direct observation of cadavers corrected many misunderstandings of the human body, such as men having fewer ribs but more teeth than women, among others. Vesalius went on to publish a set of seven books known as *De corporis humani fabrica libri septem*, or *Fabrication of the Human Body*, featuring illustrations by the pupils of Titian, one of the most influential artists of the Italian Renaissance [4]. Utilizing its expert illustrations, *De corporis humani fabrica libri septem* brought Vesalius' observations beyond the classroom and contributed to the renaissance in scientific understanding. Today the use of cadaver labs and dissection is universal in medical education (Fig. 14.1).

THE IMPORTANCE OF OBSERVATION

Much of scientific function cannot be understood until it has been observed. For instance, it was believed that digestion was a purely mechanical process until

1822 when a man named Alexis St. Martin was shot in the abdomen. The bullet shot an opening into his stomach and fused with his skin to leave a permanent open fistula, a window into the inner workings of the organ. St. Martin's doctor, William Beaumont, began a series of experiments, including tying pieces of food to a string and repeatedly inserting and removing them from the stomach to ultimately learn that digestion is primarily the chemical process we recognize today [4a]. Without the opportunity for observation, it might have been decades before the true nature of the stomach was revealed.

LIMITATIONS OF ALTERNATIVE OBSERVATIONS

Improvements in technology are giving us clearer images of the inner workings of the human body through the use of photography, microscopes, X-rays, and body scans. However, these images are still limited by singular viewpoints, narrow scope, and lack of true-to-life color.

It is currently virtually impossible to observe the inner workings of the human body as they naturally and actively occur. Endoscopy is limited to narrow lumens and a single point of view. Most other procedures would be too invasive and dangerous to perform for the sake of education. Surgical observation is difficult as surgeries are performed in the quickest and safest manner possible, while documentation and exploration are low priority compared to the health of the patient. As far as learning from cadavers goes, most preservation techniques strip tissues of their color and truth to life leaving tissues difficult to distinguish from each other and not representative of their living counterparts. With no foreseeable improvements to our abilities to document or preserve true to form life and medicine, the only available option to visualize medical techniques and procedures is to illustrate them.

ARTISTS CAN DRAW WHAT DOES NOT EXIST

Visual references are an invaluable resource when learning a subject for the first time or when putting a new subject into the context of one the reader already knows. Photographs of anatomical structures can be very hard to decipher even for expert anatomists. Direct observation is limited by access, cost, and availability of cadavers or operating rooms. Medical illustration serves as a bridge between what is observable and what can be shown. Illustrations are also not limited to clarifying the information within the text. They can serve many purposes, from attracting the attention of potential readers, conveying an emotional tone of new information, to even serving as an additional avenue for readers to retain the information of a piece. Whichever purpose they may serve, medical illustrations are a key component within medical publishing (Fig. 14.2).

Illustrations allow those without access to a cadaver lab or operating room to learn and understand the workings of the human body. They help make this information accessible everywhere from middle school biology textbooks and

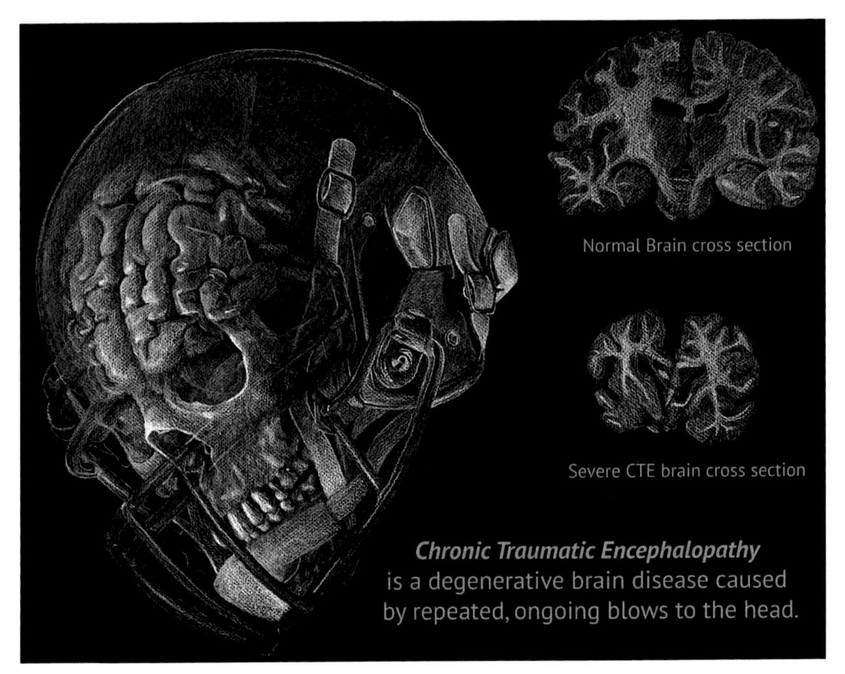

FIGURE 14.2 An example of an editorial illustration used to educate the reader and to paint the article, this one about a degenerative brain disease common among athletes, in a grave light (Evan Oxner).

YouTube videos to doctors' offices, courtrooms, and medical journals. Medical illustrations can build on a reader's existing knowledge by placing new procedures, new inventions, and new discoveries into preexisting context. Artists can also create conceptual illustrations to show invented tools and methods that do not exist yet. It is much easier to market invented tools to medical device manufacturers with a clean, clear, professional concept illustration than just a description. The extrapolated possible future of new scientific discoveries can be shown in illustrations accompanying journal publication.

THE PROFESSIONALIZATION OF MEDICAL ILLUSTRATION

To work in this industry, the majority of medical illustrators obtain a masters degree from one of the four masters programs in North America. In each of these programs students learn the current technologies and techniques pertaining to medical illustration. Students also receive a thorough education in anatomy and medicine with many of the relevant science courses taken alongside medical students. During my education as a medical illustrator, I took courses in anatomy and human dissection alongside the medical students at Michigan State University. Many medical illustrators choose to join the Association of Medical Illustrators (AMI) [5], which acts as the industry's governing body.

Whether they join or not, most choose to go through the AMI vetting process by which they become certified medical illustrators.

MEDICAL ILLUSTRATION TODAY

Medical illustration has moved beyond traditional, two-dimensional (2D) illustration. Most readers will likely know that much of the work today is done on the computer using digital image creating software such as Procreate or Adobe Photoshop. What you may be less familiar with is the rapidly evolving world of versatile three-dimensional (3D) designs. Artists can utilize programs such as Pixologic ZBrush, Maxon Cinema 4D, and Autodesk Maya to sculpt objects in a digital 3D space. Both 3D models such as these and more traditional 2D illustrations can be utilized in computer-generated animations. The newest frontier of illustration and animation is interactive virtual reality. Artists are working to create expandable anatomical facsimiles that can be shown standing in the room with you, are able to be spun around, zoomed into, or animated with the flick of a wrist.

You may not consider technology such as this immediately applicable to the publishing community; however, publishers are already incorporating interactive technology and videos into online textbooks, and we are not far from models that appear to jump off the page through a virtual reality headset. 3D models are an invaluable resource to 2D illustrations. A single 3D model can be rotated in 3D space to create multiple points of view for multiple images, e.g., anterior, posterior, medial, and lateral views. 3D rendering software is such that large-scale changes, which may have previously been cause for restarting a traditional project, can be made quickly and accurately with a few clicks of the mouse.

Medical illustrators work closely with doctors and medical authors to ensure clarity and accuracy in their illustrations. When beginning a new project a medical illustrator will discuss the purpose of the writing with each client. We will discuss what the key point of the piece is and how to quickly and effectively visually convey that idea for the viewer. It is particularly useful to start with sketches from the doctors or authors themselves to see the project from their perspective (see Fig. 14.3). Building off of this framework, the artist can prepare a number of possible designs for the illustration. Following the client's feedback and edits, the chosen design is rendered in the best medium and style to fit the project. Drawing on both the doctor or author's medical knowledge, and the artist's, an illustration is created that can effectively convey the content and tone of the piece it accompanies.

WHERE TO FIND A MEDICAL ILLUSTRATOR

So you have decided to contact a medical illustrator? Excellent choice! Now, where do you start? No matter what avenue you take you will most likely find yourself going through the AMI. The AMI has a searchable online directory making it easy to find illustrators nearby, illustrators with specific technical

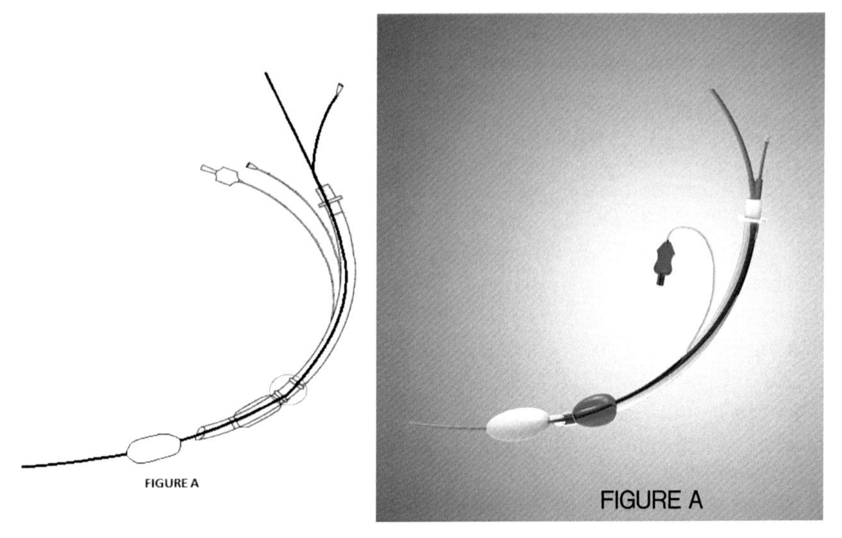

FIGURE 14.3 A proposed tool sketched on the left by inventor Noah Jentzen, MD, and rendered on the right by medical illustrator Evan Oxner.

specialties, or even illustrators with specific medical specialization. If you are a more hands-on individual, you can also request the *Medical Illustration Source Book*. The source book is an annual publication of the AMI featuring full-page promotions for individual artists and medical illustration companies. It is accessible both online and in print and is an excellent resource to browse if you are an "I'll know it when I see it" type of person [5].

Medical illustrations bring value to medical publications by putting into context, clarifying, and elaborating on the text of the author. The partnership between medical illustrators and medical authors allows the complexity of the human body and human medicine to be understood by a broader audience and give a clearer vision for learning and advancement in the medical field. I encourage all medical authors to seek qualified illustrators to ensure their words are accurately represented with the necessary skill and technical knowledge and are alluring to the intended audience.

REFERENCES

[1] Paster GK, Brown TM. Shakespeare and the four humors. U.S. National Library of Medicine. National Institutes of Health; September 19, 2013. https://www.nlm.nih.gov/.

[2] Stanford. "A history of the heart." History of the body, n.d. web.stanford.edu.

[3] Vesalius A. De corporis humani fabrica libri septem. Historical Anatomies on the Web, Page 368. US National Library of Medicine; August 26, 2016. [nlm.nih.gov].

[4] Vons, J. "André Vésale/Andreas Vesalius." Medic@ - André Vésale/Andreas Vesalius — BIU Santé, Paris. Biu Ante, n.d. www.biusante.parisdescartes.fr/.

[4a] Beaumont W. Experiments and observations on the gastric juice, and the physiology of digestion. London: Maclachlan & Stewart; 1838

[5] AMI. Access our directory. Association of Medical Illustrators, n.d. https://www.ami.org/.

Chapter 15

My Journey to Becoming an Author

Ellen R. Abramson
Medical Development, University of Michigan, Ann Arbor, MI, United States

START OF MY JOURNEY

I did not expect to become an author. I had not published journal articles and had not been involved in book projects. During my years in academic settings, I was aware of the importance of having publications, but my jobs had not involved any particular publishing requirements or expectations.

The opportunity to write *Esophageal Cancer: Real-Life Stories From Patients and Families* [1] arose during the time I was employed by the University of Michigan Health System as a major gift officer. My job was to raise funds to support medical research and education. I was responsible for several areas—Orthopaedic Surgery, Physical Medicine and Rehabilitation, Transplant Surgery, and Thoracic Surgery. I worked in partnership with physicians in these departments to connect with individuals—residency alumni, patients, and family members of patients—who were grateful for the training or care they received and had the ability and inclination to provide philanthropic support to advance research or enrich education in those areas.

The donors with whom I worked, and the gifts they provided, varied widely. In Orthopaedic Surgery, for example, I helped a residency alumnus develop a $3 million bequest intention, which was divided equally among the three areas of the University, which made a difference in his life—his undergraduate concentration, the athletic program, which had granted him a scholarship, and the Department of Orthopaedic Surgery. In Transplant Surgery, I worked with a couple who, in gratitude for care received by the husband, made a gift of $75,000 to benefit both the Transplant Patient Emergency Fund and a lectureship honoring the physician to whom they were grateful. In Physical Medicine and Rehabilitation, I worked with a patient who had been treated 30 years earlier after a devastating attack rendered him a paraplegic, to help him document a $670,000 bequest intention for the Department.

Medical and Scientific Publishing. https://doi.org/10.1016/B978-0-12-809969-8.00015-2

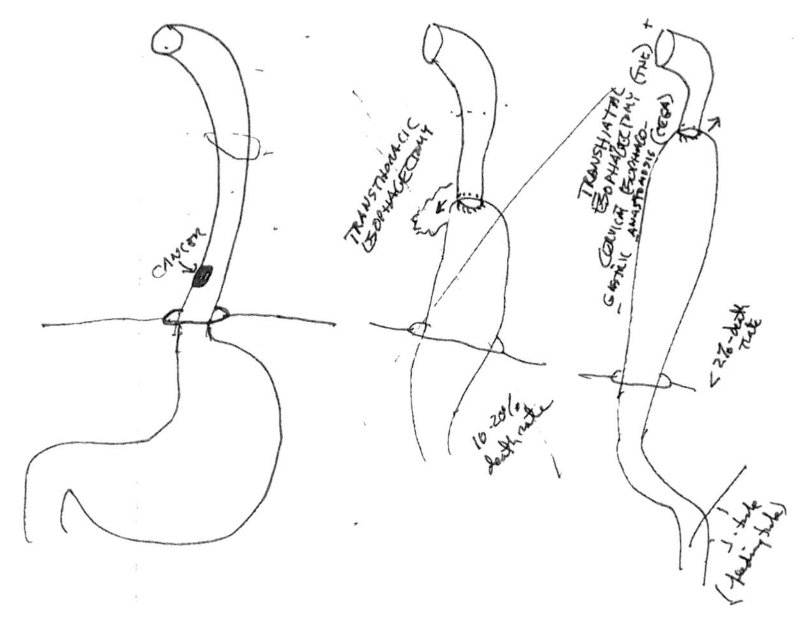

FIGURE 15.1 Transhiatal esophagectomy procedure. *(Drawing by Dr. Mark Orringer, reprinted with permission.)*

In Thoracic Surgery, I had the honor of raising funds to establish the Mark B. Orringer, MD, Research Professorship in Thoracic Surgery. Beginning in the 1970s, Dr. Orringer challenged the paradigm for treatment of esophageal cancer, for which patients had a long-term survival rate of less than 15%. He rediscovered and refined the technique of transhiatal esophagectomy without thoracotomy and a cervical esophagogastric anastomosis with the aim of reducing the complications and mortality associated with standard surgical approaches for esophagectomy, which required a thoracotomy and reconnecting the esophagus and stomach in the chest rather than the neck. Postoperative mortality dropped from nearly 20% in the 1970s to 1%. Combined with chemotherapy and radiation therapy, 3-year survival rates increased to 46% [2]. Dr. Orringer's drawing of this technique is depicted in Fig. 15.1.

OPPORTUNITY ARISES

As Dr. Orringer was preparing to retire from a remarkable academic and medical career, I began to hear stories—stories of esophageal cancer patients who survived and stories of husbands and wives whose spouses had passed away but who were grateful for the extra months or years their care at University of Michigan Thoracic Surgery had made possible. I became aware of the Esophagectomy Support Group, where patients who had successfully undergone the surgery offered support and guidance to new patients preparing for it.

I was warmly welcomed by group members and given the opportunity to attend several meetings.

I shared with a colleague who worked on a publishing team within the Medical School about the remarkable stories I was hearing. Her response to me was "Ellen - you should write a book of these stories, and I will help you." So began my career as an author.

THE PROCESS

As I met with grateful patients and shared with them the idea for the book—a compilation of stories of patients who had undergone esophagectomy surgery—I offered two purposes. One was the opportunity to have an additional helpful resource for esophageal cancer patients and their families, who were facing this surgery. The other was the opportunity to dedicate the book to Dr. Orringer, the surgeon to whom they were all so deeply grateful, on the occasion of his retirement.

In all, 11 patients and family members wrote and shared their stories with me. The two nurses who had led the Esophagectomy Support Group each wrote about the group's impact on participants and on themselves personally. Dr. Andrew Chang, Section Head of Thoracic Surgery, wrote the book's introduction, thanking the patients whose stories appeared in the book, introducing the reader to Dr. Orringer, and providing a brief explanation of esophageal cancer and the esophagectomy surgery. Rounding out the publication was a biography of Dr. Orringer, a list of resources for more information on esophageal cancer, and a preface describing what led me to write the book.

I had the pleasure of asking one of our patient-authors to provide the book's cover art. Following his surgery, this gentleman had taken up watercolor painting. He sent me a photo of himself proudly displaying one of his paintings. It occurred to me that his art was a symbol of his triumph over illness and would be the perfect cover image (Fig. 15.2). He was touched and delighted for us to include his painting in this way.

THE RESULT

The book was published in November 2016 and presented to Dr. Orringer at his retirement party. Subsequently, a copy has been provided to each new patient in the University of Michigan Thoracic Surgery section who is facing esophagectomy surgery. It has proven to be a useful educational resource for these patients and their families. It's also available for purchase on Amazon, with a portion of the proceeds benefiting the University of Michigan Thoracic Surgery Patient Education Fund.

Recently, my son-in-law's mother shared the book with her friend whose husband, diagnosed with esophageal cancer, was scheduled for the surgery. The friend telephoned me to express her gratitude for *Esophageal Cancer: Real-Life*

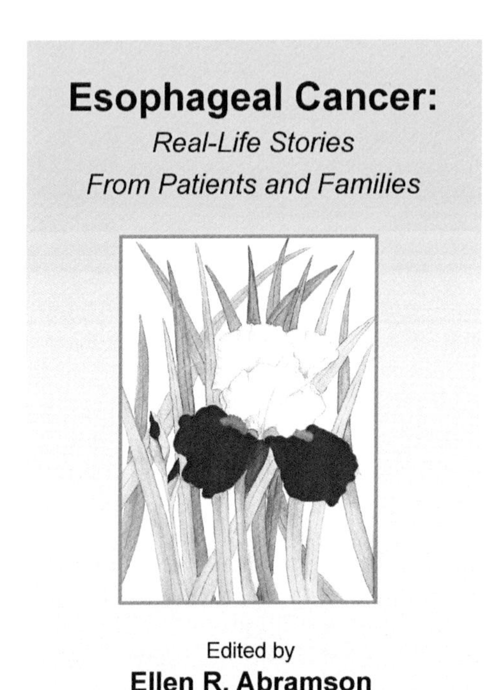

FIGURE 15.2 Front cover of book on Esophageal Cancer Survivorship. *(Published in 2015 by Michigan Publishing, Ann Arbor.)*

Stories From Patients and Families. She and her husband felt less alone knowing others had traveled this road and reading the stories of their challenges and successes.

I was delighted to hear from several of my patient-authors about the meaning they received from sharing their stories. One wrote, "This publication, I know took a great deal of your time and much effort, which produced a tremendous result. I am honored to have had my story memorialized through what you have done. This book will serve future patients and survivors well and it has aroused interest in my primary care physician here in Florida and surprisingly my dentist. I continue to lend my support to those who have been diagnosed and are recovering from the surgery and its after effects. It's hard to put into words the gratitude I feel for what you have done and honoring me as one of the patients who were asked to participate."

NEXT STEPS IN MY JOURNEY

Though I did not expect to become an author, I am grateful for having become one. Recently, I celebrated my 60th birthday. As part of that milestone, I reached out to a small group of friends, each of whom had changed my life, to

FIGURE 15.3 Author in her office working on her new book. *(Photographer, Rose Massey; reprinted with permission.)*

thank them for their profound impact on me. My publishing colleague is one of those friends. By helping me to become an author, she provided me a new way to express who I am. Currently, I am in the final stages of writing a new book (Fig. 15.3)—a personal story of how my husband and I dug ourselves out of considerable credit card debt and built a solid foundation for our future. I believe our story can be a source of hope to a great many people [3].

The experience I acquired writing *Esophageal Cancer: Real-Life Stories From Patients and Families* enabled me to easily make the move to writing about my own life. Becoming an author was transformative. It increased my capacity to tell the story of who I am.

REFERENCES

[1] Abramson E. Esophageal cancer: real-life stories from patients and families. Ann Arbor, USA: Michigan Publishing; 2015.

[2] Lerut T. Landmark article commentaries – 50th anniversary landmark commentary on Orringer MB, Forastiere AA, Perez-Tamayo C, Urba S, Takasugi BJ, Bromberg J. Chemotherapy and Radiation Therapy before transhiatal esophagectomy for esophageal Carcinoma 1990;49:348–55. Ann Thorac Surg 2015;100:1530–3.

[3] Abramson E. Winning the Money Game (in development).

Chapter 16

Writing for Fun and Profit: An Author's Longitudinal Study

Robert Marcus, MD
Department of Medicine, Stanford University, Stanford, CA, United States

INTRODUCTION

I have been asked to write a memoir of my experience with the publishing industry. As a retired academic endocrinologist, it seemed to me at first a peculiar request, but on considering the invitation I soon realized that in my 40-year medical career I have been an author, a manuscript reviewer, and an editor. I have done so as a member of an academic department and as an employee of a major pharmaceutical corporation, and this has occurred over a period of extraordinary change in medical publication. I specify *medical* publication because its nature differs from all other forms, including other types of scientific publication, thanks to the fact that in large measure it deals with the description of interventions involving human research participants.

I have organized my essay to describe my relationship to this world sequentially, first from the perspective of an author, then of a reviewer, and then of an editor. In the discussion of authorship, I will review the different requirements encountered by independent academicians contrasted with those imposed on industry scientists.

THE JOYS OF AUTHORSHIP

One of my proudest moments as a clinician investigator was the acceptance in 1973 of my first solo-authorship paper. Technology was different then as were the ways that authors communicated with publishers. Prior to submission I had typed at least five versions of that manuscript on an IBM Selectric and sent it to the journal by US Mail. I was pleased to receive my first review in about 8 weeks and was able to turn around a revision several days later. Final acceptance took another month, and the article appeared in print about a year after first submission in *Endocrinology* [1]. Even at that time I remember paying author's page charges, but these were nominal. I remember also spending a great deal of thought on how many reprints to purchase. All in all, I found the process to be very pleasant and it remained so for the next decade.

Medical and Scientific Publishing. https://doi.org/10.1016/B978-0-12-809969-8.00016-4

IMPACT OF TECHNOLOGY

Almost 10 years later my research unit acquired a word processor, a bulky, forbidding machine for use only by the secretary. At the time I was employed at a Veteran's Administration Medical Center whose director made a decision that "not just everybody" would be entitled to possess a computer (how times changed!). But at least the secretary spared me the need to type my own multiple versions.

Also at this time, fax machines appeared on the scene, adding pressure to accelerate the submission and review process. Reviewers were given shorter intervals to do their work, and authors were required to provide revisions quickly. Rapid turnaround became the goal, as journals became attentive to the parameters of publication: time to first review, time to acceptance or rejection, and time to publication—all these became marketing ammunition in a journal's efforts to attract authors. I think that was when I first became aware of the Impact Factor. This metric might be highly controversial today, but it was important for any given journal to boast a higher Impact Factor than any other journal in its field. Impact factors frequently are not only used by editorial offices to assess journal performance, but they also may have influenced authors' choice of journals to which they will submit manuscripts. Consequently, I have reviewed many manuscripts for the Journal of Bone and Mineral Research (JBMR) that were ultimately rejected for being of only parochial interest, whereas acceptance would have been far more likely had the manuscript been submitted to a more focused smaller journal.

Another useful feature of word processing was the ability to "cut and paste." In particular, when using a single set of analytical methods for multiple publications, one could simply paste the methods section into each new manuscript without having to retype it. I remember being highly amused at papers exploring the role of the adenylate cyclase-cyclic adenosine monophosphate system as the initial response to various peptide hormones. These papers seemed to be identical except that one was talking about parathyroid hormone and kidney cells instead of glucagon and liver cells. Today such a practice is frowned on, as it treads closely to self-plagiarism. It is now considered preferable to state that the analytical methods were identical to those in an earlier paper and give a citation.

The computer era quickly and completely transformed the authorial process. Programs such as PowerPoint, Statview, and End Note greatly reduced preparation time for manuscripts, and electronic submission put an end to delays in the mail and the need for expensive courier services. Consequently, journals began to experience palpable reductions in review and publication times. I cannot say that my experiences in review were always as pleasant as with my first paper, but in general I maintained a positive view of the process. I felt that the critiques were, more often than not, fair and constructive.

ACADEMIC VERSUS INDUSTRY AUTHORSHIP: A CONTRAST

During my career, I had occasion to write research papers in three distinct settings: (1) as an academic physician completely unaligned with industry; (2) as an academic participant in industry-sponsored research; and (3) as a full-time research physician for a pharmaceutical company. These settings differ significantly from each other, with particular aspects and unique challenges.

1. **Independent academic publication**. Most journals require an author to acknowledge sources of support for the work that is being reported—the National Institutes of Health (NIH), for example, or a granting society such as the American Diabetes Association, as well as any commercial entity that may have given reagents, equipment, or other support. Such acknowledgments generally appear at the end of a manuscript. Beyond that, however, the author is free to write an account of the research as he or she sees fit. Such freedom may include speculation about the relation of the findings to clinical practice, at least insofar as peer review permits. If the results of a clinical intervention suggest that a drug may be useful in settings other than those approved by the Food and Drug Administration (FDA), the author has carte blanche to propose that new utility.
   With respect to clinical trials published in a purely academic environment, there is no requirement for independent audit either of the conduct of the trial or of the reported findings. This lies in striking contrast to what is required for industry-sponsored trials, on which I will elaborate below.
2. **Participation in industry-sponsored research**. To orient the reader, I first will describe briefly the types of research publications that come from the pharmaceutical industry. The most important industry-sponsored papers are those describing the primary outcomes of trials aimed at receiving FDA approval. These papers are often pivotal to the package on which the agency bases its decision to approve a drug. The conditions of approval then appear in the product label (package insert) that defines what may be included in any promotional material for the product.
   Next in importance are reports on secondary or exploratory data from the registration trial that were not prespecified in the research plan and therefore would likely not be considered for inclusion in the product label. In addition, once FDA approval is obtained, postapproval papers may describe additional outcomes from the registration trials or may represent new nonregistration trials that the sponsor has undertaken. The findings in these papers would not become part of product label unless the manufacturer submitted a successful application for FDA to expand its approval of the given agent.

   Finally, many pharmaceutical companies sponsor investigator-initiated trials (IITs), for which independent academic scientists submit formal proposals to conduct a research study using drugs, reagents, or financial support from the company. I chaired a corporate committee to evaluate these proposals and we supported many excellent projects. The investigator's only obligation to the company was to report the progress of subject recruitment and provide a summary of the research on its completion, generally in the form of a published manuscript. The company itself assumed no responsibility for the completion of IITs, and its scientists did not appear as coauthors on the completed paper. In my view, IITs are a very useful adjunct for large pharmaceutical companies. Many creative research ideas offered potentially important insights into disease mechanism or description but were not of sufficient priority to the company to devote the time and resources necessary to conduct the research. Moreover, many of these proposals would not have reached sufficient priority by governmental funding agencies, so the IIT provided the investigator adequate support to conduct the research, and, perhaps, generate data that might be preliminary to obtaining an NIH grant on that topic.

3. **Registration trials**. Pharmaceutical companies that conduct clinical trials with a view toward achieving product registration typically undergo a rigorous sequence of regulated activity before their trials reach the publication stage. Large companies with considerable regulatory expertise generally have their research teams meet with members of the FDA to receive feedback and guidance regarding the design and conduct of their trials before a protocol is even written. This exercise is not absolutely required, and companies with smaller budgets may decide (at their peril) to forgo the opportunity. Trials in which "real" events constitute the research end points—such as the incidence of fracture or heart attack—require many-fold more participants than the smaller, less expensive protocols that a few academic physicians can undertake. The company must provide a cadre of individuals whose job it is to supervise protocol compliance in the various study centers, check on the completeness and accuracy of individual participant data and other aspects of quality control. This responsibility is often contracted to an outside Clinical Research Organization. The statistical plan for the study is almost always submitted in advance for FDA approval. None of these steps is required of purely academic research (except for government-sponsored multicenter programs, for which the scrutiny is similar to that pertaining to industry).
   Thus, in terms of the quality of research publication, such as data accuracy and statistical rigor, clinical trial papers from industry-supported research, in my experience, have been far more reliable than those from smaller, academic studies. The difference, I think, is particularly evidenced when the sponsor has built into the study an executive committee of academic physicians whose laboratories are directly involved with the research. Such a committee, along with industry representatives, periodically reviews the progress of the trial and participates in the analysis and reporting of the study findings. Usually, members of this committee undertake the manuscript writing.

   I was fortunate to be able to participate in several of these executive committees and found that they provide excellent protection against some of the reporting bias of which industry is frequently accused. That is not to say that moments of friction between the sponsor and executive committee do not occur, as each group may have a different set of priorities. For a drug that awaits FDA approval, the paper must provide a strong case to justify a positive action from the Agency, and the company may also have a timeline that is considerably shorter than the slower pace of usual academic publication. By contrast, the executive committee may be more interested in addressing an interesting point of physiology that is not directly relevant to approval. However, in my experience these conflicts were not difficult to resolve, and the industry–academic relationship appeared to be an excellent approach.

4. **Publishing as an industry physician**. During my 8 years as a physician working in the pharmaceutical industry, I had considerable responsibility for research activity in the United States surrounding a specific product. When I first joined the product group, the writing assignment for an important manuscript concerning the drug had already been contracted to an outside vendor. This vendor had no general or specific knowledge about the research program or about the drug that was under study. Not surprisingly, the draft manuscript was completely unsatisfactory. My first official act was to arrange for product-related manuscripts to be written in-house. I was, and still am, quite surprised to find the extent to which outside vendors write research papers for pharmaceutical companies. The rationale for outsourcing is to reduce the necessity for in-house writing staff, but in my experience, the tradeoff in quality is stark.

   Preparing a research manuscript in a corporate setting is not as simple as sitting down at a laptop and starting to write. Various corporate resources are indispensable. Statistical support, checking manuscript drafts for compliance with FDA regulations, corporate and legal document review, and data verification all consume corporate resources. As a result, any proposal to write a manuscript on some aspect of a clinical trial must first be reviewed by a prioritizing committee.

   Once permission to proceed with a manuscript is acquired—never a given—the next step is the selection of outside authors. Journals often are loath to accept papers for which the only authors are employees of the sponsor. Generally, the group of outside investigators includes well-established experts in the particular field who have helped guide the trial from protocol design through data analysis and interpretation. It is logical, then, for these experts to help in data interpretation and publication.

   In a major clinical trial that I headed, we included two outside investigators in the committee tasked with writing the protocol. These individuals also shared authorship on the primary study paper [2]. Other participating investigators were included in subsequent papers for various reasons, including recruitment of large numbers of patients to the trial. Each of these coauthors had full access to the trial data and statistical analysis and had multiple opportunities to contribute to the text. In general, top tier journals require the primary author to submit a list of all coauthors along with their various contributions to the paper.

I mentioned above the enviable position that nonindustry academicians enjoy by being able to speculate and propose new uses for drugs or procedures. Such a happy state of affairs does not apply to the pharmaceutical industry. Once a drug is approved, any promotional information offered by the manufacturer must adhere to the conditions of approval, i.e., the product label. Thus, if a drug is approved only for management of high cholesterol, the company might publish a paper showing that it lowered blood pressure, but the company could not legally market the drug for this "off-label" use. In fact, promotion based on an off-label publication could result in serious legal sanctions against the company, including heavy fines or even a costly consent decree for a major infraction. In my experience, the legal department of my company did not approve the inclusion of off-label efficacy claims in any of our papers, even those that were not intended for promotional activity.

Further, all publications intended for promotional use are required to show "fair balance." I learned quickly that fair balance to an academic scientist is not what the FDA means by that term. In a scholarly presentation, fair balance means that one has described and compared the benefits and risks of several drugs. For example, I would compare the benefits and risks of various osteoporosis therapies currently on the market. To the FDA, fair balance means *only* that one has described both the benefits and risks of the particular drug under consideration as listed in the product label. In fact, comparisons of one drug to other approved agents are not allowed. So, my response to the general criticism that industry presentations do not provide fair balance is that in the current system, meeting these expectations of the reader is not legally possible.

Of course, none of this means that publications emanating from industry are necessarily free of bias. Just because one is restricted from making off-label claims and comparisons to other products does not eliminate the chance of biased reporting. For example, unless the publication is intended for promotional use, authors would not necessarily include findings that might reflect badly on a drug—which may contribute to the problem of positive publication bias in biomedical publication today.

MY LIFE AS A REVIEWER

Shortly after I started to achieve recognition as a contributor to my field, I began to field requests to review submitted manuscripts. Although peer review today has taken some hits, I quite enjoyed the activity. I liked getting an early, firsthand look at interesting scientific results, although it did require me to make a conscious attempt to remain dispassionate when authors promoted ideas that I believed not to be true. After I reviewed a manuscript, the editor's decision was almost always transmitted to me accompanied by the comments of the other reviewers, so over time I was able to see many styles of reviewing. Unfortunately, I frequently encountered a style that would be completely unhelpful to the author: a simple list, by page and line number, with such perfunctory comments as "misspelling" or "change *that* to *which*," but completely lacking in thoughtful analysis of the authors' work. Such reviews are devoid of scientific insight. They fulfill the responsibility of a manuscript editor but not of a reviewer. In general I have found such reviews to be contributed by younger, inexperienced scientists, and I have occasionally proposed to journal editors that they institute a training program for prospective reviewers. This idea has never caught on, and even today, in retirement, I occasionally encounter such incompetent reviews.

To ameliorate resentment by authors for having issued a negative review, early on I developed a strategy that has served me well during my career. I begin with a brief sentence or two stating what the authors studied and why they did so. I follow that with a bare outline of the research plan, methods, and results, trying to limit this outline to the essentials and not, as too many reviewers do, reiterate the entire methods and results section of the manuscript. I then am careful to quote the authors' concluding sentence—which all too often is simply a summary and not a conclusion. That is one of my pet peeves, and so I always point out that after spending the time and energy to do the research, surely the authors can find something worthwhile to conclude from their efforts. I finish this general section by stating that the manuscript has both strengths and weaknesses. Here I delineate the strengths and say that issues of concern will be enumerated below. After this section follows a numbered list of critical comments, each giving a page and paragraph number, explaining what the problem appears to be and suggesting either a rewording or something more serious, like a different statistical approach or even an additional experiment. I consider whether the authors' conclusions are justified by the data and routinely examine the reference list to determine if recent work has been appropriately cited as well as to see if important older literature in this field has been neglected. This approach is simple to use and I highly recommend it.

I alluded in a previous section to the introduction of Statview [3] and other statistical programs that allowed individual investigators to conduct statistical analysis of their work themselves. I think these software products have proven to be a double-edged sword. They certainly made it easier and less expensive to prepare a manuscript, but in my career as a reviewer I have encountered many submissions that suffered inadequate or inappropriate statistical analysis that had been carried out by the author without expert statistical consultation. Luckily, some of the top tier journals now require independent statistical analysis, particularly of clinical trials, from their own editorial offices.

I particularly enjoyed the occasional requests I received to write an editorial on the topic addressed by the manuscript that I had reviewed. Editorial writing was always a source of great pleasure because I was able to write in a prose style less formal than for my research papers [4–7]. It also seems to me that I got many more congratulatory comments from my colleagues on these editorials than on my research publications!

I would be remiss here not to mention my indebtedness to Dr. Robert Utiger, who, as an editor for the Journal of Clinical Endocrinology and Metabolism and later the New England Journal of Medicine, showed me on my own manuscripts how a proper review should be conducted and conveyed to the author.

GRADUATION TO EDITORSHIP

My introduction to the world of editing came through an invitation to organize an issue of *Endocrinology and Metabolism Clinics of North America* around the topic of calcium and parathyroid hormone [8]. This project first involved constructing a table of contents, writing my own chapter, and finding potential authors for the remaining chapters. By that time in my career, I knew a sufficient number of colleagues for me to select the best authors for the job, so it was easy to fill out the volume. In addition, the project came during a relatively generous climate for research funding so authors had the time to take from their grant applications to write a chapter. This gratifying first experience with editorship led me to accept another opportunity, this time to edit a multiauthored book on osteoporosis, to be published in 1994 by Blackwell Scientific. At that time, the only other book devoted to osteoporosis was edited by Drs. Lawrence Riggs and Joseph Melton at the Mayo Clinic, so Blackwell saw a reasonable market for another volume on that topic. This book was also a great pleasure to prepare, as the publisher was extremely helpful and agreeable to my approach, and the authors, most of whom were my good friends, were enthusiastic about the project [9].

Also in 1994, I was approached by Dr. John Bilezikian to work on a comprehensive textbook on the parathyroid glands, "The Parathyroids," along with another old friend, Michael Levine. John was to be Editor-in-Chief and Michael and I were associate editors. The three of us had in common the fact that we had all trained at the NIH in the laboratory of Gerald Aurbach. Dr. Aurbach was a titan in the parathyroid field who died in 1991 in an act of violence, and this book was created as a tribute and memorial to him and his work. The publisher of the first edition was Raven Press, and a senior publishing editor who had been trained as a scientist was our primary contact with that company. She and Dr. Bilezikian met at a funeral for DeWitt Goodman, another wonderful scientist who died the same day as Dr. Aurbach, and casual conversation eventuated rapidly into the concept for this memorial volume.

We instantly formed a strong bond with and affection with our publishing editor, who continued to work with me and with John in subsequent efforts over the next 20 years as she moved from Raven to Academic Press and then to Elsevier. The Parathyroids was a magnificent achievement—the first comprehensive research text in this area whose contributing authors were outstanding members of the scientific community [10]. By virtue of its size and the busy calendars of its contributors, however, bringing the book to publication readiness was far more complex and occasionally frustrating than my earlier editing projects. Most

FIGURE 16.1 The Parathyroids, first edition and the most recent third edition.

frustrating was dealing with recalcitrant authors, each of whom had been given many months to complete their chapters and who had received periodic reminders. We learned that some of the authors had not even started their chapters by their due date, and we even had to impose drop-dead dates after which the chapter would not be accepted. In one case our publishing editor camped out at an author's office, refusing to leave until he finished writing his chapter! The Parathyroids also was a financial success and is currently in its third edition, now published by Elsevier. Fig. 16.1 shows both the first and third editions of The Parathyroids.

Following completion of The Parathyroids, the same publishing editor invited me to edit a comprehensive textbook devoted to the field of osteoporosis. The intended audience for this book was to be basic and clinical scientists conducting research in this field. The book was designed to contain sections on basic bone biology, skeletal acquisition during growth, bone maintenance in adult life, the impact of physical activity and exercise, nutrient status, and hormonal sufficiency on bone health, as well as a large section on the various hygienic and pharmacologic interventions that could treat established osteoporosis. To manage this enormous workload, I recruited two other members of the faculty at Stanford University, Drs. David Feldman and Jennifer Kelsey, to join me as editors. In all, approximately 80 chapters were submitted and the book, published in 1995 under the title OSTEOPOROSIS, has been an enormous success [11]. In 1996 its first edition won the prestigious PROSE award in the Medical Science category from the Association of American Publishers for excellence in professional and scholarly publishing. Twenty years after its first appearance, Osteoporosis is now a two-volume set and is in its fourth edition (Fig. 16.2).

Publication success often depends a great deal on timing. The first edition of OSTEOPOROSIS appeared about the time that the first blockbuster drug for

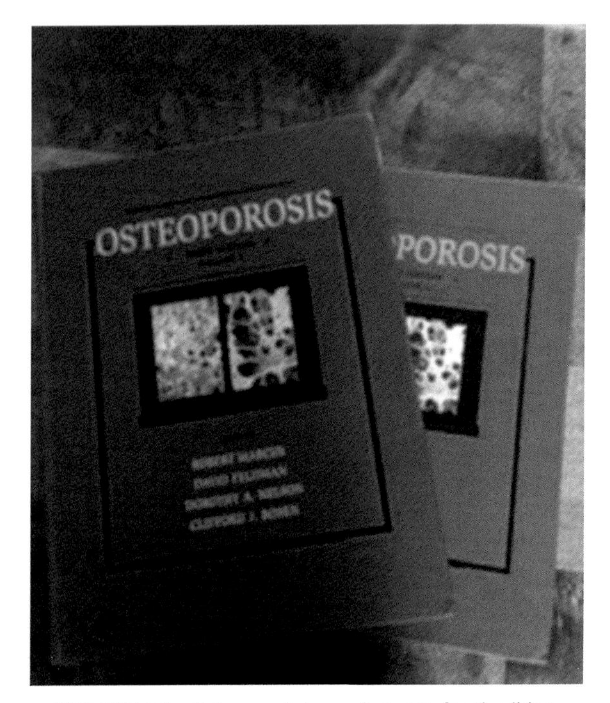

FIGURE 16.2 Osteoporosis two-volume set, fourth edition.

osteoporosis, Fosamax (Merck & Co.), was released, and the second edition appeared in close proximity to release of the first anabolic agent for this disease, Forteo (Eli Lilly & Co.). I am sure that interest in these books was catalyzed as a result of the increased exposure to osteoporosis and its myriad aspects that physicians received from pharmaceutical sales personnel.

Between the appearance of its first and fourth editions, the publication environment dramatically changed. Potential authors, having to devote more time and resources to fundraising, are now more reluctant to accept invitations to write a chapter, and among those who do accept, tardiness in manuscript submission is much more common. Even those authors who have been asked to prepare minor updates of their previous work are reluctant to do so. Moreover, the price of large textbooks is now out of reach for many scientists and students. Although many libraries still see a need to purchase major textbooks, many readers are more interested in acquiring only those sections or chapters most relevant to their work rather than purchasing an entire book. Therefore, some publishers have begun to respond to this environment by putting their books online and permitting individual chapters to be downloaded at a much reduced price. I think this approach is eminently sensible, and I suspect it is the wave of the future, so I hope the publishing industry will fully embrace it.

MY TIME AS CHAIR OF A PUBLICATIONS COMMITTEE

One other facet of academic publication that faced me deserves mention, and that concerns the question "who should publish scholarly journals?" In the 1990s I became chairman of the publications committee of my primary organization, the American Society for Bone and Mineral Research. At that time we were considering a change of publisher for society's publications, the JBMR. This journal, which began as a mom-and-pop venture under the sole direction of Dr. Lawrence Raisz, had been published by a small publishing house, Mary Ann Liebert, Inc., for more than a decade. Although the journal made great strides over this time, it was becoming clear that a larger publishing house would be more effective in generating revenue and handling the logistics of managing what had become a major scientific journal. Thus, the Society switched to Blackwell. For several years, our relationship with Blackwell was mutually satisfactory, but eventually the Society realized that another change was needed, one that would increase the advertising revenues to the Journal and would ultimately be less costly. At the time, the idea to self-publish seemed very attractive, so under the capable editorial leadership of Dr. Marc Drezner, my committee supported a very successful self-publication operation over the next decade. In recent years, decreases in research funding and industrial support for the bone field reached the point that self-publication was no longer the most propitious model for our journal, and so publication once again reverted to a commercial publishing house, this time to Wiley (who, interestingly, had acquired Blackwell some years ago). At present, this relationship appears to be satisfactory to all concerned. The lesson here is that there is no hard and fast rule on whether self-publication is the best choice for scientific societies. This decision requires consideration of the size and maturation of the society, whether it is in a growth or stabilization phase, what the advertising environment may be, and what the future trends in a particular area of research may portend.

POSTSCRIPT

The opportunity to write this memoir led me to look back on an immensely satisfying career in academic medicine. Until now I have viewed that career only in respect to the training I received from Gerald Aurbach, the research accomplishments I achieved, the students whose careers I influenced, the validation I received through elective office in my organizations, and the exhilaration of helping to spearhead the development of an exciting new drug at Eli Lilly & Co. I realize now that my writing, reviewing, and editing activities form a shadow career that is gratifying to contemplate. In closing I thank my dear friends Jasna Markovac, John Bilezikian, and Hunter Heath for their long-standing professional and personal encouragement that made this career thread a reality.

REFERENCES

[1] Marcus R. Cyclic nucleotide phosphodiesterase from bone. Characterization of the enzyme and studies of inhibition by thyroid hormones. Endocrinology 1975;96:400–9.

[2] Saag KG, Shane E, Boonen S, Marin F, Donley DW, Taylor KA, Dalsky GP, Marcus R. Teriparatide or alendronate in glucocorticoid-induced osteoporosis. N Engl J Med 2007;357:2028–39.

[3] SAS Campus Drive. Statview Software, Cary, North Carolina 275, 1985.

[4] Marcus R. Cyclic etidronate – has the rose lost its bloom? Am J Med 1993;95:555–6.

[5] Marcus R. Bones of contention: the problem of mild hyperparathyroidism. J Clin Endocrinol Metab 1995;80:720–2.

[6] Marcus R. Exercise: moving in the right direction. J Bone Miner Res 1998;13:1793–5.

[7] Marcus R. New perspectives on the skeletal role of estrogen. J Clin Endocrinol Metab 1998;83:2236–8.

[8] Marcus R, editor. Endocrinology & metabolism clinics of North America. Hypercalcemia. WB Saunders; September 1989.

[9] Marcus R, editor. Osteoporosis. Contemporary issues in endocrinology. Blackwell Scientific; 1994.

[10] Bilezikian JP, Marcus R, Levine M, editors. The parathyroids: basic & clinical concepts. NY: Raven Press; 1994.

[11] Marcus R, Feldman D, Kelsey J, editors. Osteoporosis. San Diego: Academic Press; 1996.

Part III

Legal and Ethical Issues

Chapter 17

Rights and Publishing Contracts: What Authors Need to Know

Melissa Levine, JD[1], Karen Kost[2]
[1]*Copyrights Office, University Library, University of Michigan, Ann Arbor, MI, United States;*
[2]*Health Information Technology and Services, University of Michigan, Ann Arbor, MI, United States*

Copyright scholar Jessica Litman describes copyright law, which determines who gets to control works of scholarship like academic papers and textbooks, as "peculiarly counterintuitive," completely baffling to lay people, and designed by copyright lawyers for copyright lawyers [1]. Unfortunately, most physicians and researchers do not have a copyright lawyer on hand to help them review their publishing agreements and make decisions about permissions and licensing. The following chapter provides a brief overview for academic authors to help them understand the basic rules of copyright and how to understand and negotiate their publishing contracts.

COPYRIGHT

The Statute of Anne was passed in England in 1710. Considered the first copyright statute, the full title of the law was "An Act for the Encouragement of Learning, by vesting the Copies of Printed Books in the Authors or purchasers of such Copies, during the Times therein mentioned." From around 1440 when Johannes Gutenberg invented the printing press, the number of printing presses increased rapidly. Along with the growing number of books, more people were literate. This in turn raised concerns by governments, authors, and publishers about unregulated copying of books. This led to the development of intellectual property rights by the time of the Statute of Anne. From its earliest concept, copyright law today has grown to address the rights and ownership of not only written materials but also other types of creative works including music, choreography, films, visual arts such as photographs, architectural works, software, and more. US copyright law does not protect ideas, facts, titles, data, or useful articles, the latter of which is protected by patent laws.

The United States Constitution deals with rights of the author in Article 1, Section 8, Clause 8, known as the Copyright Clause, stating that the US

Medical and Scientific Publishing. https://doi.org/10.1016/B978-0-12-809969-8.00017-6

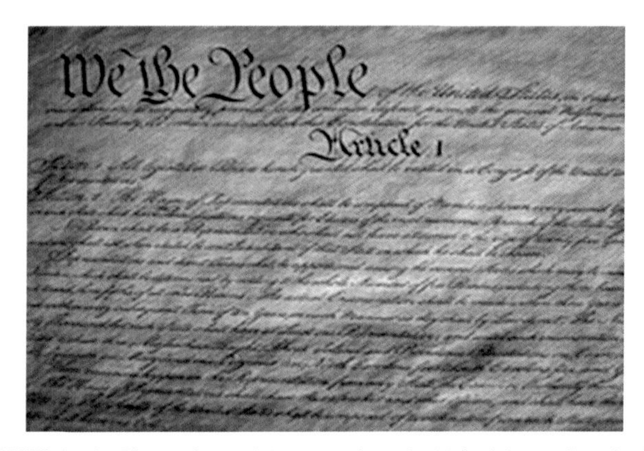

FIGURE 17.1 Photo of copyright clause from the United States Constitution.

Congress shall have the power "To promote the Progress of Science and useful Arts, by securing for limited Times to Authors and Inventors the exclusive Right to their respective Writings and Discoveries" (Fig. 17.1). This represents the foundation for US intellectual property law, specifically that having to do with patents and copyright.

The term of copyright protection generally begins from the moment of creation and continues for the life of the author, plus 70 years. Copyright protection is automatic; it is not necessary to register a work with the US Copyright Office to be eligible for copyright protection. However, there are legal benefits to registration and most publishers will register your work in your name as the author or in their name as publisher, depending on the terms of your contract. Creative works are sometimes registered to formally assert the author's rights and as a basis to claim statutory damages in the case of infringement. Notice of copyright (whether registered or not) is labeled with the name of the work, the copyright symbol "©," and the year the work is created or published. (Registration and notice are formalities that were required under earlier copyright law in the United States, so they may be particularly relevant for works you may have published in the past.)

It is important for today's editors and authors to have a basic understand what copyright is and what it is not. Copyright is often referred to as a bundle of legal rights. Examples of those rights include the following:

- The right to reproduce the work,
- The right to distribute the work,
- The right to prepare derivative works,
- The right to perform the work, and,
- The right to license any of the above to third parties.

A copyright holder can assign (license) or transfer all or just some of these rights, thus the concept of the bundle.

To be eligible for copyright, a work must be original, creative, and fixed in a tangible form. This may include anything from a jotting on a paper napkin (works that can be perceived directly) to those that require the aid of a machine or device. The creator is typically the copyright holder unless the work was produced within the scope of one's employment. Under US law, work created within the scope of employment is typically owned by the employer. (Note that many universities have policies that specifically provide that their faculty and scholars hold copyright in their scholarly works such as journal articles or monographs.) Creators who retain their copyrights may transfer or assign their copyrights by contract or license. Copyright is held jointly for works that have multiple creators, i.e., each creator shares equally in the copyright unless there is something in writing among the creators to the contrary. For administrative purposes, it may be easiest to authorize one copyright holder/author to negotiate for the group, for example.

COPYRIGHTABILITY

Copyright law gives the copyright holder the right to control certain uses of works that are protected by copyright. It also gives users the right to make certain uses of those works without permission.

In the United States, to be eligible for copyright protection, a work must be as follows:

- Original
  - To qualify as original, the work must be created independently and must have "at least a modicum" of creativity. Information in charts, graphs, and tables may not be subject to copyright protection because they are purely factual; copyright might exist if there is something original about the way data are expressed. They do not meet the originality requirement unless there is something creative about the way they are presented (e.g., a special design or a novel way to show data). Copyright may protect a database or other compilation of otherwise unprotectable elements, but only if it meets the originality requirement. Faithful reproductions of two-dimensional works of art generally are not considered to have a copyright beyond the copyright in the original work. They lack the required "modicum of creativity." If the original work of art is in the public domain, the reproduction is typically treated as being in the public domain, too. If the original work of art is protected by copyright, it remains protected, but the person who made the reproduction does not get any new rights.
- A work of authorship
  - Works of authorship include literary works, musical works, pictorial, graphic, and sculptural works, audiovisual works, and sound recordings, as well as many other types of creative works. If two or more people make copyrightable contributions to a work with the intent that their contributions

  be merged into one whole, they are joint authors under US law and hold equal shares of the copyright from the time the work is created. To transfer the copyright or grant an exclusive license, all joint rightsholders must agree. As mentioned earlier regarding employees and "works made for hire," the copyright holder is the person who employed or commissioned the creator of the work, rather than the creator of the work him- or herself.
- Fixed
  - A work must also be "fixed in a tangible medium of expression" by or under the authorization of the author. Writing a work on paper or on a computer, recording a work on tape, and sculpting a work out of marble all satisfy this requirement. An unrecorded improvisation (e.g., in music or dance) would not satisfy this requirement. An ice sculpture would probably not be sufficiently fixed to be eligible for copyright.

US copyright does not protect "any idea, procedure, process, system, method of operation, concept, principle, or discovery." It also does not protect works prepared by an officer or employee of the US Government as part of that person's official duties. In the United States, copyright protection lasts for a limited time only. All copyrightable works eventually lose copyright protection.

RIGHTS AND PERMISSIONS

When using copyrighted work in their own content, authors must seek permission from the copyright holder. This step can be very time- and labor-intensive, but it is necessary to assure that no infringement occurs. Authors and publishers must decide if they are willing to pay a licensing fee or royalty for using material copyrighted by others. In cases when it seems too onerous a task to seek or pay for permission from a copyright holder, they may look for and use alternative material or limit their use of the material to that portion that seems eligible for fair use.

The concept of fair use grants exceptions to copyright protection for purposes such as critiques, comments, news reporting, and teaching (including multiple copies for classroom use, scholarship, or research) and thus is not considered copyright infringement. Fair use is a limitation that balances the otherwise exclusive rights of copyright holders in a manner consistent with the First Amendment of the US Constitution regarding speech.

The factors considered when determining whether usage qualifies as fair use include the following:

1. Purpose and character of the use, i.e., is it for nonprofit educational use and is the current author/creator's work noncommercial in nature? It is important to remember that if an author's work is going to be sold as a book, for example, then the work becomes commercial.
2. Nature of the copyrighted work, i.e., is the use for a factual or creative expression?

3. Amount and substantiality of the portion used in relation to the copyrighted work as a whole, i.e., is the creator using a little or a lot of the copyrighted material and which part(s) is being used?
4. The effect of fair use on the potential market for or value of the copyrighted work, i.e., might fair use result in a substantial decrease in the potential market and value of the original work?

There are occasions when materials are no longer covered under copyright and are considered to be in the public domain. Works in the public domain may be used without the permission of the copyright holder and may be used for free. These include the following:

- Works published before 1923 in the US
- Some works published between 1923 and 1963, but "publication" complicates determination (e.g., was it published, registered, renewed?)
- Works by the US government

TERMS OF USE

In addition to copyright, it is important to define the license terms under which a work can be used. When there is no specific indication of the terms of use, one may assume a work is subject to copyright—unless fair use or one of the other exceptions to copyright is applicable to your use (more on that in a moment). This means that while the work can be read, it cannot be reused, copied, or distributed in any way without the express permission of the rights holder in the absence of an applicable exception.

More liberal licenses are available for authors and copyright holders to enable readers to reuse the work. The most commonly used suite of licenses is Creative Commons (CC) [2]. CC is a nonprofit organization that offers free licenses and related legal tools that help authors mark their creative work with specific licenses that indicate how others may use their work. It facilitates sharing one's work while retaining legal and practical control over one's work product while making it available for some level of reuse. It is designed to address the fluid nature of uses via the Internet—but it can be and is used for all kinds of media. Open Access books and journals typically use some form of a CC license.

CC licenses are well accepted and useful. This chapter draws from materials prepared by the University of Michigan Copyrights Office which are distributed under a CC BY license [3]. The CC BY license indicates that the copyright holder grants permission for others to reuse the material provided proper attribution is made. By following the terms of the license, no additional permission is needed. Some people still pursue permissions as a courtesy even with an attribution license. As a practical matter, this is not necessary.

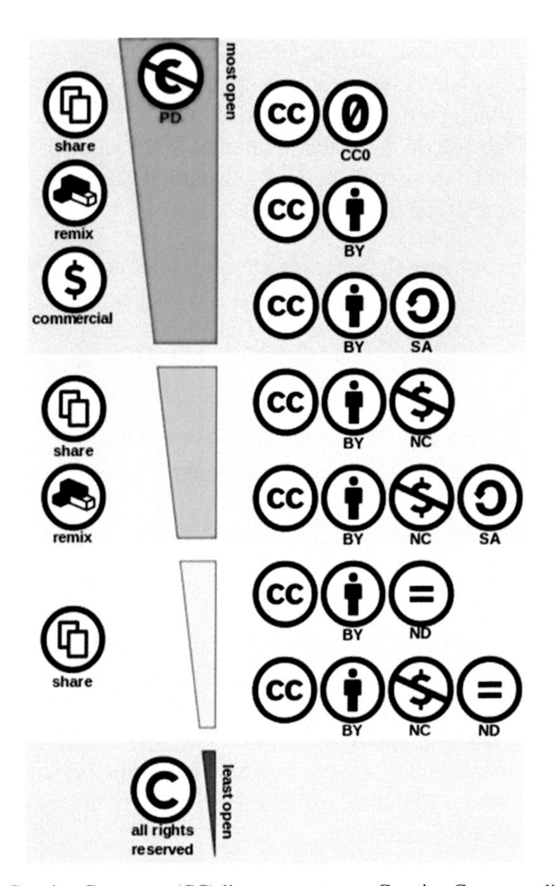

FIGURE 17.2 Creative Commons (CC) license spectrum. Creative Commons license spectrum between *public domain* (top) and *all rights reserved* (bottom). Left side indicates the use cases allowed, right side indicates the license components. The dark green area indicates *Free Cultural Works* compatible licenses, the two green areas indicate compatibility with the *Remix culture*. The CC licenses present authors with a range of options to license—and communicate their license effectively—to encourage sharing of their work. *(https://creativecommons.org/policies. Creative commons (the original CC license symbols), combined work by Shaddim cc-by-4.0 licensed. From https://en.wikipedia.org/wiki/File:Creative_commons_license_spectrum.svg.)*

Fig. 17.2 illustrates the spectrum of licenses available from CC, and Fig. 17.3 shows the symbols for each type of CC license along with the rights associated with that license.

PUBLISHING AND CONTRACTS

Authors who are interested in getting their work published will need to sign a contract with a publishing company or a university press. Author agreements

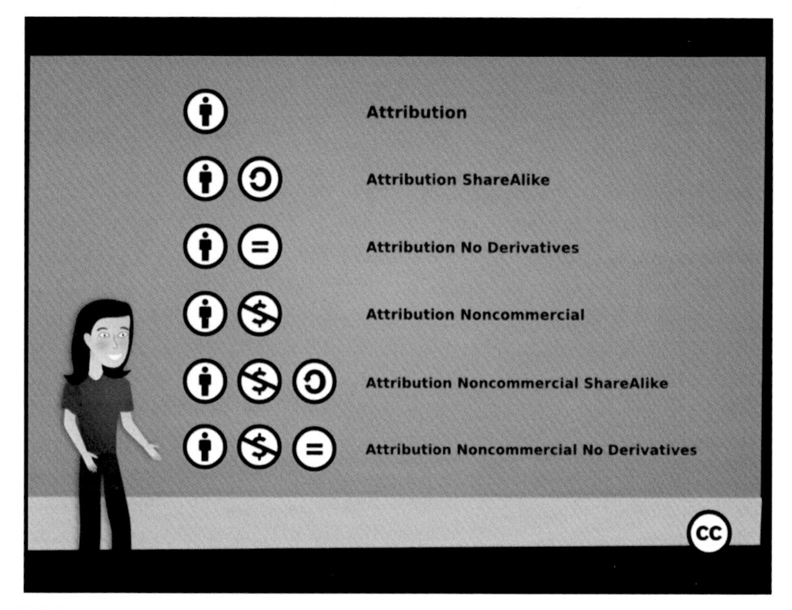

FIGURE 17.3 Creative Commons licenses, symbols, and associated rights. *(From https://wiki.creativecommons.org/wiki/Sharing_Creative_Works_14. Creative Commons CC BY.)*

are issued after the publisher has evaluated and approved the project as submitted by the author. The author can present a well thoughtout idea for a publishing project or a few sample chapters or even the complete manuscript for evaluation. Publishers typically provide information on their websites for prospective authors.

Publishing contracts address at least the following kinds of concerns:

- Parties—who will sign the contract (author, publisher, publishing partners)
- Grant of rights—which rights will be transferred and under what sort of exclusivity
- Copyright
- Distribution—where and how the work will be distributed
- Warranty and Indemnity—promises made to protect each party
- Royalties—calculated usually on net receipts (the amount the publisher actually receives, taking into account discounts and commissions)
- Share of profit—calculated based on the actual profit of a publishing project, after all the direct (and sometimes indirect) costs are deducted
- Distribution of author payments—if there are multiple coauthors or publishing partners, the way the royalties (or share of profit) is distributed needs to be clearly stated
- Out of print provisions—specifies what happens to the rights if the publication is declared out of print (rights often revert back to the author)

- Term and Termination—under what circumstances each party may terminate a publishing contract

  Publishers need certain elements to be stipulated in a publishing contract to be able to publish an author's work as follows:

- Transfer of rights to the publisher to legally publish the work
  The contract will define which rights are being transferred and/or which rights will be retained by the original creator. It will also specify the exclusivity of those rights. Most commercial publishers and university presses require exclusive rights to publish, in part to prevent other parties (including the author) from distributing and selling the work independently.
- Roles and responsibilities of signing parties
  The publishing agreement will clearly define what the publisher will do and what the author will do. The publisher will generally assume all responsibilities, including financial responsibility, for producing, marketing, distributing, and selling the work. The author assumes all responsibilities for creating the content and delivering it to the publisher.
- Original work
  Publishers need assurance that the work is entirely original, except where attributed to others (e.g., attributed quotations or third-party content with permission). This includes work from all the coauthors and co-contributors.
- Schedules, deadlines
  By signing the contract, the author commits to delivering the manuscript (and any related components) to the publisher by a specific date. In addition, the author agrees to work with the publisher in a timely manner during the publishing process so that the project does not encounter unnecessary delays (e.g., reviewing proofs).
- Defamation, indemnification
  Publishers need assurance that no defamation to others is included in the work to reduce the possibility of legal action. Likewise, should a situation arise where either the author or the publisher becomes involved with any third-party disputes, the contract's indemnity clause (as agreed on by the author and publisher) would go into effect.

AUTHOR CONTRACTS FOR JOURNALS AND BOOKS

When a paper is accepted for publication in a journal, the author will receive an agreement from the journal publisher to sign that gives the publisher the right to publish that paper. Likewise, when an author submits an invited chapter for a book, a similar agreement will be sent before the manuscript can proceed to publication. Traditionally publishers (and professional societies with publishing programs) have required that authors transfer copyright and all other rights exclusively to the publisher. However, more recently, with the rapid growth in Open Access publishing, authors are able to retain copyright for their contributions and can retain certain rights regarding the use of the work.

These include the right to

- post the manuscript (or even the published article) on the author's personal website or laboratory website
- post the manuscript (or published article) to an institutional repository
- distribute a specified number of offprints (electronically) to colleagues and collaborators

Ultimately, the publisher may be willing to give the author the option of retaining certain rights as well as copyright for his or her paper. But, some journals are quite rigid in their policies and the author may not have much room for negotiation. Then the author will need to decide whether to publish the paper in that particular journal or perhaps consider alternatives. Similarly, some book publishers are not willing to negotiate terms for author contracts (in part because it can cause problems with the other authors in the same book). Again, the author has a choice whether to accept the terms as they are presented or to withdraw the chapter from the book.

PROTECTING YOUR INTELLECTUAL PROPERTY

Publishing research and other scholarly information is an essential part of science and medicine. Working with publishing organizations, be they large companies, university presses, professional societies, or related vendors, is usually inevitable. Authors will receive contracts and those contracts will need to be signed to proceed with publication. But, it is important to be aware of author rights, what may or may not be negotiable, and what it means for the way the work will ultimately be distributed and used.

One of the most important and most commonly overlooked considerations for managing your intellectual property is also the simplest: retain copies of your signed contracts. These are important documents about one's career. Knowing what your rights are and what you have agreed to over the years is vitally important to later use of your work and managing your intellectual legacy. It is never too early to start. The Authors Alliance promotes authorship with particular attention to the scholarly community and is an excellent source of information (http://www.authorsalliance.org/).

> *Be a responsible steward of your intellectual property. Retain vital rights for you and your readers while authorizing publishing activities that benefit everyone by making scholarship more widely available [4].*

Many research libraries increasingly are hiring people with copyright and publishing expertise to help faculty, staff, and students with concerns related to publishing contracts and copyright. The University of Michigan Library has an established Copyright Office with knowledgeable staff available to assist the University community with questions about any of the topics covered in this chapter.

http://lib.umich.edu/copyright.

copyright@umich.edu.

REFERENCES

[1] Litman J. Digital copyright. Prometheus Books; 2006.

[2] https://creativecommons.org/.

[3] Portions of the copyrightability and rights and permissions sections are adapted from the University of Michigan Library.

[4] www.arl.org/sparc.

Chapter 18

Publication Ethics in Medical Publishing

Christina N. Bennett, PhD
American Physiological Society, Bethesda, MD, United States

INTRODUCTION

Preparing manuscripts for publication requires considerations for details that extend beyond the science. Questions such as the following should be answered prior to submitting the manuscript:

Who qualifies for authorship?
Should my consulting work be declared in the disclosures section?
Is it appropriate to present my data this way?
Can I publish this piece of data again?

Failure to do so may result in editors or reviewers raising questions about the ethics of the work that may either delay or, ultimately, prevent publication of the work.

To support authors in answering these types of publishing ethics questions, scholarly medical journals have publication ethics policies included as part of the Information for Authors or Ethics Policy on the journal website. Many of the journals reference the recommendations from the International Committee of Medical Journal Editors (ICMJE) [1] as their source for setting publication ethics policy.

The following sections highlight the main ethics issues that are addressed in journal editorial offices and describe ways to keep your manuscripts ethically sound.

AUTHORSHIP

Disputes over authorship are perhaps the most difficult ethics issue to resolve as unlike most components of a manuscript, decisions regarding authorship often raise feelings of ownership that make negotiating more complex. According the

Medical and Scientific Publishing. https://doi.org/10.1016/B978-0-12-809969-8.00018-8

ICMJE recommendations, the following four criteria must be met for a contributor to qualify as an author. A contributor should have

- Made "substantial contributions to the conception or design of the work; or the acquisition, analysis, or interpretation of data for the work; AND"
- Drafted "the work or revising it critically for important intellectual content; AND"
- Approved "the final version to be published; AND"
- Agreed "to be accountable for all aspects of the work in ensuring that questions related to the accuracy or integrity of any part of the work are appropriately investigated and resolved."

These recommendations are very useful for beginning the conversation with potential authors about who qualifies for authorship. Discussions about authorship should occur with every transition in the study as contributions to the work change and those who you initially thought would be authors ultimately may not make a significant contribution and, likewise, those who you had not considered initially may be the ones making significant contributions.

Not only may the author list change but also often the order of the authors in the list changes as the project develops. For biological sciences, the first author is the person who develops the study, performs the experiments, and drafts the manuscript. The last author is the person who obtains the funding, manages the study, and provides critical evaluation of the work throughout the study and during manuscript preparation. The middle authors often contribute to a portion of the work and support the intellectual development of some, or all, aspects of the study.

The first-author designation is often the most coveted as it is considered to represent the person who provided the most significant amount of work on the study and deserves the credit for the discovery. Authorship disputes often arise between would-be first and second authors, especially when two contributors have made equally significant contributions to the work. One way to resolve these types of concerns is to have two first authors by acknowledging their equal contributions in the author byline.

The other major ethical issue regarding authorship in medical publishing involves guest and ghost authors. Guest authors are defined as those listed as an author who did not contribute to the study in any significant way. Authorship should not be given simply because someone provided a reagent, read the paper, married you, plays golf with you, or is your department chair.

Ghost authors are those who made a significant contribution to the manuscript but do not want to be listed as an author. Why would someone NOT want to be an author? Perhaps, you collaborate with a company that asked you to test their reagents in your model system. Not only do they provide the reagents but they also assign one of their scientists to work with you on the study. Their support on the study and in drafting the manuscript qualifies for authorship. However, the scientist requests to remain off the author list to avoid the perception that the company influenced the findings. Now consider that 3 months later both the guest- and ghost-authored papers are called into question for plagiarism. It is likely that the guest authors will immediately claim that they had nothing to do with the study and ask to be removed from the author list. Likewise, the authors on the manuscript with the ghost authors will be angry that some of the "real" authors are not being held accountable. The authorship list cannot be changed once concerns are raised, that is why authors should be limited to those who meet all four of the ICMJE criteria.

Medical Journals try to confirm that the authorship list is correct by informing all authors that the manuscript has been submitted for review after the initial submission and after every revision. In addition, any changes to authorship require that ALL authors sign a change of authorship form. In this way, the journal is assured that all authors are aware of the additions, deletions, or reordering of authors before the manuscript is published.

How can you avoid authorship issues? Be sure to seek approval of the draft manuscript from all authors before you submit it to the journal. In fact, save the emails from your coauthors that states their agreement with the submission. Be sure to have current contact information for all authors and load it into the manuscript submission system so that all authors will be informed about the manuscript progress. Lastly, be sure to resolve any authorship disputes before submitting the manuscript.

CONFLICTS OF INTEREST

Researchers use their expertise in many activities, both academic and private. That is, researchers may run an academic lab and work as a consultant for a biotech company that has similar research interests. Some academic researchers also run small private companies to translate their research discoveries into a marketable product. When authors have these dual interests there is concern that the findings reported in their research articles are biased to promote their private interests. These types of situations are defined as conflicts of interest. To help address this concern, medical journals require authors to declare their nonacademic activities at submission and within the manuscript in the Disclosures section. The Conflict of Interest disclosure form (posted here: http://www.icmje.org/conflicts-of-interest/), developed by the ICMJE, is used by many journals to obtain information from authors about their other interests. These interests include financial activities outside of the submitted work (i.e., consulting fees, speaker fees, funding from nonacademic companies), intellectual property (i.e., patents and copyrights), and any other interests not covered by the first two sections.

Declaring your interests does not mean that you have a conflict of interest. It simply allows editors, reviewers, and ultimately readers to be aware of your interests and make their own conclusions about whether there may be a bias in the interpretation with the work. In fact, it is when you fail to disclose your interests that someone may say that you are hiding a conflict of interest.

How can you avoid concerns about conflicts of interest? Be sure to disclose all of your nonacademic activities that could be perceived to influence your study. Even if you do not think that your consulting activities or speaker engagements for a private research or pharmaceutical company are of consequence, they show that you have a relationship with the company. Your name and pictures of your activities may be posted on company websites, issued in press releases, or declared on their financial records. It is better for you to declare your nonacademic interests, than to have a reviewer or reader raise concerns about your relationships to the journal.

Also, take the time to consider your nonacademic relationships and assess whether they have influenced your thinking or perception of a particular study or area of research. It could be that by being immersed in a new and exciting initiative, you are motivated to show that a new drug works as intended or that a new device is superior to the ones currently available. To help avoid this, be sure that you have seen all the data acquired and analyses performed, and interpret the data according to the standards that make you an expert in the field. If you are not assured that the study was performed and analyzed as you had planned, then you should seek clarification about the data before including them in a manuscript.

DUPLICATE SUBMISSION

Once the study has been completed, the paper drafted, and authorship assigned, you are ready to submit your manuscript to a journal for review. Specifically, your manuscript should be submitted to one journal for review at a time. This means that you submit your work to your journal of choice and wait for them to give you a decision (i.e., accept, revise, or reject) before you consider sending your work to any other journal. The reason why there is a publication ethics guideline prohibiting submission of the same manuscript to more than one journal, a duplicate submission, is twofold. First, it protects the efforts of the reviewers who donate their time to review manuscripts by not having the same pool of experts review the same paper. Second, it protects the publishers from having to determine rights to the paper if both journals accept the work.

Authors may think that they are increasing their chances to have their paper accepted by submitting the manuscript to more than one journal at a time. However, when experts in the field are invited to review the same manuscript by two or more different journals, chances are that the reviewer will refuse to review the work for any journal. Moreover, the journals will likely reject the manuscript when the reviewer informs them of the duplicate submission.

How can you avoid concerns about duplicate submission? Be sure to submit your work to one journal at a time. Also, be sure you have designated one person to submit the work, often this is the corresponding author. Particularly, when the author list is composed of authors from different institution or locations, be sure that all authors are aware of where the work will be submitted and who will submit it. Lastly, inform the journal if your work was submitted to more than one journal. It is much better for you to alert them to the error rather than a reviewer raising the concern.

DUPLICATE PUBLICATION AND DUPLICATE DATA

Publishing ethics issues are not always detected during peer review. Often, readers alert authors and journals to potential errors in the literature, including duplicate publications. According to the ICMJE, a duplicate publication is a "publication of a paper that overlaps substantially with one already published without clear, visible reference to the previous publication." Medical journals review manuscripts for publication with the expectation that the work is new and original. As such, publishing the same information, in whole or in part, in more than one manuscript violates acceptable standards of practice. First, publishing the same study more than once can make a singular conclusion more influential than it actually is. For example, if a metaanalysis study to determine which drug is the most effective treatment for a particular disease includes results from two articles that present the exact same data, then the same results will count as two "independent" findings. As such, the metaanalysis may identify a treatment that may in fact have not been rigorously evaluated in the literature and influence recommendations for patient treatment or care. Second, duplicate publication is seen as a means to "game the system." By publishing the same work more than once, perhaps with a slightly different title or figure set, it shows that the authors are trying to increase the appearance of productivity without doing the work. Of course, once it is discovered by a reader, a tenure committee, or a journal, the likely outcome is that one, if not both articles, will be retracted.

There are instances when publishing the same data in more than one article is appropriate. For example, if data from one study result in multiple publications, then the baseline data for the human subjects in the study should be reported in each publication. Another example would be if there is a longitudinal study and there is value in comparing the data from year 1, which is already published, with the new data from year 3. In such cases, the original data could be included in the new manuscript. However, any reuse requires full disclosure within the manuscript. The original article should be referenced, and the reuse should be declared within the text and the legends so that the reviewers, and ultimately the readers, are aware that portions of the data have been previously published. Note that duplicate data have their limits. Portions of a figure or table may be reused; but if a large portion of the data has been previously published, likely reviewers will not consider the work to be original.

How can you avoid issues regarding duplicate publication? Be sure to publish your manuscripts and data sets only once. If there is a reason to reshare previous results be sure that the reuse is disclosed within the text, and it is a great idea to inform the editor within the cover letter at submission. Again, be proactive in declaring the reuse so that the editor is not taken by surprise when a reviewer asks about why previously published information is included in the manuscript.

SELF-PLAGIARISM

As you begin to publish manuscripts in a particular research field, it is likely that you will develop a writing style that clearly articulates the discoveries in your area of research, the methods used in the lab, and the questions that remain unanswered. The descriptions of these details may be perfect as written and seem reasonable to use in subsequent manuscripts. However, reuse of text, even if it is your own words, is not acceptable practice. Every manuscript should be an original work with original thoughts. Moreover, publishing portions of a manuscript that have already been previously published could violate the copyright of the original publisher and constitute a legal infringement of the original work.

The one section of the manuscript that often contains previously published text is the Methods section as research groups tend to use similar methods in multiple papers. However, even if the experimental method is similar to prior publications, it does not mean that it is the exact same method. For example, the reagents that were used in the study that the lab published 5 years ago may be discontinued or supplied by another company. By copying and pasting the old methods into the new manuscript, the information is outdated and it will make it difficult for a reader, or even you, to repeat the work if you use those methods as your protocol. In cases, where the methods really are the same from one paper to the next, it is good practice to refer readers to the prior publication rather than copying and pasting the same information into the new publication. This practice also allows you to have more space in the manuscript to write a detailed methodology for the experiments that are new to the field and increases the likelihood that readers will use, and cite, your manuscript as the source for their experimental methods. That being said, no reader likes to follow "as previously described" references that do not lead to the original and definitive method. Be sure that what you reference describes the experiment you performed.

Reusing text from prior publications in other sections of the manuscript is much less acceptable. Introduction and Discussion sections should set the context for the current paper and contain original interpretations rather than repeating the same information from a prior publication. Likewise, the text included in a Review article should be your own words, not a compilation of text from your prior publications. Medical journals want you to write a manuscript with your own thoughts, own voice, and own interpretations.

Many medical journals use plagiarism detection software to screen articles for textual overlap before publication. The plagiarism software is designed to compare your manuscript to those already publishing in the scholarly literature. Thus, reuse from prior publications is likely to be detected by the software. When it is detected, depending on the degree of reuse, journals will require authors to rewrite the manuscript before considering it for publication. If the reuse is significant, editors may prefer to reject the manuscript due to a lack of originality.

How can you avoid self-plagiarism? First, be sure that your contribution to the manuscript is original. Even if you obtain facts from your prior publications, write them in your own words rather than copying and pasting from one document to the next. One may have good intentions to copy text and rewrite it later but there is always the possibility that you will forget this step and retain verbatim thoughts from prior publications in your new manuscript. Second, be sure to remind your coauthors to write their contributions to the manuscript in their own words. In particular, the coauthors with an extensive list of publications may find it easier to copy and paste technical and factual descriptions from their prior publications rather than write it "again" in their own words. They should be reminded that this is a new work that requires a new way to present the information. Lastly, consider screening the manuscript for text reuse before submitting it to the journal. By doing so, the work can be corrected before the journal detects it.

PLAGIARISM

Unlike self-plagiarism, plagiarism is a much more serious offense. Plagiarism is when an author takes text from another source and uses it as if it was his or her own. This is not acceptable practice as there is an expectation in the academic and professional community that the work you present represents your own thoughts and ideas. Most institutions and funding agencies consider plagiarism to be research misconduct that could lead to termination of employment in addition to a loss of confidence in your work by your research community.

When preparing the text of a manuscript, it is good practice to write facts and concepts in your own words and cite the original source. By writing facts in your own words, you lessen the likelihood of including verbatim text in your new work. Likewise, by citing the original source, you are sure to acknowledge the original contributor and give them credit for their contribution to the research field. When

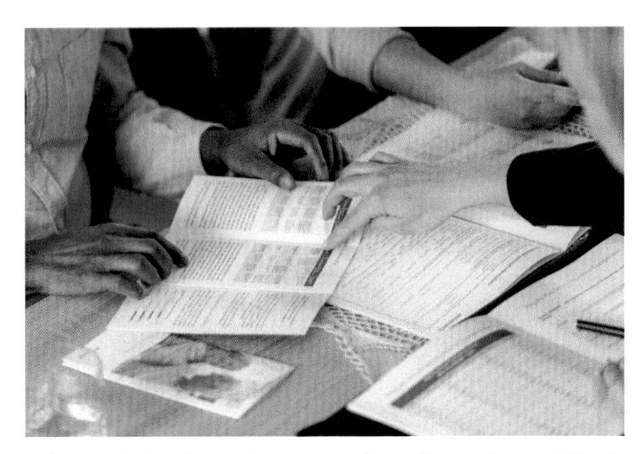

FIGURE 18.1 Photo depicting data and text comparisons from various publications (iStock.com/simonker).

working with multiple authors, it is important to remind them to write their portion of the manuscript in their own words and cite all relevant primary sources. Moreover, every author has a particular writing style that has been influenced by different education backgrounds and cultures. Thus, what may be considered plagiarism to you may not be plagiarism to a coauthor. Be sure that there is a defined expectation for text preparation for all authors on the manuscript.

Medical journals screen for plagiarism using the same plagiarism detection software as described above. If plagiarism is detected, it is possible that you will be allowed to submit corrections, particularly if there is an indication that education on proper writing practices was not provided to a junior author. Senior authors, however, may not get the opportunity to rewrite text that has been plagiarized. Rather, the medical journal could reject the manuscript and inform author's institution about the concern. It is up to the institution to determine if the plagiarism is a one-time mistake or a common practice that requires immediate correction.

How can you avoid having plagiarized text in your publications? As noted above, all coauthors must use writing practices that support writing text in their own words and citing the original sources. Also, try to work on writing projects in a stress-free environment and give yourself time to think and write because if you are working with a tight deadline under high stress, it may increase the likelihood of taking shortcuts and making mistakes. Even a few simple copy-paste entries into your text may lead to rejection of your work. If you need more time, ask for it. It is far better to write an original work and preserve your academic reputation (Fig. 18.1).

ANIMAL AND HUMAN STUDIES ETHICS

Many research studies involve the use of animals or humans as subjects. Reporting these studies requires special consideration, as medical journals need to be assured that the work was performed according to international, national, and institutional guidelines and that the animals or human subjects were treated with the highest standards of care and treatment. Therefore, medical journals provide specific guidance in the Information for Authors on how to report animal and human subject studies.

Medical journal require that the Methods section include specific statements of approval for conducting ethical and humane animal and human subjects studies. For animal studies, statements of IACUC approval and details about humane treatment of the animals throughout the study should be declared. For human subject studies, statements of IRB approval and receipt of written informed consent from all subjects must be declared. Likewise, the protocol included in the Methods section should be detailed enough for readers to know how the experiment was performed. For animal studies, experimental details should include species, strain, age, and sex of the animals. It should also describe how animals were selected for the various treatments and the number of animals used for each intervention or treatment. Likewise, the description should describe any adverse events or changes to the protocol to support the humane treatment of the animals. For human subject studies, the experimental details should provide a description of their baseline health and characteristics such as age, weight, sex, and any other parameters that are relevant to the study. It should also describe how many subjects were studied in each intervention and whether any dropped out of the study or experienced adverse events.

Medical journals rely on reviewers and journal staff to ensure that this information is reported in the manuscript. Reviewers and editors will raise concerns even if statements are included if the study sounds unethical or would not be approved by most academic IACUC's. Authors would have to justify their reasoning for performing the experiment as described and confirm that their institution approved the work. If the concerns are addressed, then the editor may consider publishing it with modifications to the manuscript that includes a more complete description of the study and rationale for the work. Manuscripts will not be published if a study was not approved.

How can you avoid concerns about animal and human subject study ethics? Be sure to seek approval from your institution for your studies prior to starting. Also, be sure to run the study that was approved and seek approval for any major changes to the protocol. Likewise, be sure that your manuscript describes the studies completely and declares the steps you took to run a humane study. Failure to run studies or report the findings following appropriate institution standards leads to more regulations for everyone.

FABRICATION AND FALSIFICATION

The data that are presented in the figures of your manuscript are often the most viewed portion of your work. Readers can get a lot of information from the figures without having to read much more than the legends that accompany them. There is an expectation that the data presented in the figures represent actual results obtained during the course of the study. The data may be derived from one experiment or from several repeat experiments. In either case, what is presented in the figure should have an associated data set, original photograph, trace, or other source data that can confirm that the information is real. If there are no original source data, then it should not be presented as a figure. Making up results for experiments that never took place is defined as data fabrication. Manipulating data to reach conclusions that are different than the source data is defined as data falsification. Falsifying and fabricating data is considered research misconduct by most academic institutions and funding agencies and can result in termination of employment or loss of good standing in your research community.

Medical journals have strong statements in their ethics policies that prohibit the submission of falsified or fabricated data for publication in a research articles. They are becoming increasingly savvy at detecting manipulated images using forensic software and data analysis programs. Moreover, more and more journals are requiring authors to submit the source data for review or to store the source data associated with the manuscript on a public repository. In this way, reviewers can confirm that the information reported in the figure is consistent with the source data.

When inconsistencies are detected, authors must explain how the figures were prepared and show the data that support the figure. In many cases, inconsistencies are correctable and not considered fabrication or falsification. However, when data are inappropriately manipulated, the manuscript will be rejected and most likely the author's institution will be informed. It is the institution's responsibility to determine whether inappropriately prepared figures are isolated to one manuscript or are common practice for the author and/or lab group. It is also the institution's responsibility to find the best course of action to prevent it from happening again.

How can you avoid including falsified or fabricated data in your manuscript? Be sure to prepare your figures based on actual results and confirm that the information reported in the legends is an accurate description of the data presented. If you did not prepare the figure, ask to see the raw data along with the results presented in the figure. Many journals require authors to declare that they have seen the raw data and take responsibility for the integrity of the data presented. Thus, this request to your coauthors is not a sign of mistrust. Rather, it is an acknowledgment that you take your responsibility as an author seriously. Of course, there are collaborations where an expert has provided data that you could not interpret even if were provided. In those cases, be sure that they are included as authors on the work and that the list of contributions states that they provided a particular piece of data. In this way, if concerns arise, it is clear that the piece of data in question was the responsibility of a particular person.

Second, label and store all of your raw data, especially the data that were used to create the figures. The data should be accessible to all coauthors including the corresponding author as they may be contacted by the journal to submit the original data for a figure even after you leave the lab for another position.

FIGURE 18.2 Photo showing a stack of scientific journals. *(Photo courtesy Marc Stephens, Ann Arbor, MI.)*

Third, avoid editing photographs with digital editing software. Even small edits, such as adjusting contrast, rearranging the order of the lanes, or erasing imperfections can give the impression that the data have been manipulated. Moreover, numerous small edits can add up and alter the original image to a point where it is no longer consistent with the original capture. Likewise data points used to make graphs and charts should not be excluded, added, or adjusted to make the differences more convincing. If the results are not definitive, then it should be presented as is and interpreted as such.

Last, if you are experiencing pressure to produce a certain result or repeat a particular outcome, step back and remind yourself that that is not how science works. You ask a question and interpret the results. If the results are not what you expected, then you adjust your interpretation, not the data.

CONCLUSION

These examples represent the most common ethics issues in publishing. However, publication ethics issues are rarely black and white and outcomes will vary based on the journal. Thus, it is important to be your own best advocate for publishing an ethically sound research article. The more you declare to the editor up-front regarding authorship, potential conflicts of interest, data reuse, and manuscript preparation concerns, the more likely you will be successful in publishing a sound paper. And that is the goal, to have a publication that is honest, transparent, and reproducible (Fig. 18.2).

REFERENCE

[1] International Committee of Medical Journal Editors. http://www.icmje.org/.

Chapter 19

Publishing Ethics: An Interview With the Founders of Retraction Watch

This was a Q&A with Adam Marcus and Ivan Oransky, co-founders of Retraction Watch and The Center for Scientific Integrity, conducted by Jasna Markovac, University of Michigan.

WHAT IS "FAKE PEER REVIEW?"

To answer that question, let us first consider what constitutes "real" peer review. Simply put, peer review is the process by which journals vet the scientific quality and content of a submitted manuscript. Although the systems they use to do so can vary, they all rely in some fashion on the willingness of experts to volunteer their time to read and critique submissions and offer guidance about whether and under what circumstances—modest revisions, major overhauls, etc.—to accept the papers. Generally, journal editors ask two or three experts to peer review each paper that makes it past a first-pass look.

Here is where fakery comes into play. Some journals rely on authors of submitted manuscripts to provide the names of and contact information for experts to serve as potential reviewers. While that is not considered best practice in many fields, some editors frequently find themselves short of reviewers and need to enlarge the pool with suggestions.

Most of the time, the process works smoothly. However, unscrupulous researchers have been known to game the system by giving editors the names of real scholars paired with bogus email addresses linked to accounts they control [1]. How often does this deception occur? At the time of this writing, Retraction Watch has reported on more than 500 such cases, including the removal of 107

Medical and Scientific Publishing. https://doi.org/10.1016/B978-0-12-809969-8.00019-X

papers from a single journal believed to be tainted by fake peer reviews. In one of the earliest cases, a researcher in South Korea was discovered to be cheating the review system after he submitted reviews of 28 of his papers—using someone else's name—in less than 24 h [2].

In a variation of fake peer review, the *Journal of Vibration and Control* had to retract more than 60 papers in a scandal involving a "peer review" ring [3]. In that scheme, a group of authors agree to give favorable reviews to one anothers' papers.

Sometimes the authors are not to blame. In some countries, particularly China, it is not uncommon for researchers to hire third-party companies to help them get their papers into shape, particularly if English is not their first language. However, sometimes unbeknownst to these authors, the companies are doing more than that, namely arranging for fake peer review.

Meanwhile, many authors fall victim to so-called "predatory publishers" with legitimate-sounding journal names that promise conventional peer review but deliver none at all.

We will note that unethical scientists have tried other ways to make their work look more impressive than it might be, including by creating citation "cartels" to inflate the impact of their articles. In other words, if gaming an aspect of science is possible, people will try to cheat.

I HEAR THERE IS NOW SOFTWARE THAT LETS PEOPLE AVOID PLAGIARISM DETECTION. IF I AM ESSENTIALLY REWRITING SOMETHING, WHY IS THAT PLAGIARISM?

The US Office of Research Integrity (ORI)—and many scientific codes of conduct around the world—considers plagiarism one of the three cardinal sins for scientists in the United States, along with fabrication and falsification of data [4]. "Plagiarism-detection software" does exist and is widely used by journals.

However, the term is something of a misnomer. Although these engines, including iThenticate, DejaVu, and even Google, can find passages of overlapping text by comparing documents, they do not on their own identify plagiarism. Rather, they produce percentages of similarity between articles—say, an unpublished manuscript and a previously published doctoral dissertation. The amount of overlap serves as warnings to editors, who then must compare by hand the suspicious parts to see if in fact copying has occurred and whether it rises to the level of plagiarism.

In some instances, the offense is obvious—verbatim copying of entire sections, or even entire papers, without attribution, quotation marks, or other attempts to properly cite the original material. Other times the lines are blurrier.

For example, a group of authors that publishes several papers about work using a particular model may have a good reason for writing their methods section in a precise and therefore repetitive way.

Depending on the amount of copying, this sort of repetition can rise to the level of what is clumsily called "self-plagiarism," which the ORI does not consider an offense but publishers might find to be a copyright violation. At the extreme sits duplicate publication of entire articles, in which an author or groups of authors publishes highly similar, or even identical, papers in different journals. Again, this is not typically an issue for regulators but a problem for publishers and the scientific record, in the sense that duplicate articles can artificially inflate the importance in the literature of a particular finding.

The ORI and the National Science Foundation (NSF), a major funder of research in the United States, are quite clear about how authors can protect themselves against claims of plagiarism: Cite, quote, attribute. If you feel that you need to write something in a way that is similar to or perhaps even identical to the way someone else has stated it, quoting that source will keep you on the side of the angels [5].

WE HEAR A LOT ABOUT "SCIENTIFIC MISCONDUCT." WHAT EXACTLY IS IT AND DOES IT ONLY APPLY TO PUBLISHING?

"Scientific misconduct" is a broad and not particularly illuminating term to describe the range of transgressions researchers can make as researchers (as opposed to private citizens). The ORI specifies fabrication of data, falsification of results, and plagiarism as the three behaviors for which it will consider sanctions against scientists who receive federal funding [4].

A carefully done 2012 study suggested that roughly two-thirds of retracted papers are removed for misconduct as opposed to honest error [6]. Yoshitaka Fujii, a Japanese anesthesiologist who holds the record for most retractions by a single author, with 183, is believed to have fabricated his data [7]. Joachim Boldt, of Germany, who sits in second place with 96 retractions, lost his articles for falsifying his data and for failing to obtain ethics approval in some of his studies [8].

And the notion of "questionable" research practices that do not quite meet the threshold for misconduct is gaining attention. For example, Don Poldermans, a noted cardiovascular researcher in The Netherlands, lost his faculty position at Erasmus Medical Center for a variety of transgressions including being "careless" in his collection of data, according to the institution [9]. And in Sweden, two scientists at Uppsala University were reprimanded after a panel concluded that they had failed to adequately supervise a doctoral student who was found to have manipulated images in four published papers [10].

According to Robert Nerem, who led a panel for the U.S. National Academy of Sciences looking at scientific misconduct: "Research institutions need to have a much broader focus than fabrication, falsification, and plagiarism — one that includes detrimental research practices — if they are to really foster research integrity" [11].

The panelists also urged institutions, universities, publishers, journals, etc., to help create a nonprofit Research Integrity Advisory Board that would help promote best practices for research ethics [12].

And no, misconduct does not only apply to published papers. The US NSF says it has seen a dramatic increase in the number of funding proposals with evidence of misconduct, particularly plagiarism [13]. According to the agency, had it applied plagiarism-detection software to all 8000-odd grants it made in Fiscal Year 2011, between 1% and 1.5%, or roughly 800–1200, would have triggered alarms.

I HEARD THAT PEOPLE WHO HAVE BEEN ACCUSED OF SCIENTIFIC MISCONDUCT, SOMETIMES EVEN MULTIPLE TIMES, CONTINUE TO RECEIVE FEDERAL FUNDING. CAN THIS BE TRUE?

The ORI, which oversees recipients of federally funded research, seeks sanctions against scientists who commit misconduct with taxpayer funding [14]. A typical penalty might involve a 3-year proscription on receiving another government grant plus a similarly long period of supervised research [15]. Occasionally the punishment will last 5 years and in only the rarest cases will the penalty involve criminal prosecution [16].

However, the answer to the question is, Indeed it can. A 2017 study in the Journal of Empirical Research on Human Research Ethics found that 17 researchers who received sanctions from the ORI for committing misconduct on federal grants subsequently received a total of more than $100 million in additional government funding once their sanctions were over [17]. "Clearly, misconduct is not the career-killer one might have expected," Kyle Galbraith, of the University of Illinois, who conducted the study, told *Science* [18].

DO PUBLISHERS OR JOURNAL EDITORS KEEP LISTS OF PEOPLE WHO HAVE COMMITTED VARIOUS TYPES OF PUBLISHING MISCONDUCT (OR PLAGIARISM)?

Perhaps editors keep the names of known miscreants in a file somewhere the way delis post the names of check bouncers for cashiers to see, but journals and publishers tend to say little about whether they keep black lists. Elsevier, for example, states only that "Decision by the editorial board on future submissions by the author or author group" is among the possible sanctions resulting from misconduct [19].

The most aggressive reaction to deceitful authors is the publishing ban. The Committee on Publication Ethics (COPE), which advises editors, authors, and journals about best practices, discourages publishing bans against individual authors [20]. One reason COPE gives for this stance is that retraction is intended to clean up the literature, not punish wayward authors. Another reason: the group has been warned that bans might constitute restraint of trade.

That said, publishers have levied bans of varying terms on certain authors [21]. For example, in 2010 the American Society of Microbiology issued a 10-year publishing and reviewing ban on Naoki Mori, a Japanese virologist whose questionable use of images led to the retraction of 30 papers [22]. And in 2012, the Indian Journal of Dermatology banned a group of authors from Tunisia who turned out to be serial plagiarists [23]. We also have seen 5-year bans from journals for the misuse of a student's research and double submission (when authors submit the same paper simultaneously to different journals) [24].

WHAT ABOUT COAUTHORS ON A PAPER WHERE THERE IS SUSPECTED SCIENTIFIC MISCONDUCT? WHAT IF THERE IS AN INVESTIGATION? IS EVERYONE RESPONSIBLE OR JUST THE SENIOR AUTHOR OR JUST THE CORRESPONDING AUTHOR?

This question has no short answer. Sometimes coauthors are involving in misconduct, other times not. And sometimes the involvement does not emerge until later. Perhaps the most spectacular such case involves a Japanese anesthesiologist named Yuhji Saitoh, who was a frequent coauthor of Yoshitaka Fujii (see Question 3). Saitoh initially was not implicated in the scandal that led to the retraction of 183 of Fujii's papers. But a recent statistical analysis concluded that the odds were overwhelming that Saitoh had tampered with his data, if not outright fabricated his findings, in many of his papers [25]. Saitoh has received a permanent ban from the Japanese Society of Anesthesiologists [26].

On the other hand, retraction notices frequently single out one author and state that coauthors were unwitting of any misconduct. In those instances, it's safe to assume that the editors—and any institutional investigators—were satisfied that no conspiracy to commit fraud occurred. Such was the case with Scott Reuben, a pain specialist in Massachusetts who fabricated data in roughly 20 clinical trials [27]. As an investigator in multisite studies, Reuben was easily able to deceive his coauthors because they had no way of knowing that the results he was contributing were not legitimate.

However, it is worth pointing out that every named author on a paper in theory bears at least some responsibility for the content of the article. Journals often now require attestations of contributions from each person who participated in the writing of the manuscript, and no one who did not take part in the work should be listed among the authors [28].

WE OCCASIONALLY GET EMAIL INVITATIONS TO SUBMIT OUR WORK TO VARIOUS JOURNALS. HOW DO WE KNOW THESE ARE REAL JOURNALS AND NOT A SCAM? I HEARD OF "PREDATORY JOURNALS." CAN YOU EXPLAIN?

So-called "predatory journals"—the term is a bit controversial—promise all the trappings of legitimate scholarly publishing, peer review, editing, etc., but offer little or none of those services. Meanwhile, they charge fees to researchers who get effectively nothing in return for their money except an article in a journal that few, if any, of their colleagues will read and no reputable scientists will take seriously.

Predatory publishers often choose names for their companies and journals that sound legitimate. Outfits such as the *American Society of Science and Engineering*, *Engineering and Technology Publishing*, and the *Global Science Publishing Group* appear at first glance to be above board but on inspection are simply paper factories. They also have been known to spam authors to solicit submissions, play games with their editorial boards, charge royalty fees for citations, and otherwise behave badly [29].

Jeffrey Beall, a research librarian at the University of Colorado, Denver, identified roughly 1000 publishers and journals he considered predatory. His blog, ScholarlyOA, cataloged these unscrupulous companies and explored the many and often creative ways they attempted to deceive unsuspecting authors. (Beall's list was not without critics, including publishers who objected to his characterization of their business model, some of whom threatened lawsuits against him, as well as advocates of open-access publishing, which Mr. Beall scorned. Many of those advocates raised serious questions about his methodology.)

Mr. Beall has abandoned his black-list project, so although his catalog can still be found online, it is no longer being updated. Some groups are compiling "white lists" of legitimate publishers and journals, but these have the disadvantage of not being comprehensive. A group of major publishers, along with COPE, has launched a campaign, called Think.Check.Submit, to offer guidance for authors hoping to steer clear of predatory journals [30].

Why predatory publishers are able to thrive involves a systemic problem with science itself. The heavy emphasis on the paper as the most important output in research—above even robust and reproducible results—drives scientists to churn out papers at a dizzying and, ultimately, preposterous pace: at least a million per year and growing. Until science comes to terms with the demand problem—by deprioritizing the published article—the situation is unlikely to improve.

REFERENCES

[1] Ferguson C, Marcus A, Oransky I. Publishing: the peer-review scam. Nature 2014;515(7528):480–2.

[2] Oransky I. Retraction count grows to 35 for scientist who faked emails to do his own peer review. Retraction Watch; 2012. Available at: http://retractionwatch.com/2012/09/17/retraction-count-for-scientist-who-faked-emails-to-do-his-own-peer-review-grows-to-35/.

[3] Oransky I. SAGE Publications busts "peer review and citation ring," 60 papers retracted. Retraction Watch; 2014. Available at: http://retractionwatch.com/2014/07/08/sage-publications-busts-peer-review-and-citation-ring-60-papers-retracted/.

[4] Code of Federal Regulations (CFR). Research misconduct. Public Health Serv Policies Res Misconduct 2005:93–103. 42 CFR §.

[5] Roig M. Avoiding plagiarism, self-plagiarism, and other questionable writing practices: a guide to ethical writing. 2015. Available at: https://ori.hhs.gov/avoiding-plagiarism-self-plagiarism-and-other-questionable-writing-practices-guide-ethical-writing.

[6] Fang FC, Steen RG, Casadevall A. Misconduct accounts for the majority of retracted scientific publications. Proc Natl Acad Sci USA 2012;109(42):17027–33.

[7] Marcus A, Oransky I. How the biggest fabricator in science got caught. Nautilus; 2015. Available at: http://nautil.us/issue/24/error/how-the-biggest-fabricator-in-science-got-caught.

[8] Oransky I. The retraction watch leaderboard. Retraction Watch; 2015. Available at: http://retractionwatch.com/the-retraction-watch-leaderboard/.

[9] Marcus A. Breaking news: prolific Dutch heart researcher fired over misconduct concerns. Retraction Watch; 2011. Available at: http://retractionwatch.com/2011/11/17/breaking-news-prolific-dutch-heart-researcher-fired-over-misconduct-concerns/.

[10] Palus S. Biologists earn 5th retraction following Swedish investigation. Retraction Watch; 2016. Available at: http://retractionwatch.com/2016/05/05/biologists-earn-5th-retraction-following-swedish-investigation/.

[11] McCook A. U.S. panel sounds alarm on "detrimental" research practices, calls for new body to help tackle misconduct. Retraction Watch; 2017. Available at: http://retractionwatch.com/2017/04/11/u-s-panel-sounds-alarm-detrimental-research-practices-calls-new-body-help-tackle-misconduct/.

[12] National Academies of Sciences, Engineering, and Medicine (NASEM). In: Fostering integrity in research. Washington (DC): The National Academies Press; 2017.

[13] Mervis J. NSF audit of successful proposals finds numerous cases of alleged plagiarism. Science 2013. Available at: http://www.sciencemag.org/news/2013/03/nsf-audit-successful-proposals-finds-numerous-cases-alleged-plagiarism.

[14] Office of Research of Research Integrity (ORI). Administrative actions. 2017. Available at: https://ori.hhs.gov/administrative-actions.

[15] Office of Research of Research Integrity (ORI). Case summaries. 2017. Available at: https://ori.hhs.gov/case_summary.

[16] Oransky I, Abritis A. Who faces criminal sanctions for scientific misconduct? In: 5th world conference on research integrity. 2017. Available at: https://www.eventure-online.com/eventure/public/publicAbstractView.form?id=310802&congressId=10578&from=session&fromId=377128.

[17] Galbraith KL. Life after research misconduct: punishments and the pursuits of second chances. J Empir Res Hum Res Ethics 2017;12(1):26–32. http://dx.doi.org/10.1177/1556264616682568.

[18] McCook A. U.S. researchers guilty of misconduct later won more than $100 million in NIH grants, study finds. Science 2017. Available at: http://www.sciencemag.org/news/2017/02/us-researchers-guilty-misconduct-later-won-more-100-million-nih-grants-study-finds.

[19] Elsevier. Questions and answers. Elsevier Website; 2017. Available at: https://www.elsevier.com/editors/perk/questions-and-answers.

[20] Committee on Publication Ethics (COPE). Researcher banned for 10 years. Comm Publ Ethics Website 2017. Available at: https://publicationethics.org/blogs/researcher-banned-10-years.

[21] Marcus A, Oransky I. What's behind paper retractions? (5): Banned!. Lab Times 2011;6:49.

[22] Marcus A. Japanese virologist hit with publishing ban after widespread data manipulation. Retraction Watch; 2010. Available at: http://retractionwatch.com/2010/12/24/japanese-virologist-hit-with-publishing-ban-after-widespread-data-manipulation/.

[23] Oransky I. Serial plagiarizers banned from dermatology journal forever. Retraction Watch; 2012. Available at: http://retractionwatch.com/2012/06/20/serial-plagiarizers-banned-from-dermatology-journal-forever/.

[24] Oransky I. Heads up: "Borrowing" your student's work will earn you a partial retraction — and a five-year publishing ban. Retraction Watch; 2014. Available at: http://retractionwatch.com/2014/01/29/heads-up-borrowing-your-students-work-will-earn-you-a-partial-retraction-and-a-five-year-publishing-ban/.

[25] Carlisle JB, Loadsman JA. Evidence for non-random sampling in randomised, controlled trials by Yuhji Saitoh. Anaestheisa 2016;72(1):17–27.

[26] McCook A. Anesthesiology society bans co-author of researcher with record-number of retractions. Retraction Watch; 2017. Available at: http://retractionwatch.com/2017/06/06/anesthesiology-society-bans-co-author-researcher-record-number-retractions/.

[27] Palus S. Scott Reuben notches 25th retraction, for a letter to the editor. Retraction Watch; 2015. Available at: http://retractionwatch.com/2015/11/23/scott-reuben-notches-25th-retraction-for-a-letter-to-the-editor/.

[28] International Committee of Medical Journal Editors (ICMJE). Defining the role of authors and contributors. Int Comm Med J Ed 2017. Available at: http://www.icmje.org/recommendations/browse/roles-and-responsibilities/defining-the-role-of-authors-and-contributors.html.

[29] Oransky I, Marcus A. Science sting exposes how corrupt some journal publishers are. STAT; 2017. Available at: https://www.statnews.com/2017/03/22/science-journal-publishers-sting/.

[30] Oransky I, Marcus A. A famed journal blacklist is dead. Long live a blacklist!. STAT; 2017. Available at: https://www.statnews.com/2017/01/27/journal-predatory-blacklist/.

Part IV

Expanding Access and Increasing Impact

Chapter 20

The Digital Age of Academic Medicine: The Role of Social Media

Todd A. Jaffe[1], David C. Cron[1], Joseph R. Linzey[1], Vahagn C. Nikolian, MD[2], Andrew M. Ibrahim, MD, MSc[2]
[1]*University of Michigan Medical School, Ann Arbor, MI, United States;* [2]*Department of Surgery, University of Michigan, Ann Arbor, MI, United States*

The growth of digital technologies has provided researchers with novel strategies to help disseminate their work. Medical journals and investigators have begun to leverage the connected nature of a digital world to expand their reach. Specifically, the recent upsurge of social media usage in the medical community has provided a new, valuable tool for many involved in health care. Social media connects researchers and learners globally—breaking down walls that were once barriers to rapid and wide dissemination of medical research. Younger generations of learners and physicians have grown up entrenched in social media and are rapidly adopting this technology to advance their education and academic lives. In fact, recent studies have indicated that social media use by both doctors and medical students has increased to over 90% [1].

Although the use of social media in medicine may still be nascent, many benefits have already been demonstrated. The augmentation of social media for medical journals has been shown to improve the reach of published research [2,3]. Social media channels have also facilitated increased collaboration and debate among investigators. Trainees can more easily engage with mentors and stay informed of the latest research through social networks. These functions can provide advantages throughout the research process for researchers and readers, alike.

Medical and Scientific Publishing. https://doi.org/10.1016/B978-0-12-809969-8.00020-6

This chapter provides an overview of popular social media platforms, case studies of successful adoption within academic medicine, overall benefits and drawbacks, and anticipated future directions.

OVERVIEW OF POPULAR SOCIAL MEDIA PLATFORMS

Multiple social media platforms have found utility within the field of medical research and education. An overview of the popular platforms currently utilized in medical research is included in Table 20.1 and described in the following paragraphs. Many specific applications for well-recognized platforms, such as Facebook and Twitter, have already been established. Other social strategies for content dissemination, including blogs and podcasts, are finding their niche in the market. Blogs and podcasts have even partnered with medical journals to improve the reach of content and enhance engagement with readers [4]. Other emerging platforms, including Instagram and Snapchat, may also find their advantages with regard to medical research. However, the scope of these platforms is much less determined at the time this chapter is written.

Twitter

The utilization of Twitter as a social media platform in academic medicine is well established. In total, Twitter has 328 million monthly active users as of April, 2017 [5]. As a "microblogging" platform, Twitter limits users to posts ("tweets") of 140 characters or less. Users can post text, links, pictures, videos, and Twitter-specific polls, and new functions are introduced periodically. Individuals (such as researchers, physicians, patients, lay public) and organizations (such as medical journals) utilize the platform. A user's tweets can be searchable and viewable to the public, unless one chooses to make his or her account private. Users "follow" each other, meaning they can more easily see the posts of those they follow, and their own tweets can be seen by their followers. For example, the Journal of the American Medical Association (JAMA; @JAMA_current) has over 202,000 followers at the time of writing this chapter. These followers are able to see JAMA tweets, which often includes new research articles, and JAMA's followers can then "retweet" these posts to their own followers. Other medical journals and their associated number of followers are included in Table 20.2. The culture of Twitter encourages sharing others' posts, which combined with its simplicity of use and concise posts, has made it a successful platform for journals and researcher to disseminate their research. Users also have the capability to "tag" any other user on Twitter, allowing users to connect with colleagues, strangers, and well-known public figures alike. This aspect is valuable for researchers, physicians, and trainees to form connections and communicate across institutions. Such connections can also be facilitated

TABLE 20.1 Common Social Media Platforms

Social Media Platform	Description	Examples
Twitter	Message broadcasting system that allows users to "tweet" messages and engage with other users in 140 characters or less	• Visual Abstract • Tweet Chats • Journal Clubs
Facebook	Large social network where users can interact with their connections, "like" and share content, create events, and form communities for private engagement and discussion	• Hernia consortium • Research organizations
Podcasts	Audio media platform where content creators can develop episodes that are typically released incrementally; users can subscribe for notifications of content	• Behind the Knife • Healthcare Triage
LinkedIn	Social network designed for professionals, which provides a medium for connections, content sharing, job postings, and occupational support	• LinkedIn groups • Research recruitment
WordPress tumblr Blogger TypePad LiveJournal SQUARESPACE Blogs	Informal channel for content sharing and engagement with readers and has expanded to include video content and live messaging	• Skeptical Scalpel • The Incidental Economist

through the use of "hashtags," which entails placing a "#" before a phrase. These hashtags are searchable through the platform and allows users to search for tweets related to specific topics. For example, using the hashtag #MedEd will allow users to join together in a common conversation related to medical education topics.

Facebook

Facebook is the largest social network platform, with 1.9 billion monthly active users as of December, 2016 [6]. Facebook posts are typically longer and are

TABLE 20.2 Representative Medical Journals on Twitter

Journal	Handle	Followers[a]	Created
Annals of Surgery	@AnnalsofSurgery	18,324	October 2011
Annals of Surgical Oncology	@AnnSurgOncol	2,331	March 2014
Annals of Thoracic Surgery	@annalsthorsurg	1,528	February 2015
ANZ Journal of Surgery	@ANZJSurg	869	March 2010
BJS	@BJSurgery	9,846	March 2011
International Journal of Surgery	@IJSurgery	2,740	March 2013
JAMA	@JAMA_current	202,455	May 2009
JAMA Surgery	@JAMASurgery	17,128	July 2009
Journal of Surgical Education	@JSurgEduc	660	April 2016
Journal of Surgical Research	@JSurgRes	1,621	June 2015
Journal of American College of Surgeons	@JAmCollSurg	5,838	April 2013
Journal of Trauma and Acute Care Surgery	@ JTraumAcuteSurg	8,929	June 2011
New England Journal of Medicine	@NEJM	419,901	March 2009
Plastic and Reconstructive Surgery: PRS Global Open	@prsjournal	15,563	April 2009
Surgery	@SurgJournal	1,553	May 2015
The American Journal of Surgery	@AmJSurgery	1,686	April 2015
The Lancet	@TheLancet	263,389	March 2009
World Journal of Surgery	@worldjsurg	1,641	October 2014

[a]*Follower numbers as of 5/3/17.*
Adapted from Logghe, H.J., Boeck, M.A., Atallah, S.B., 2016. Decoding Twitter: Understanding the History, Instruments, and Techniques for Success. Ann Surg 264, 904–908.

often focused on social communities and visual media including pictures and videos. This factor limits the platforms utility for individuals to widely disseminate research articles. Nevertheless, the large user base of Facebook makes it a valuable platform for researchers and journals. For example, JAMA's Facebook account is followed by 404,000 people as of April, 2017. Since Facebook does not have a character limit, users tend to be more likely to comment on posts, facilitating more in-depth dialogues compared to Twitter, albeit with less broad reach. Facebook can therefore be more valuable than Twitter for enabling conversation within smaller groups and communities. Individuals most commonly communicate with one another by writing comments on each other's "walls" or profile pages. Though users can privately message any other user, individuals often cannot write on another's wall without being "friends" with that person. Social circles tend to be tighter on Facebook, often limited to family, friends, and close colleagues. As such, it is more difficult for researchers, physicians, and trainees to form connections across institutions on Facebook. A recent and unique addition to Facebook is the implementation of live video streams, which can enable live broadcasting of events such as medical conferences. The benefits of these novel functionalities for the medical community are still being explored.

Blogs and Podcasts

Blogs and podcasts are becoming more prevalent and finding their niche in medical education and research. Many medical journals use these media to improve the reach of content and enhance engagement with readers. A blog, short for web log, is a forum for sharing content and allowing public commentary. Medical journals can share posts about recent publications. Researchers, physicians, and the lay public can engage in a discussion via comments on these posts. Podcasts are audio or visual versions of blogs, and they enable more convenient consumption of medical literature. However, they generally have less opportunity for the listener or viewer to engage in discussion. Podcasts may feature authors or other experts discussing a recent publication, and many are educational in nature. For example, the journal *Diseases of the Colon and Rectum* produces podcasts 20–30 min in length that feature an article's author and a guest discussant. This format delivers an engaging and memorable discussion of important articles. *Behind the Knife* is an example of a podcast that targets surgical trainees, covering clinical and academic topics in the portable form of an audio podcast. Podcasts such as this offer a convenient medium that allows medical research to be consumed on the go.

LinkedIn

LinkedIn is a social media platform that is unique when compared to other social media outlets. It operates primarily as an online method of displaying the user's accomplishments in the form of an online resume or *curriculum vitae* (CV) to allow for professional networking. Similar to other social media outlets, individuals can "like" and "follow" organizations such as the "University of Michigan Medical School" or the "Barrow Neurological Institute." These organizations will post articles or content that resembles content found on Facebook. However, for the most part, academic journals have not attempted to utilize LinkedIn to disseminate academic research in the same way they have used Facebook and Twitter. LinkedIn can be a powerful tool to display accomplishments and network professionally, so providing a website link to a well-maintained LinkedIn page on a professional Twitter or Facebook account can be beneficial.

Other Emerging Platforms

Other emerging platforms, including Instagram and Snapchat, may also find their advantages with regard to medical research. Instagram is a social image-sharing network in which the primary posts are photos. Users can engage with one another through comments on shared photos. Given that Instagram is limited to photos, it has not yet been widely adapted by journals. However, like Twitter, Instagram can broadly connect users and enable wide dissemination of content. With the advent of visual abstracts (described below), Instagram may soon gain popularity with journals. Another emerging form of social media, Snapchat, is a mobile phone application for sharing photos and short (10 s maximum) video clips. The benefits of Snapchat include it is easy to use interface that allows rapid and efficient browsing. News sources have taken advantage of this usability, producing multimedia that links to stories and websites. Medical journals have not yet adopted this technology, likely due to the younger user base of Snapchat.

The remainder of this chapter will explore specific examples of successful utilization of social media in medical research, examine the professional and ethical implications of its use, and consider emerging evidence with a look to the future for the technology.

THREE CASE STUDIES OF SUCCESSFUL SOCIAL MEDIA INITIATIVES IN ACADEMIC MEDICINE

Visual Abstracts to Disseminate Research

In July of 2016, *Annals of Surgery* introduced the "Visual Abstract." Simply put, a visual abstract is a visual representation of the information typically found in the abstract section of the article. It is not part of the article itself, but an adjunct used by journals to help disseminate the work on social media so that readers can quickly find articles they are interested in reading. It is also not a substitute for the article, but instead acts like a "movie trailer" to preview the

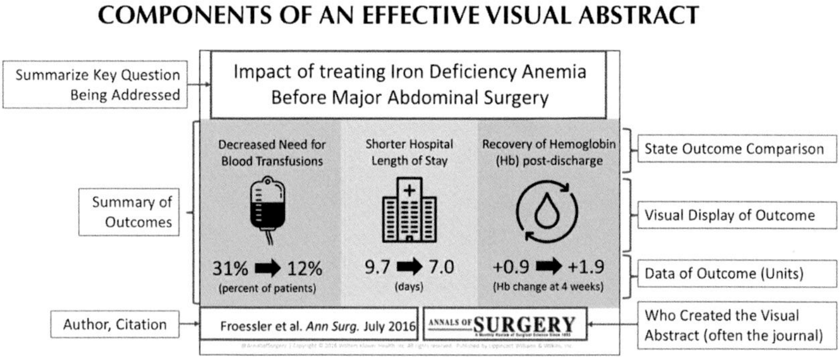

COMPONENTS OF AN EFFECTIVE VISUAL ABSTRACT

Summarize Key Question Being Addressed
Impact of treating Iron Deficiency Anemia Before Major Abdominal Surgery
Decreased Need for Blood Transfusions
Shorter Hospital Length of Stay
Recovery of Hemoglobin (Hb) post-discharge
State Outcome Comparison
Summary of Outcomes
Visual Display of Outcome
31% ➡ 12% (percent of patients)
9.7 ➡ 7.0 (days)
+0.9 ➡ +1.9 (Hb change at 4 weeks)
Data of Outcome (Units)
Author, Citation
Froessler et al. *Ann Surg.* July 2016
ANNALS OF SURGERY
Who Created the Visual Abstract (often the journal)

FIGURE 20.1 Components of an Effective Visual Abstract. *(Reproduced with permission from Andrew M. Ibrahim. A Primer on How to Create a Visual Abstract; April 2017. www.SurgeryRedesign.com/resources.)*

key findings [7]. Components of a typical visual abstract are seen in Fig. 20.1. To date, more than 20 journals have adopted this visual graphic into their social media platforms [8].

The use of a visual abstract on social media has several benefits. A recent case control crossover study conducted by the *Annals of Surgery* demonstrated that tweets including a visual abstract had a three times higher rate of click through to article on the publisher website compared to tweets without a visual abstract [9]. The same images have been downloaded by presenters and used in their slides to reference the study or to facilitate discussion during journal clubs. Because visual abstracts often clarify the key message of the article in a more accessible format, it has also been used by institutions to share information with lay media and press.

Despite its rapid adoption, use of the visual abstract has potential drawbacks. Because a visual abstract contains only a subset of the information in the full article, it has potential to be biased in selectively highlighted specific findings over others. Similarly, the visual and focused nature of the format may "over simplify" or "dumb-down" the research. Finally, although not the intent, many time-crunched readers may use the visual abstract as *substitutes* for reading the article and miss important points about the study's design and limitations. Journals have taken several steps to mitigate these potential drawbacks by established safeguards of external review and always linking the visual abstract to the full text article.

Online Journal Clubs

For many medical professionals, journal club is a ubiquitous component of academic medicine. The modern journal club, which many have attributed to William Osler, serves a dual role: allowing participants to develop skills necessary for critical appraisal of literature, while simultaneously disseminating important information [10]. The Internet has significantly altered the manner in which individuals access information and communicate findings. As such, effective use of social media may allow more efficient sharing and discussion of high-impact literature with a broad audience.

The principles of establishing and maintaining an effective journal club have previously been described by Deenadayalan et al. [11]. Although established for the traditional "in-person" setting, these principles serve as the pillars for "web-based" journal clubs as well. Specifically, an effective journal club is dependent on consistent attendance, a well-defined purpose, strong leadership, thoughtful discussion, appropriate selection of articles, and regular assessment of effectiveness. An additional principle, efficient moderation of discussion, is paramount for a successful Internet-based journal club.

Online journal clubs have evolved with the widespread utilization of the Internet. Journals, such as *Kidney International*, first introduced the concept of online discussion of selected articles [12]. Experts would summarize the strengths and weaknesses of a variety of articles to provide readers with a different perspective of the manuscripts [13]. Unfortunately, these initial ventures, though informative, lacked the interactive qualities, which are quintessential to a strong journal club.

The development of Twitter provided a potential solution to the static nature of previous journal clubs. Twitter allows for real-time communication between users, unlocking geographic limitations associated with traditional journal clubs, and allowing for the breakdown of hierarchy encountered in academic settings. Even novices in a particular realm can assess the thoughts of experts, while testing the waters with statements that may result in further discussion. The first use of Twitter for academic journal clubs was described by Reich in 2011 [14]. This journal club utilized a hashtag to identify tweets associated with the conversation and facilitated the processes of summarizing the discussion. Since this initial venture into Twitter journal clubs, a variety of other journal clubs have emerged. Below we will highlight the use of Twitter-based journal clubs at one institution.

The #UMichSurgJC Experience

Given the success of online-based journal clubs, authors from this article have developed a journal club to supplement residency education. These interactive discussions, though in their infancy, have allowed residents within the institution to connect with authors of the articles being discussed and engage with other faculty and residents around the world. The journal club is promoted through the departmental account @UMichSurgery. Residents and faculty involved in the program simultaneously contact featured guests (article authors) and experts in the field to maximize the content discussed.

To date, topics from emergency general surgery to colorectal surgery have been discussed. These 2-h journal clubs, which follow the traditional journal club, have generated significant attention. Metrics associated with these events have attracted more than 200 surgeons, trainees, and students to actively discuss topics through more than 1000 tweets. The majority of the discussions happen over designated 2-h blocks on a monthly basis. Beyond content of the tweets, which often will include references to supplemental reading, links to educational resources and provide insights that are not in print, these online journal clubs have provided exposure for all involved. Further, these events not only provide content for those who actively engage in that form of communication but can also provide content for others within our department. For these events, a transcript of the discussion is developed and disseminated to all individuals within the department, allowing non-Twitter users to learn from the experience and encourage future involvement.

Although these innovative approaches to resident education are reassuring, limitations are well recognized. Specifically, these events require significant coordination and planning to be successful. Active recruitment of key participants and development of topics in advance of the event are necessary to maintain engagement of the audience. Additionally, these events may exclude a large segment of surgeons who may not utilize social media regularly. In recognition of these concerns, it is likely that future journal clubs will have a hybrid approach, incorporating real-time discussions between intramural participants with Internet-based discussions from extramural participants.

Facebook's International Hernia Collaboration

Facebook's International Hernia Collaboration (IHC) is an example of an innovative and successful use of Facebook for sharing clinical best practices and research. The IHC is a group that allows surgeons to post difficult cases and solicit opinions from other surgeons across the world. This creates a platform for sharing best practices that, at the time of writing this chapter, has 3618 practicing surgeons or trainee members from over 42 different countries. This is a closed group, meaning members must be vetted and approved prior to acceptance, and no protected health information is posted. Surgeons may post a case vignette, often accompanied by preoperative imaging or intraoperative photographs or videos. Surgeons may also share video recordings of their laparoscopic cases to teach clinical skills, and the IHC has recently begun posting live streams of cases. Other members may comment on posted content to share their opinion or stimulate a dialogue among the collaborative. The IHC page receives dozens of clinical posts per week, and posts commonly have 20 to 40 comments each. The success of this social media platform has since led to the creation of similar Facebook groups in other surgical subspecialties.

Facebook is uniquely suited to hosting a collaborative such as the IHC. Facebook is free and accessible in most countries, enabling such a collaborative to have international reach. The creation of groups is a core functionality on Facebook, and without character restrictions to posts, users can have lengthy dialogues about difficult clinical cases. Although a main goal of the group is to share clinical experience and expert opinion to inform clinical decision-making, discussions in this group can serve as an impetus to stimulate new research ideas. Group members may also upload content to the group, thus enabling a repository of literature relevant to the group's medical specialty. Surgeons in practice can benefit by crowd-sourcing difficult cases, and trainees can benefit from the searchable log of clinical vignettes and surgical skill recordings. This concept is applicable to other medical specialties and can also be used as a tool to bring together researchers with similar academic interests.

There are drawbacks to the IHC model that are important to consider. Protecting patient data is a paramount concern. The IHC has administrators who monitor the content to ensure no protected health information is shared, but this remains a concern whenever clinical cases are shared on social media. The closed nature of the group and the member vetting process further ensure content protection and quality control; however, this also limits the reach of any content produced in the group. This is a necessary restriction for groups such as IHC focused on sharing clinical cases and expert opinion. If the goal is to disseminate medical research, an open group with less membership restrictions would provide a more scalable model and enable sharing content in a manner similar to Twitter.

THE BENEFITS OF SOCIAL MEDIA

Multiple benefits of social media in medical research have been documented throughout this chapter. This includes connecting investigators, expanding the reach of published content, and engaging with readers. The advantages of these functions are shared by the various stakeholders, which includes researchers, journals, and readers directly and indirectly involved in the medical profession.

Benefits to Researchers

Social media benefits medical investigators by providing a link to key collaborators and a direct medium with which to interact with readers. As demonstrated by the Facebook International Hernia Collaborative, social media provides a virtual forum with which individuals in the medical community can engage with potential experts. It is not uncommon to witness productive debate, and the opportunity for novel ideas is highly apparent. This benefit is not limited to Facebook as the direct engagement provided by Twitter has paved the way for constructive discussions and the development of multiple collaboratives. Examples include a monthly obesity social media chat, Twitter journal clubs, and direct engagement with national organizations including the CDC and NIH [15]. These forums connect investigators with readers, thus enabling directed feedback and discussion.

Benefits to Journals

For medical journals, social media can help expand the reach of content and improve engagement with readers. Journals have searched for digital methods to expand readership for years. Initial strategies included email and web-based content, which are now core components of new medical journals [16]. Social media has further helped journals reach a broader audience, which includes younger readers. As demonstrated by the success of the Visual Abstract, these novel strategies are having tangible benefits in both reach and engagement. Furthermore, journals can continue to address their diverse audience preferences with dissemination of content through other digital channels including podcasting and video channels.

Benefits to Readers

Readers of medical research can draw significant benefits from social media as well. This includes not only those involved in the medical profession but also patients and their families. As stated above, medical researchers may have more visibility into relevant content through social channels, which can improve collaboration. Others in the medical profession may benefit from prompt knowledge of new practice guidelines or novel studies as published in medical journals. Patients and families may appreciate additional benefit as they have a convenient strategy for both knowledge acquisition and direct engagement with providers.

POSSIBLE DRAWBACKS OF SOCIAL MEDIA

Social Media Use and Professionalism

The public nature of social media lends itself to professionalism concerns regarding its users and its content. Although still a nascent technology, there have been multiple instances in which individuals within the medical profession have faced repercussions due to content they have posted on social media [17]. Content posted on social media can be lasting. Although social networks have implemented safeguards to keep specific material private, it can be challenging to control the dissemination of content. Furthermore, published content invites comments and reader interactions, which can be linked to original authors. This lends further opportunity for professionalism concerns, which may be out of a given publisher's control. Employers frequently use online searches to investigate potential candidates, and these searches often link to social media content. It remains both essential and challenging to maintain one's professional image in a social world.

TABLE 20.3 Guidelines for Social Media Usage and Privacy in Health Care

Agency	Title of Guideline	Representative Sample
AMA AMERICAN MEDICAL ASSOCIATION	AMA Policy: Professionalism in the Use of Social Media	*"Physicians should be cognizant of standards of patient privacy and confidentiality that must be maintained in all environments, including online, and must refrain from posting identifiable patient information online."*[a]
ACP American College of Physicians Leading Internal Medicine, Improving Lives	Online Medical Professionalism: Patient and Public Relationships	*"Use of online media can bring significant educational benefits to patients and physicians, but may also pose ethical challenges. Maintaining trust in the profession and in patient–physician relationships requires that physicians consistently apply ethical principles for preserving the relationship, confidentiality, privacy, & respect for persons to online settings & communications."*[b]
DEPARTMENT OF HEALTH & HUMAN SERVICES · USA	Health Insurance Portability and Accountability Act Privacy Rule	*"The HIPAA Administrative Simplification provisions provided for the establishment of national standards for the electronic transmission of certain health information, such as standards for certain health care transactions conducted electronically and code sets and unique identifiers for health care providers and employers."*[c]

[a]*Association AM. Oberserving Professional Boundaries and Meeting Professional Responsibilities. Am Med Assoc J Ethics May 2015;17(5):432–34.*
[b]*Farnan JM, Snyder Sulmasy L, Worster BK et al. Online medical professionalism: patient and public relationships: policy statement from the American College of Physicians and the Federation of State Medical Boards. Ann Intern Med 2013;158(8):620–7 https://doi.org/10.7326/0003-4819-158-8-201304160-00100 [published Online First: Epub Date].*
[c]*Modifications to the HIPAA Privacy, Security, Enforcement, and Breach Notification rules under the Health Information Technology for Economic and Clinical Health Act and the Genetic Information Nondiscrimination Act; other modifications to the HIPAA rules. Fed Regist 2013;78(17):5565–702.*

Social Media and Privacy

Regarding privacy, the use of social media is largely unexplored territory, and thus content ownership and protecting the dissemination of published work can present a challenge. Health Insurance Portability and Accountability Act and other governing guidelines contain many specifiers related to the appropriate and inappropriate use of content in the public realm. A sample of representative guidelines regarding the appropriate dissemination of information on social media is included in Table 20.3. The concept of deidentifying patient data is one measure that has been widely supported to promote patient privacy. Further adapting these guidelines remains challenging as social media presents a forum for providers to interact directly with patients in a public setting. Appropriate regulation of social media content as it relates to medical research is in flux, and it can be hypothesized that further guidelines will be implemented as the use of the technology matures.

FUTURE DIRECTIONS OF SOCIAL MEDIA IN ACADEMIC MEDICINE

As academic physicians have become more comfortable with utilizing social media, it has played an increasingly significant role in the dissemination of academic research. For example, in 2012 the Twitter account for the *New England Journal of Medicine* (@nejm) had only 66,000 followers, by October 2014 it had 218,000, and at the time of writing this chapter it has over 420,000 followers [18,19]. The exponential increase in followers demonstrates an elevated desire of physicians and the public to engage with academic medical literature. It is imperative for the medical community to adopt social media outlets to not only elevate abilities to disseminate academic research but also to begin engaging with a community that desires more information in a convenient, digital manner. Social media allows academic journals, health organizations, and physicians to connect with each other and the lay public to ultimately increase the awareness and impact of health care research.

Users of social media need to be continually cognizant of modern technologies and additional methods for disseminating their message. Both Instagram and Snapchat, while not currently used to disseminate academic research, have the potential to reach an even broader and younger population. Live video chats, demonstrations, and discussions provide a medium with which master educators and researchers can explain difficult concepts and inspire junior colleagues, medical students, and the public.

In conclusion, social media has provided a long-desired method for quickly and effectively disseminating important research. While Twitter, Facebook, podcasts, and blogs are currently the most popular methods of research dissemination, there are multiple other social media outlets that could prove to be useful in the future.

REFERENCES

[1] George DR, Rovniak LS, Kraschnewski JL. Dangers and opportunities for social media in medicine. Clin Obstet Gynecol 2013;56:453–62.

[2] Baan CC, Dor FJ. The transplantation journal on social media: the @TransplantJrnl journey from impact factor to Klout score. Transplantation 2017;101:8–10.
[3] Peoples BK, Midway SR, Sackett D, Lynch A, Cooney PB. Twitter predicts citation rates of ecological research. PLoS One 2016;11:e0166570.
[4] Khan K. Best of BJSM online: highlights from the blog, podcasts, YouTube page, Google+ and more!. Br J Sports Med 2016;50:258.
[5] Twitter. Selected company metrics and financials, Q1 2017 earnings, Q1 2017 ed. Twitter, Inc.; 2017. www.investor.twitterinc.com.
[6] Facebook. Company Info and Stats. In: Facebook, editor. Facebook Newsroom; 2017.
[7] Ibrahim AM, Bradley SM. Adoption of visual abstracts at circulation CQO: why and how we're doing it. Circ Cardiovasc Qual Outcomes 2017;10.
[8] Ibrahim AM. A primer on How to create a visual abstract. 2016. December 2016 (1st ed.), March 2017 (2nd ed.).
[9] Ibrahim AM, Lillemoe KD, Klingensmith ME, Dimick JB. Visual abstracts to disseminate research on social media: a prospective, case-control crossover study. Ann Surg 2017.
[10] Linzer M. The journal club and medical education: over one hundred years of unrecorded history. Postgrad Med J 1987;63:475–8.
[11] Deenadayalan Y, Grimmer-Somers K, Prior M, Kumar S. How to run an effective journal club: a systematic review. J Eval Clin Pract 2008;14:898–911.
[12] De Broe M. Journal club. Methylation determines fibroblast activation and fibrogenesis in the kidney. Kidney Int 2010;78:430.
[13] Honey CP, Baker JA. Exploring the impact of journal clubs: a systematic review. Nurse Educ Today 2011;31:825–31.
[14] Reich ES. Researchers tweet technical talk. Nature 2011;474:431.
[15] Bhattacharya S, Srinivasan P, Polgreen P. Social media engagement analysis of U.S. federal health agencies on Facebook. BMC Med Inf Decis Mak 2017;17:49.
[16] Colbert JA, Steinbrook R, Redberg RF. Announcing a new JAMA internal medicine website. JAMA Intern Med 2016;176:1749.
[17] Ventola CL. Social media and health care professionals: benefits, risks, and best practices. PT 2014;39:491–520.
[18] Alotaibi NM, Badhiwala JH, Nassiri F, Guha D, Ibrahim GM, Shamji MF, Lozano AM. The current use of social media in neurosurgery. World Neurosurg 2016;88. 619–624 e617.
[19] Micieli JA, Tsui E. Ophthalmology on social networking sites: an observational study of Facebook, Twitter, and LinkedIn. Clin Ophthalmol 2015;9:285–90.

Chapter 21

Caring Through Conversation: Communication in Health Care

Nathan Houchens, MD[1,2]
[1]*Department of Internal Medicine, University of Michigan, Ann Arbor, MI, United States;*
[2]*Department of Medicine, Inpatient Care, Veterans Affairs Ann Arbor Healthcare System, Ann Arbor, MI, United States*

INTRODUCTION

There is no greater agony than bearing an untold story *inside you.*

Maya Angelou

"The best." Time and again, health-care providers have heard these words in answer to the question of what patients would want for themselves while receiving care. And many times, the assumption underlying this phrase is the best doctor, the best test, and the best medication. Time and again, however, it becomes clear that the meaning of these words changes during the patient's journey. "The best," in patients' and families' viewpoints, shifts from the most knowledgeable physician, the most precise computed tomography scanner, the lowest mortality rate to something very different. It becomes the people who care, who can communicate effectively about a range of important topics, who demonstrate compassion and empathy, and who see the patient as a person and not a collection of symptoms and laboratory values. Patients and providers are most satisfied (and indeed have improved health outcomes) when they share a mutual respect and a therapeutic relationship that is based on caring and compassionate people (Fig. 21.1), not things. These values and qualities are shared through communication.

This chapter will focus on the following topics regarding communication in health-care settings:

- The fundamental purposes, complexities, and challenges of communication in health-care settings
- Patient and provider outcomes of effective (and not so effective) communication
- Various models for teaching, learning, and assessing communication
- Novel dissemination methods for scholarship, techniques, and ideas in this field

Medical and Scientific Publishing. https://doi.org/10.1016/B978-0-12-809969-8.00021-8

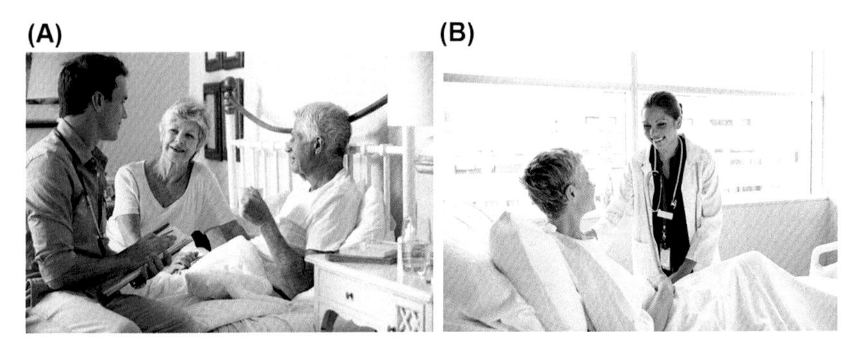

FIGURE 21.1 Photos depict meaningful communications between health-care providers and patients and their families (A photo: iStock.com/bowdenimages) (B photo: iStock.comJohnnyGreig).

It will discuss the importance of learning the principles of effective communication skills so that health-care providers may better elicit the untold story in their patients, their colleagues, and themselves.

COMMUNICATION PURPOSES, COMPLEXITIES, AND CHALLENGES

Communication, the exchange of ideas among individuals in any format, is paramount for the safe and effective practice of medicine. Health-care conversations occur over 2.7 million times each day [1]. Each physician will perform 140,000–160,000 conversations during his or her career [2]. Communication is the foundation for the patient–provider relationship, the method by which the patient's myriad data are acquired and assimilated into diagnostic and therapeutic treatment plans, and the way empathy and humanism are shared in challenging times. Indeed, talk has been referred to as "… the main ingredient in medical care and it is the fundamental instrument by which the doctor-patient relationship is crafted and by which therapeutic goals are achieved" [3].

There are different purposes for communication between physicians and patients. Broadly speaking, these purposes include (1) creating a good interpersonal relationship, (2) exchanging information, and (3) making treatment-related decisions [4]. Yet, these simplified purposes do not adequately capture the complexity inherent in health-care communication. For instance, these conversations are conducted among individuals with nonequal positions that lead to power dynamics and the potential for patriarchal decision-making. They are also usually not voluntary conversations, they are often concerning issues of great personal significance to the patient and are therefore rich with emotion, and they require close understanding and adherence [5].

Unfortunately, physicians play a role in providing ineffective communication, which adds to the challenge. Fifty percent of psychosocial and psychiatric concerns from patients are missed or go unaddressed [6]. Physicians interrupt patients after an average of 18–23 s of patient description of their chief concerns

[7,8]. Fifty four percent of patients' problems and 45% of patients' concerns are neither obtained by the physician nor provided by the patient [9]. Physicians and patients disagree on the presenting problem in half of all cases [10]. Patients are often dissatisfied with physician-provided information [11]. Clearly, there is a need for continued attention at elevating health-care conversations, and it begins with investment in people and relationships.

COMMUNICATION AND OUTCOMES

Safety

Poor patient satisfaction and disagreement are not the only concerns resulting from ineffective communication. Patients are put at risk. Approximately 1 out of every 5 inpatient admissions to the hospital involves inaccurate medication reconciliation (the process of identifying a patient's home medications and reconciling them with the ones to be prescribed in the hospital) with potential to cause meaningful harm to the patient [12]. Medication reconciliation relies significantly on and is intertwined with effective communication. Insufficient communication among house officers caring for patients in the hospital is associated with patient harm during transitions in care (Fig. 21.2) [13,14]. Inadequate history gathering by providers may result in patients undergoing unnecessary tests or procedures, which increase both cost and risk to the patients themselves [15].

It is worth noting that medical communication failures do not occur simply as a result of faulty or inaccurate exchanges of information. Rather, communication failures often occur in contexts of unequal power relations (e.g., between house officer/attending, house officer/specialty service, and house officer/nurse). For instance, house officers' concerns of appearing inadequate in front of their superiors and their uncertainty of their role within decision-making

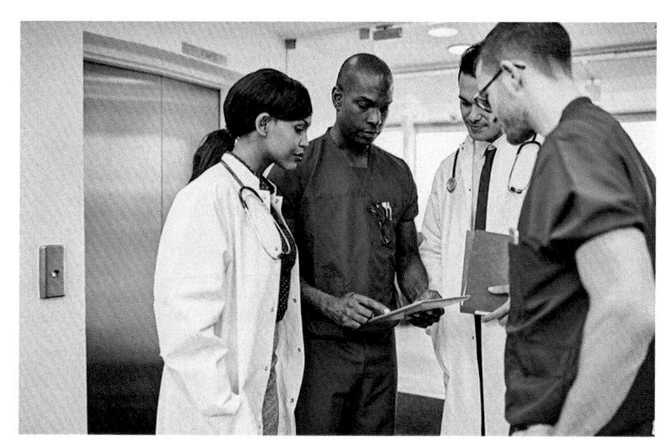

FIGURE 21.2 Photo illustrating physicians discussing ways to optimize patient care (iStock.com/UberImages).

processes can prevent them from fully communicating with attending physicians and patients [16–18]. These findings point to the complexity of clinical communication and to the necessity of communication training for all levels of health-care providers.

Outcomes for Patients

Effective communication between patients and their providers, founded in relationships, leads to a plethora of beneficial effects for both parties. This has been empirically demonstrated in numerous scholarly articles over decades. Interestingly, the benefits of high-quality communication between patients and providers are not relegated only to subjective assessments of well-being.

Indeed, patients enjoy improved health outcomes. Some of these include the following:

- improvement in blood pressure readings [19,20]
- weight loss [21]
- control of blood sugar in patients with diabetes mellitus [22,23]

Patients demonstrate the following:

- improved recall and understanding of medical information [20]
- increased trust with their health-care providers [24,25]
- enhanced sense of self-efficacy and support [26]
- satisfaction with their care [24,27,28]
- adherence and self-management of chronic disease [22,27]
- improvement or resolution of symptoms and improvement in functional abilities [29,30]
- improvement in overall health status and quality of life [27,31].

According to one metaanalysis, the patient–provider relationship's effect on a patient's health outcomes is estimated to be larger than that of aspirin in the primary prevention of myocardial infarction (heart attack) for 5 years, a well-documented and thoroughly studied effective intervention for cardiovascular health [32].

Outcomes for Providers

The benefits of effective relationship-centered communication are not limited only to patients. Providers who focus on this form of communication experience a myriad of beneficial outcomes as well, which also have been well-documented in the literature. Providers demonstrate improvements in accuracy of diagnostic decision-making and problem identification [33], a particularly important outcome given the findings of the Institute of Medicine's report on diagnostic error and its deleterious effects on patient safety [34].

Providers have shown the following:

- improved efficiency [8,35,36]
- confidence in their own abilities [37]
- satisfaction and engagement in their profession [35]
- decreased costs of health care [32]
- decreased malpractice claims [38–40].

With the challenges of professional resilience and compassion fatigue at the forefront of the field's dialogue, it is important to note that relationship-centered communication and the training of these skills among health-care providers lead to decreased rates of professional burnout [41,42] and enhanced expressions of empathy [41,43]. With incredible costs, waning provider resilience, and increased scrutiny of quality and efficiency of care, never has there been a more critical time to reinvest in high-quality communication at all levels.

TEACHING, LEARNING, AND ASSESSING COMMUNICATION

Throughout the last several decades, volumes of scholarship have been amassed that have closely examined relationship-centered interviewing and broken it into discrete series of skills and techniques. These specific skills have been used in establishing models of health-care communication, which may therefore be used to facilitate teaching, learning, and assessment [44]. In 1991, many experts in the field as well as architects of various communication models coalesced to discuss and identify ideas and conceptualizations of communication in this arena. At that time, a consensus was reached known as the Kalamazoo consensus statement [45]. This overarching framework of crucial skills in the field of health-care communication was seen as a turning point, which allowed standardization of communication skills in patient–provider interactions and uniformity that, prior to this, was not present. This newfound common vernacular was foundational to cultivating patient–provider relationships and allowed for more sophisticated educational and research pursuits. And it was clear that nearly all of the skills and techniques within the various models were similar, indicating key themes that underlie effective interchanges. Regardless of the branding, the methods were nearly the same [43].

Effective interpersonal communication is an imperative skill for all practicing health-care providers. This is evidenced by its assessment in medical training and practice at all levels. More and more medical schools are identifying specific entrustable professional activities and milestones for students that revolve around building rapport and high-quality interpersonal communication with their patients and their colleagues [44,46]. As another example, the Accreditation Council for Graduate Medical Education lists interpersonal communication skills as a domain of core competencies among house officers who continue to hone their clinical

FIGURE 21.3 Centers for Medicare and Medicaid Services Hospital CAHPS survey information page. (From *https://www.cms.gov/Research-Statistics-Data-and-Systems/Research/CAHPS/hcahps1.html.*)

and relational skills [47]. Though assessment of these behaviors in action has proven to be elusive on a wide scale, suggestions for direct observation, evaluation methods, and feedback have been put forth [48].

Finally, patient satisfaction with providers is increasingly important on the national landscape as can be seen by its relevance in the value-based purchasing system from the Centers for Medicare and Medicaid. As an example, patients who are admitted to the hospital are given a survey called HCAHPS (Fig. 21.3), and several elements of this survey involve patients' perceptions of physician communication and interaction [49]. Organizations with higher scores on this type of survey will receive increased funding from federal regulatory agencies, whereas those that score poorly face financial penalty [50]. This shift in focus, from volume of care to quality of communication and patient satisfaction, underscores recognition by regulatory agencies of the significant role communication plays in patient safety, satisfaction, and quality of care.

DISSEMINATION OF SCHOLARSHIP, TECHNIQUES, AND IDEAS

Health care represents one of the most complex and sizable of all the scientific communities. As the field expands, new approaches to disease and new techniques for delivering safe, effective, and efficient health care are being instituted. Indeed, the volume of new knowledge and discovery is continually and exponentially increasing. This is true also of scholarship in communication. As a result, many different resources and venues have been established to disseminate new advances, techniques, and theories in the arena of health-care communication.

Training Efficacy and Domains

In the not so distant past, the belief that talent in communication was innate within individuals prevailed. That those who could communicate effectively were born with those skills and those who were not were destined to struggle through conversations. However, it is now known that dedicated training with coaching, practice, observation, reflection, and specific feedback can improve communication skills at all learning levels [51,52]. The good can become great, and the great can become even better.

Various types of programs and interventions have been established to specifically train health-care providers in communication skills as well as enhance the systems of communication in various settings. Faculty development programs and training opportunities for house officers have been created and implemented. These involve dedicated observation and feedback with encouraging results. Providers feel more comfortable and confident in their skills and also report reduced levels of burnout and enhanced empathy toward their patients and colleagues [41]. Consultation programs have been established in which communication experts travel to other facilities to train interested health-care providers in their own authentic clinical environments [53]. So-called "train the trainer" programs have been developed to foster groups of like-minded facilitators interested in elevating communication at their own organizations [43].

Finally, different subdomains of conversations have been the subject of additional scholarship and training opportunities above and beyond foundational skills. These include delivering serious news, motivational interviewing, challenging relationships, utilizing the computer with patients present, disclosing unanticipated outcomes, sexual health topics, and pain management [54,55]. Much time, energy, and forethought have been placed into these resources and training opportunities to assist providers in building relationships, often within nuanced and challenging situations.

Scholarly Journals

Scientific literacy refers to specialized knowledge—not only of the subject matter within a particular discipline but also of the reading and writing practices, typified genres and audiences, rhetorical moves, characteristic vocabulary, and syntactical structures that comprise the textual and communicative activities of a discipline. Much research and literature exists to guide health-care providers in the discussion of effective health-care communication. Indeed, the number of journals dedicated specifically to communication in health-care settings is in the double digits and includes the Journal of Health Communication, Journal of Communication in Healthcare, Patient Education and Counseling, Medical Encounter, Communication and Medicine, and International Journal of Communication and Health, to name a few.

Organizations, Events, Books, Online Offerings, and Workshops

Many organizations dedicated to improving communication in health care have been founded and cultivated over the last few decades. Several of these organizations including the American Academy on Communication in Healthcare (AACH) [56] and the Institute for Healthcare Communication [57] have thorough and up-to-date bibliographies on a variety of topics. Several health-care organizations have founded centers dedicated to improving communication skills locally, regionally, nationally, and internationally. Some of these include the Center for Enhancement of Communication in Healthcare at the University of California, San Francisco [58], the Center for Excellence in Healthcare Communication at the Cleveland Clinic Foundation [59], and the Center for Communication and Medicine at Northwestern University Feinberg School of Medicine [60]. A search for books related to communication in health care yields numerous references by many renowned leaders in the field. There are frequent events held to discuss up-to-date and novel techniques, behaviors, scholarship, and best practices in communication including the International Conference on Communication in Healthcare and Health Literacy Annual Research Conference [61], ENRICH and AACH Research Forum [62], the annual Patient Experience Empathy and Innovation Summit [63], and the Beryl Institute [64], the latter two of which focuses on the patient experience of health care. There is a myriad of online curricula and modules such as DocCom [65] (a production of the AACH) as well as webinars, videos, social media campaigns, and online faculty development opportunities and workshops. Many of these offer continuing medical education or maintenance of certification credit for professionals to demonstrate ongoing professional learning.

CONCLUSIONS

Communication in health-care settings is complex and critically important to ensure patient safety, satisfaction, and resilience of both providers and patients. This is achieved through building meaningful relationships. Cultivating and honing communication skills is not only possible but it is also imperative to train the next generations of effective physicians and to achieve high-quality care. To meet this challenge, many organizations, events, and resources have been created to help disseminate the newest techniques, behaviors, and advances in health-care communication. Other industries striving to achieve the goals of high reliability may view health care as a field that has embraced the need for training in this arena.

REFERENCES

[1] Hsaio C, Cherry DK, Beatty PC, Rechtsteiner EA. In: Johnson G, Hyattsville V, editors. National ambulatory medical care survey: 2007 summary. Maryland: National Center for Health Statistics; 2010. Available from: http://www.cdc.gov/nchs/data/nhsr/nhsr027.pdf.
[2] Lipkin M, Putnam SM, Lazare A, editors. The medical interview: clinical care, education, and research. New York: Springer-Verlag; 1995.
[3] Roter DL, Hall JA. Doctors talking with patients, patients talking with doctors. Westport: Auburn House; 1992.
[4] Ong LM, de Haes JC, Hoos AM, Lammes FB. Doctor-patient communication: a review of the literature. Soc Sci Med April 1995;40(7):903–18.
[5] Chaitchik S, Kreitler S, Shaked S, Schwartz I, Rosin R. Doctor-patient communication in a cancer ward. J Cancer Educ 1992;7:41.
[6] Davenport S, Goldberg D, Millar T. How psychiatric disorders are missed during medical consultations. Lancet 1987;2:439–40.
[7] Beckman HB, Frankel RM. The effect of physician behavior on the collection of data. Ann Intern Med November 1984;101(5):692–6.
[8] Marvel MK, Epstein RM, Flowers K, Beckman HB. Soliciting the patient's agenda: have we improved? JAMA January 20, 1999;281(3):283–7.
[9] Stewart MA, McWhinney IR, Buck CW. The doctor/patient relationship and its effect upon outcome. J R Coll Gen Pract 1979;29:77–82.
[10] Starfield B, Wray C, Hess K, et al. The influence of patient-practitioner agreement on outcome of care. Am J Public Health 1981;71:127–31.
[11] Haug M, Lavin B. Consumerism in medicine: challenging physician authority. Beverly Hills (CA): Sage Publications; 1983.
[12] Cornish PL, Knowles SR, Marchesano R, Tam V, Shadowitz S, Juurlink DN, Etchells EE. Unintended medication discrepancies at the time of hospital admission. Arch Intern Med 2005;165(4):424–9.
[13] Horwitz LI, Moin T, Krumholz HM, Wang LL, Bradley EH. Consequences of inadequate sign-out for patient care. Arch Intern Med 2008;168(16):1755–60.
[14] Kitch BT, Cooper JB, Zapol WM, Marder JE, Karson A, Hutter M, Campbell EG. Handoffs causing patient harm: a survey of medical and surgical house staff. Jt Comm J Qual Patient Saf 2008;34(10):563–70.
[15] Hasnain M, Bordage G, Connell KJ, Sinacore JM. History-taking behaviors associated with diagnostic competence of clerks: an exploratory study. Acad Med 2001;76(10 Suppl.):S14–7.
[16] Sutcliffe KM, Lewton E, Rosenthal M. Communication failures: an insidious contributor to medical mishaps. Am Med 2004;79:186–94.
[17] Danswereau F, Markham SE. Superior-subordinate communication: multiple levels of analysis. In: Jablin FM, Putnam LL, Roberts KH, Porter LW, editors. Handbook of organizational communication. Newbury Park (CA): Sage; 1987. p. 343–88.
[18] McCue JD, Beach KJ. Communication barriers between attending physicians and residents. J Gen Intern Med 1994;9:158–61.
[19] Cooper LA, Roter DL, Carson KA, Bone LR, Larson SM, Miller III ER, Barr MS, Levine DM. A randomized trial to improve patient-centered care and hypertension control in underserved primary care patients. J Gen Intern Med 2011;26:1297–304.

[20] Orth JE, Stiles WB, Scherwitz L, et al. Interviews and hypertensive patients' blood pressure control. Health Psychol 1987;6:29–42.

[21] Pollak KI, Alexander SC, Coffman CJ, Tulsky JA, Lyna P, Dolor RJ, James IE, Brouwer RJN, Manusov JRE, Ostbye T. Physician communication techniques and weight loss in adults: project CHAT. Am J Prev Med 2010;39:321–8.

[22] Heisler M, Bouknight RR, Hayward RA, Smith DM, Kerr EA. The relative importance of physician communication, participatory decision making, and patient understanding in diabetes self-management. J Gen Intern Med April 2002;17(4):243–52.

[23] Hojat M, Louis DZ, Markham FW, Wender R, Rabinowitz C, Gonnella JS. Physicians' empathy and clinical outcomes for diabetic patients. Acad Med 2011;86:359–64.

[24] Safran DG, Taira DA, Rogers WH, Kosinski M, Ware JE, Tarlov AR. Linking primary care performance to outcomes of care. J Fam Pract September 1998;47(3):213–20.

[25] Sherbourne CD, Sturm R, Wells KB. What outcomes matter to patients? J Gen Intern Med June 1999;14(6):357–63.

[26] Street Jr RL, Makoul G, Arora NK, Epstein RM. How does communication heal? Pathways linking clinician-patient communication to health outcomes. Patient Educ Couns March 2009;74(3):295–301. http://dx.doi.org/10.1016/j.pec.2008.11.015. Epub 2009 Jan.

[27] Renzi C, Abeni D, Picardi A, Agostini E, Melchi CF, Pasquini P, Puddu P, Braga M. Factors associated with patient satisfaction with care among dermatological outpatients. Br J Dermatol October 2001;145(4):617–23.

[28] Derksen F, Bensing J, Lagro-Janssen A. Effectiveness of empathy in general practice: a systematic review. Br J Gen Pract January 2013;63(606):e76–84. http://dx.doi.org/10.3399/bjgp13X660814.

[29] Ambady N, Koo J, Rosenthal R, Winograd CH. Physical therapists' nonverbal communication predicts geriatric patients' health outcomes. Psychol Aging September 2002;17(3):443–52.

[30] Stewart MA. Effective physician-patient communication and health outcomes: a review. CMAJ May 1, 1995;152(9):1423–33.

[31] Stewart M, Brown JB, Donner A, McWhinney IR, Oates J, Weston WW, Jordan J. The impact of patient-centered care on outcomes. J Fam Pract 2000;49:796–804.

[32] Kelley JM, Kraft-Todd G, Schapira L, Kossowsky J, Riess H. The influence of the patient-clinician relationship on healthcare outcomes: a systematic review and meta-analysis of randomized controlled trials. PLoS One 2014;9:e94207.

[33] Peterson MC, Holbrook JH, Von Hales D, Smith NL, Staker LV. Contributions of the history, physical examination, and laboratory investigation in making medical diagnoses. West J Med February 1992;156(2):163–5.

[34] National Academies of Sciences, Engineering, and Medicine. Improving diagnosis in health care. Washington (DC): The National Academies Press; 2015.

[35] Tulsky JA. Interventions to enhance communication among patients, providers, and families. J Palliat Med 2005;8(Suppl. 1):S95–102.

[36] Mauksch LB, Dugdale DC, Dodson S, Epstein R. Relationship, communication, and efficiency in the medical encounter: creating a clinical model from a literature review. Arch Intern Med July 14, 2008;168(13):1387–95. http://dx.doi.org/10.1001/archinte.168.13.1387.

[37] Nørgaard B, Ammentorp J, Ohm Kyvik K, Kofoed PE. Communication skills training increases self-efficacy of health care professionals. J Contin Educ Health Prof 2012 Spring;32(2):90–7. http://dx.doi.org/10.1002/chp.21131.

[38] Beckman H, Markakis K, Suchman A, Frankel R. Getting the most from a 20-minute visit. Am J Gastroenterol May 1994;89(5):662–4.

[39] Levinson W, Roter DL, Mullooly JP, Dull VT, Frankel RM. Physician-patient communication. The relationship with malpractice claims among primary care physicians and surgeons. JAMA February 19, 1997;277(7):553–9.

[40] Ambady N, Laplante D, Nguyen T, Rosenthal R, Chaumeton N, Levinson W. Surgeons' tone of voice: a clue to malpractice history. Surgery July 2002;132(1):5–9.
[41] Weng HC, Hung CM, Liu YT, Cheng YJ, Yen CY, Chang CC, Huang CK. Associations between emotional intelligence and doctor burnout, job satisfaction and patient satisfaction. Med Educ August 2011;45(8):835–42. http://dx.doi.org/10.1111/j.1365-2923.2011.03985.x.
[42] Boissy A, Windover AK, Bokar D, Karafa M, Neuendorf K, Frankel RM, Merlino J, Rothberg MB. Communication skills training for physicians improves patient satisfaction. J Gen Intern Med July 2016;31(7):755–61. http://dx.doi.org/10.1007/s11606-016-3597-2. Epub 2016 Feb 26.
[43] Riess H, Kelley JM, Bailey RW, Dunn EJ, Phillips M. Empathy training for resident physicians: a randomized controlled trial of a neuroscience informed curriculum. J Gen Intern Med 2012;27:1280–6.
[44] Chou CL, Cooley L, Pearlman E, White MK. Enhancing patient experience by training local trainers in fundamental communication skills. Patient Exp J 2014;1(2). Article 8. Available at: http://pxjournal.org/journal/vol1/iss2/8.
[45] Makoul G. Essential elements of communication in medical encounters: the Kalamazoo consensus statement. Acad Med April 2001;76(4):390–3.
[46] Weiss BD. Health literacy and patient safety: help patients understand: manual for clinicians. 2007.
[47] Benson BJ. Domain of competence: interpersonal and communication skills. Acad Pediatr 2014 Mar–April;14(2 Suppl.):S55–65. http://dx.doi.org/10.1016/j.acap.2013.11.016.
[48] Henry SG, Holmboe ES, Frankel RM. Evidence-based competencies for improving communication skills in graduate medical education: a review with suggestions for implementation. Med Teach May 2013;35(5):395–403. http://dx.doi.org/10.3109/0142159X.2013.769677. Epub 2013 Feb 27.
[49] https://www.cms.gov/Medicare/Quality-Initiatives-Patient-Assessment-Instruments/Hospital-Value-Based-Purchasing/.
[50] https://www.cms.gov/Outreach-and-Education/Medicare-Learning-Network-MLN/MLNProducts/downloads/Hospital_VBPurchasing_Fact_Sheet_ICN907664.pdf.
[51] Maguire P, Fairbairn S, Fletcher C. Consultation skills of young doctors. Part I: benefits of feedback training in interviewing as students persist. BMJ 1986;292:1573–6.
[52] Cegala DJ, Lenzmeier Broz S. Physician communication skills training: a review of theoretical backgrounds, objectives and skills. Med Educ 2002;36(11):1004–16.
[53] http://aachonline.org/Programs/Workshops.
[54] http://www.aachonline.org/dnn/Resources/Academic-Articles.
[55] http://healthcarecomm.org/annotated-bibliographies/.
[56] http://aachonline.org/.
[57] http://healthcarecomm.org/.
[58] http://cech.ucsf.edu/.
[59] http://clevelandclinicexperiencepartners.com/.
[60] https://www.communication.northwestern.edu/centers/communication-and-health.
[61] http://www.aachonline.org/dnn/Events/International-Conference-ICCH.
[62] http://www.aachonline.org/dnn/Events/ENRICH.
[63] https://my.clevelandclinic.org/departments/clinical-transformation/depts/patient-experience/summit.
[64] http://www.theberylinstitute.org/.
[65] http://www.doccom.org/.

Chapter 22

Measuring Impact

Tyler Nix, MLIS, Judith Smith, MLIS, Jean Song, MSI
Taubman Health Sciences Library, University of Michigan, Ann Arbor, MI, United States

WHAT IS RESEARCH IMPACT

Historically rooted in scientific communication, measuring impact has become increasingly important to key stakeholders such as funding bodies, institutional administrators, researchers, clinicians, and even the general public who are asking for accountability, return on investment, and transparency in the dissemination of scientific information. Impact is defined by the Oxford English Dictionary [51] as "...commonly the effective action of one thing or person upon another; the effect of such action; influence; impression." Focusing on impact in research, Greenhalgh, et al. summarize different definitions of research impact, including the UK 2014 Research Excellence Framework definition where research impact "...benefits to one or more areas of the economy, society, culture, public policy and services, health, production, environment, international development or quality of life…" [1]. Major funding bodies are also targeting impact and some of their definitions are displayed in Table 22.1. While often having common elements in the definition including broad scope and results over time, impact definitions do vary and are nuanced depending on the organization, group, or individual involved.

WHY MEASURE IMPACT

Measuring impact can be important for decision-making in many situations including promotion and tenure, recruitment for professional positions, reporting and dissemination, selection of dissemination venue such as targeting journals for manuscript submission, and research funding. The Robert Wood Johnson Foundation's publication, "Impact Capital Measurement: Approaches to Measuring the Social Impact of Program-Related Investments" [6] lists the following purposes for measuring the impact of their program-related investments (PRIs):

1. To demonstrate the value of an individual investment post-investment
2. To inform foundation resource allocations pre-investment

Medical and Scientific Publishing. https://doi.org/10.1016/B978-0-12-809969-8.00022-X

TABLE 22.1 Funding Body Impact Definitions

Funding Body	Term Definition
Centers for Disease Control (CDC)	Impact: "...effect that interventions have on people, organizations, or systems to influence health." [2]
National Institutes of Health (NIH)	Overall Impact: "....exert a sustained, powerful influence on the research field(s) involved..." [3]
National Science Foundation (NSF)	Broader Impacts: "Advance discovery and understanding while promoting teaching, training, and learning; Broaden participation of under-represented groups; Enhance infrastructure for research and education; Broaden dissemination to enhance scientific and technological understanding; Benefits to society" [4]
Wellcome Trust	Impact: "...is usually taken to mean the longer-term sustainable change attributable to a project or intervention that remains after the project has finished..." [5]

3. To demonstrate the value of PRIs (program related investments) as a valid foundation investment strategy
4. To enable comparisons between PRIs and charitable grants
5. To demonstrate the value of PRIs as a proof of concept for the philanthropic field
6. To assess whether PRIs are promoting desired policy changes at the systemic level
7. To increase dollars invested in impact investment by creating measurement standards for the entire impact investing field

Per the Bill & Melinda Gates Foundation, "[e]valuation is best used to answer questions about what actions work best to achieve outcomes, how and why they are or are not achieved, what the unintended consequences have been, and what needs to be adjusted to improve execution" [7]. The reasons for measuring impact listed here are not just limited to these foundations. The STAR METRICS (Science and Technology for America's Reinvestment Measuring the Effects of Research on Innovation, Competitiveness, and Science) project from the Office of Science and Technology Policy (OSTP) of the United States government was created to "assess the impact of federal R & D [research and development] investments" to answer the public's "questions about the impact of federal investment in sciences, particularly with respect to job creation and economic growth" [8]. The significance of measuring impact continues to grow as funding bodies are demanding accountability, transparency, and demonstrable benefits as a result of their funding decisions.

Situations in which impact is being measured are often high stakes for individuals and organizations whether it is the hiring of new professionals or the successful grant or program investment. Therefore, knowing the strengths and weaknesses of impact indicators is paramount.

IMPACT INDICATORS

Much literature has been produced discussing research impact and research productivity with no single agreed on method of measurement [9]. The most likely reason for this seemingly fundamental flaw in research evaluation is the highly complex nature of the research environment and the difficulties of having accurate indicators that measure long-term change, which can also be directly attributable to the research work. Research is conducted on individual, group, and institutional levels, and its effect can take a significant amount of time to manifest as quantifiable results. How research is conducted, communicated, and is perceived of as being successful is highly dependent on the context. Therefore, there is neither little consistency in the language used to describe impact nor are there systematic approaches to classing outcomes or uniform systems to keep track of these outcomes.

Keeping these limitations in mind, commonly used impact indicators are typically divided into two categories: quantitative measures and qualitative measures. Historically, peer-reviewed publications have served as a surrogate for measuring impact and with the publishing pressures that ensued in the 20th century, the use of citation analysis as an impact measure was solidified with the development of the Science Citation Index (SCI) and the impact factor by Eugene Garfield and Irving Sher in the 1950 and 60s [10].

Commonly used traditional quantitative measures are as follows:

- Numbers of research grants awarded
- Grant dollars awarded
- Numbers of publications (articles, books, chapters)
- Numbers of citations to a work
- Numbers of awards or honors received
- Numbers of patents granted

Additional quantitative indicators include but are not limited to numbers of publication downloads, prototype creation, mentions about the work in social media such as Twitter and Facebook, and computer programs written. These non-traditional indicators are typically grouped into a class called "alternative metrics." Alternative metrics are an attempt to expand the measures of impact and complement citation analysis as methods of scientific communication have also broadened and impact is no longer limited to the peer-reviewed literature proxy.

Commonly cited qualitative measures are as follows:

- Quality of the journal in which an article was published as determined by surveys or impact factor
- Prestige of the granting institution

- Researcher morale—how much a researcher perceived they were being supported in dollars and time given for research by their institution, department, and peers
- Changes in behavior—for example, how a clinician may deliver care, how a patient may manage his or her weight
- Application of the work having an influence on policy

The most commonly used quantitative measure of research impact is the analysis of citation and publication data, commonly known as bibliometrics. It has been greatly researched and analyzed for its merits and weaknesses [11].

The strengths of bibliometrics include the following:

- Concrete, quantifiable units of measurement
- Broad view of citation activity—who is citing whom and how many times
- History, validity, and usage for at least several decades

The weaknesses of bibliometrics include the following:

- Homographs—the consolidation of different researchers' works due to name similarity
- Cronyism—the self-citing by individuals or mutual colleagues
- Negative citations—the high ranking of highly argued works, cited because of its many flaws or significant disagreement
- Lack of tools—historically relying on the SCI database as the sole source of cited works data though additional tools such as Scopus and Google Scholar have more recently emerged

Limiting the analysis of impact to quantitative indicators can lead to the gross inflation of seemingly arbitrary statistics in the overall evaluation of a program, individual, or work. Because of the use of these quantitative indicators without the necessary evaluative information of qualitative indicators, the Leiden Manifesto was published in 2015 to codify the best practices for "metrics-based research assessment" [36]. The ten principles are included here:

1. Quantitative evaluation should support qualitative, expert assessment
2. Measure performance against the research missions of the institution, group or researcher
3. Protect excellence in locally relevant research
4. Keep data collection and analytical processes open, transparent and simple
5. Allow those evaluated to verify data and analysis
6. Account for variation by field in publication and citation practices
7. Base assessment of individual researchers on a qualitative judgment of their portfolio
8. Avoid misplaced concreteness and false precision
9. Recognize the systemic effects of assessment and indicators
10. Scrutinize indicators regularly and update them

Even publishers have concerns about quantitative indicators. At the Annual Meeting of The American Society for Cell Biology (ASCB) in 2012, individuals, groups, and publishers put forth the San Francisco Declaration on Research Assessment (DORA) [12], which in general recommends the following:

- Eliminating the use of journal-based metrics for funding, appointment, and promotion decisions
- Assessing research for its own merits regardless of where the research is published
- Capitalizing on opportunities available through online publication

HOLISTIC EVALUATION

To address the difficulties in measuring impact, many combined approaches that use multiple indicators have been developed as frameworks or models. For example, the Centers for Disease Control (CDC) has developed the Science Impact Framework [13] based on the Institute of Medicine's (IOM) Degrees of Impact description [14]. As described by the CDC, the "...framework gauges broader societal, environmental, cultural and economic impact using a combination of narrative, quantitative and qualitative indicators. Since it takes time for the impact of a body of work to be apparent, key is finding those short term indicators that were predictive of long term impact. In tracing the link there is more emphasis on contribution rather than attribution (it is difficult to assign credit to any single entity)." The framework is built on five domains of influence (disseminating science, creating awareness, catalyzing action, effecting change, and shaping the future) in which CDC lists key indicators in each domain (e.g., trade publications, continuing education, congressional hearings, social change, new hypotheses), and these five domains can be explicitly linked to each other in multidimensional ways to address the often nonlinear way that impact may manifest itself.

There has been much literature such as Milat et al. and Greenhalgh et al. [1,15] that summarize some commonly used frameworks and models. These include the Payback Framework, the Research Impact Framework, the Canadian Academy of Health Sciences (CAHS) Framework, monetization models, societal impact assessment, and the UK Research Excellence Framework. Many of these models and frameworks such as the Science Impact Framework use a combination of qualitative and quantitative indicators, gathering data using bibliometric analysis, interviews, assessments, and surveys. Not unexpectedly, these models are labor-intensive and gathering data for the studies that employed these techniques required significant resources.

The strength of research is the ability to investigate without any limitation and, as such, the ability to measure the impact of conducting research must be tailored and customized to best reflect the individuality of the work. Knowing that impact can take a long time to become apparent and the resource-intensive nature of impact assessment, a holistic and individualized approach to measuring impact will be the most effective in providing an approachable and unbiased accounting of impact.

BIBLIOMETRICS

Bibliometrics warrant further discussion, as they are one of the most commonly used quantitative indicators for measuring productivity and developing a narrative for scholarly and societal impact. As stated earlier, these citation impact indicators have their strengths and limitations. Accessibility of tools generating these metrics has rapidly increased use, making these concrete numbers and citation-based indicators easily downloadable—some with ready-made visualizations. This access has been a boon to administrators, librarians, researchers, and other stakeholders needing quick, automated access to this content. However, there can be misunderstandings, confusion, and potential misuse.

Numerous literature reviews have described in significant detail the strengths and challenges of these indicators [16]. The indicators are often used for hiring and promotion, yet the numbers are sometimes used without intentional regard to institutional context. Bornman describes what he calls "ambivalences" surrounding the data and their use, and how it is not "the intrinsic quality of publications but the schemes for measuring this quality that have become central to the discussion...the very notion of scientific quality makes sense only in the context of quality control and quality measurement systems" [17]. Additional challenges and confusion arise if researchers and administrators proceed without a secure grasp of tools' methodology. There are nuances to the numbers since the metrics are tool driven, and numbers calculated reflect documents indexed in the tool and that tool's particular methodology. A desire for a quickly obtained, absolute number can hinder the recognition of the source and methodology. Recognition of the complexity of these tools enables informed choices about the metrics chosen to demonstrate impact. Finally, literature also points to circumstances where the desire or pressure for increased citation counts can lead to self-citation, cronyism, and other means to artificially increase numbers. Overall, an overreliance on these indicators, or focusing on just one quantitative indicator, can steer researchers and administrators away from factoring in qualitative evaluations of impact.

Despite challenges, frameworks exist for responsible and transparent metrics, including a more nuanced approach to using metrics, advocating for the term "indicators" as recognition that "data may lack specific relevance," even if they are useful overall [18]. Quantitative indicators also are outlined in Snowball metrics, university-agreed on global standards to help and use and create fair and consistent metrics [19]. While the metrics below are inextricably linked as they are all derived from citations, they can be grouped into broad categories of author level, article level, and journal level. The section below describes these key indicators and a selection of core tools available for access.

Tools for Providing Quantitative Metrics

No one tool exists that covers all content and can therefore provide consistent numbers that can be used under all circumstances. Features of tools change over time, so examining the tools with a lens of understanding their core features is crucial.

Features to scrutinize when choosing a tool include the following:

- **Content coverage**: What types of documents does the tool index? Strictly peer-reviewed journals? Gray literature? Books? Conference proceedings?
- **Dates of coverage**: What years of content are covered? How can the years be filtered in a search?
- **Metric methodology**: When calculating the particular metric, what types of documents are included? For example, if calculating *h*-index, are self-citations included? If calculating a journal impact factor (JIF), what are the details of the calculation?
- **Document-level metrics**: Does the tool offer ways to compare documents in the same field of the same type (e.g., field-weighted)? Does it offer a percentile analysis? Are alternative metrics available?
- **Filtering mechanisms and output**: In what ways can the data be broken down and exported? By document type? Discipline? Funder?

Table 22.2 displays some key indicators available in major citation-based metric tools. The table does not provide details of these indicators but is presented to illustrate variety among tools.

Article-Level Metrics

Article-level metrics can be defined in various ways, but essentially refer to metrics describing the impact of individual articles. For the purposes of this chapter, article-level metrics measure the impact of researcher's work at the article level through both traditional citation counts and alternative metrics. SPARC describes how article-level metrics seek to capture impact at the article level through various data points and provide an immediate, granular way to measure impact [21].

Traditional citation counts refer to the number of times that a particular document, such as a book or peer-reviewed article, has been cited by another researcher since its publication. This count can be seen as a measure of influence, yet the amount of times a document has been cited can be indicative of the level of importance of the document, or it can be a measure of how controversial the document is. A good example is Wakefield's [22] Lancet article on the relationship between vaccination and autism. While now retracted, the article received thousands of citations (depending on the tool) because of its controversial findings. Thus, the context of the article must be considered, and not just the citation count itself [22].

TABLE 22.2 Citation Tool Key Indicators

	Scopus	Web of Science and Journal Citation Reports	Google Scholar
Vendor	Elsevier	Clarivate Analytics	Google
Content Coverage	Largest interdisciplinary indexing and abstracting database Indexes peer-reviewed journals, books Over 21,500 titles from more than 5000 international publishers	Science Citation Index Social Science Citation Index Arts and Humanities Citation Index Over 33,000 journals and indexes journal articles, books, conference proceedings, data sets, and patents	Mixture of peer-reviewed content and gray literature. No transparency in what is indexed
Author level	*H*-Index Can include books in calculation	*H*-Index	*H*-Index
Article level	Traditional Citation counts based on documents indexed and alternative metrics offered through altmetrics.com Incomplete citation information for documents published prior to 1996 but in process of back filling	Traditional Citation Counts based on documents indexed	Crawls the web for peer-reviewed and nonpeer-reviewed documents
Journal	Citescore: https://journalmetrics.scopus.com/ SJR: SCImago Journal Rank SNIP: Source Normalized Impact per Paper	Impact Factor through Journal Citation Reports (JCR) Eigenfactor (JCR) Article Influence (JCR)	*H*-index, h-core, h-median
Output and filtering	Downloads, visualizations, and filtering available	Downloads, visualizations, and filtering available	Must use Publish or Perish software to download and analyze [20]

Alternative metrics refer to attention that an article receives from social media, such as blogs, Facebook mentions, Twitter, or downloads from publisher websites or citation management and sharing tools such as Mendeley or CiteuLike. There is growing interest in these metrics and their contribution to analyzing research impact. Studies have examined the relationship between alternative metrics and standard citation counts [23], including whether alternative metrics correlate with traditional citations [24]. Alternative metrics offer a way to provide a holistic view of a document, by measuring attention quickly since traditional citations counts take longer to accumulate. Overall, this type of attention can be included in grant proposals or other projects that require evidence of impact. Traditional citation counts are still the mainstay of many projects that require demonstration of research impact such as tenure packages. However, this type of attention provided by alternative metrics has quickly gained interest as researchers and institutions seek to analyze their impact and influence [25].

Main providers of Alternative metrics are Altmetrics.com and Plum Analytics. Additionally, the alternative metrics data can be viewed through a variety of subscription databases, such as Scopus and ProQuest databases. The Public Library of Science (PLoS) offers article-level metrics for authors publishing in its journals, offering both traditional citation data and other indicators, such as article views and downloads. Some publisher websites also display alternative metrics. Fig. 22.1 shows how Altmetrics are displayed for a May

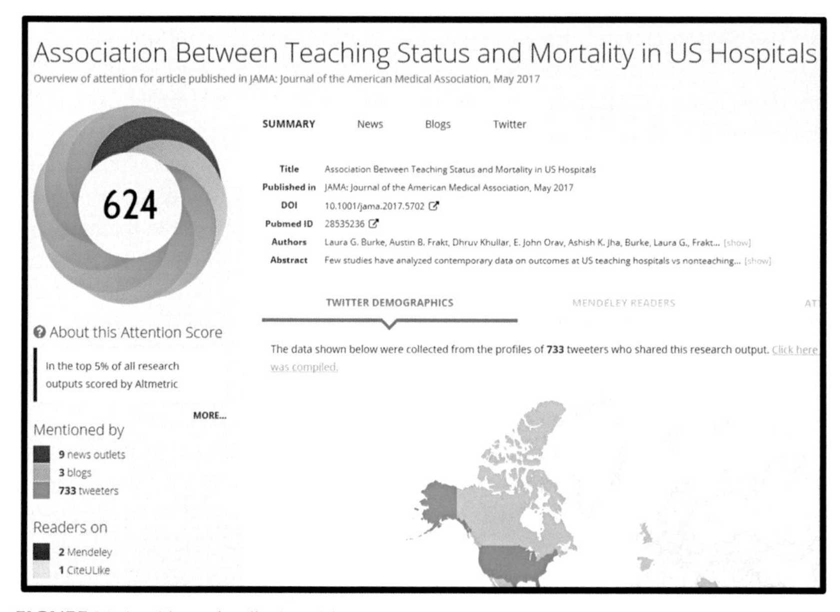

FIGURE 22.1 Altmetrics displayed for a May, 2017, Journal of the American Medical Association article. *(From https://jamanetwork.altmetric.com/details/20412887.)*

2017 Journal of the American Medical Association (*JAMA*) article. While traditional citation can take years to accumulate due to the scholarly publishing cycle, one can see the attention the author is getting within the same month published through news outlets, blogs, and tweeters.

Tools such as Scopus and Web of Science offer sophisticated ways to view document level metrics and can categorize, among other categories, by document type, field, coauthors, and funding agency. A relatively new tool, iCITE from the National Institutes of Health [26] solely covers journals indexed in PubMed. One may enter a single or multiple PubMed identifiers which are then analyzed using the NIH developed measure Relative Citation Ratio (RCR). RCR provides a benchmarking process which "represents a citation-based measure of scientific influence of one or more articles. It is calculated as the cites/year of each paper, normalized to the citations per year received by NIH-funded papers in the same field and year."

Author Level

Developed in 2005 by J.E. Hirsch, the most commonly used and well-known citation-based author metric is the *h*-index, which attempts to measure faculty members productivity and impact. A scientist has an *h*-index of h if h of their publications each have at least h citations and their remaining publications each have fewer than h+1 citations [27]. Using the tools described above, the measure is easy to capture. The metric is calculated on a researcher's citation count and the point at which the citation count and number of publications intersect. For example, an author with 20 papers that have at least 20 citations will have an *h*-index of 20. That author may have published 60 papers with varying numbers of citations for each paper, but the *h*-index is interested in that point of intersection.

Although there is ease of access for *h*-indexes, there is confusion and ongoing academic debates about the utility and consistency of this citation impact indicator [28]. First, the *h*-index is based on publications indexed in the particular tool used, as well as documents uses for calculation. For example, Scopus includes books in its *h*-index calculation—but Web of Science does not. Often a researcher's *h*-index is higher in Google Scholar because it pulls from a larger body documents, leading to confusion when their number is lower in Web of Science or Scopus. Second, issues arise when comparing *h*-indexes of researchers in different fields since citation rates vary. Researchers in the humanities and other disciplines accumulate citations at a much slower rate than researchers in the health sciences. Third, a researcher's *h*-index is impacted by their career stage. Early career researchers will have a lower number than researchers well into their career with many publications and the time needed for citation accumulation. Finally, the *h*-index calculation does not take into account the level of contribution to the article. That is, there is no "extra weight" given for being first author.

Journal Level

Various journal-level impact measures exist to demonstrate the prestige of a journal and provide a way to help researchers know where to publish [29]. These metrics also help librarians determine the direction of their collections. The most commonly used journal impact metric has been the JIF, provided by Clarivate Analytics, which is now in competition with Scopus's 2016 released CiteScore [30]. Google Scholar also offers journal metrics through its *h*-index, h-core, and h-median [31].

Below are definitions of the two prominent impact measures and how they are calculated:

Journal Impact Factor (JIF): Defined as "a measure of the frequency with which the 'average article' in a journal has cited in a particular year or period." The annual JCR impact factor is a ratio between citations and recent citable items published. Thus, the impact factor of a journal is calculated by dividing the number of current year citations to the source items published in that journal during the previous 2 years [32].

CiteScore (SCOPUS): CiteScore is a "family" of eight indicators. CiteScore itself is "the number of citations received by a journal in one year to documents published in the three previous years, divided by the number of documents indexed in Scopus published in those same three years. For example, the 2015 CiteScore counts the citations received in 2015 to documents published in 2012, 2013 or 2014, and divides this by the number of documents indexed in Scopus published in 2012, 2013 and 2014" [33].

Similar to the *h*-index, there is controversy and debate surrounding journal rankings [34]. For example, individual articles are sometimes evaluated based on the prestige of the journal rather than looking at the article itself to gauge quality. Being published in a journal with a high impact factor does not necessarily guarantee quality of research. However, journals are sometimes used as a proxy for individual article quality. Additionally, there can be an overreliance on using the journal rankings for journal choice, eclipsing other publishing interests and needs, and causing researchers to bypass options that may be better. An example is Open Access publishing, where a journal may have a lower ranking, yet studies have shown that Open Access publishing provides visibility that relates to higher citation counts [35].

Institutional Impact and Benchmarking Tools

Additional categories of tools exist that allow for monitoring the impact of individuals and groups and benchmarking across institutions. Research Information Management Systems (RIMS), such as Symplectic's Elements, Clarivate Analytics Converis, and Elsevier's PURE provide ways for institutions and individual researchers to highlight expertise and retrieve data that can then be analyzed and visualized for understanding and demonstrating research impact.

They ingest bibliographic data and often other data (e.g., grants, patents, etc...) to demonstrate productivity and influence. Benchmarking tools such as InCites from Clarivate Analytics, or SciVal from Elsevier, compare data and impact across institutions. Another Clarivate Analytics Tool, Essential Science Indicators, "reveals emerging science trends as well as influential individuals, institutions, papers, journals, and countries in your field of research."

ENHANCE PUBLICATION IMPACT

Having so far introduced key publication impact metrics and some of the tools used to identify them, the following section will present four practices researchers can consider to enhance the impact of their published works: (1) establish and maintain a consistent identity in traditional literature and social media spheres, (2) make publications discoverable in search engines, (3) share publications widely using multiple means of dissemination, and (4) use evaluation frameworks beyond quantitative citation metrics alone to demonstrate scholarly impact.

Before exploring these four considerations, it is important to clarify the goal of this section: namely, to dedicate appropriate attention to ensuring publications are optimized for engagement with researchers' target audiences using the tools available in the publishing landscape. Per Hicks et al. [36], citation metrics have the potential to establish publication incentives that lead to goal displacement, wherein the measurement becomes the goal. The intent of this section, therefore, is in opposition to "gaming" the system for metrics' sake (e.g., through excessive self-citation) and instead attempts to highlight methods to improve a researcher's publication impact considered holistically, as defined previously in this chapter.

ESTABLISH AND MAINTAIN RESEARCH IDENTITY

Although massive hyperlinked literature databases have increased researchers' access to publication and citation data, correct author/article attribution represents a major unsolved problem for information science [37]. The challenges are many: one person may publish using multiple names or variations of their name over the course of a career; many individuals have the same name (ambiguity); necessary author metadata are often incomplete (e.g., lack of publisher records related to authors' full name or geographic location); and complications related to attribution in multiauthored, multiinstitution, and multidisciplinary publications. The need for author disambiguation is commonplace in scientific communities, where investigators search for new collaborators in multidisciplinary fields and funding agencies follow grant recipients throughout their careers. As such, the ability to ensure proper article attribution, and distinction among authors, is essential for researchers to receive appropriate credit for publications across multiple platforms.

There are several tools gaining wider use in the publishing landscape that address author disambiguation and help authors to manage their research identity. One such tool, ORCID (https://orcid.org), provides researchers the ability to consolidate publications and other professional information under a unique, persistent identifier number. By registering with ORCID, a researcher can use the identification number in an expanding number of publication-related platforms, including publisher submission platforms, bibliographic databases, and grant tracking systems to seamlessly port their publication history and affiliation information.

As of May, 2017, ORCID works with over 600 member organizations, including several US and international research universities, publishers and funding organizations to integrate ORCID functionalities in systems and workflows including RIMs, manuscript tracking systems, and grant application platforms [38]. A few of the participating organizations relevant to biomedical and health sciences researchers include the American Association for the Advancement of Science (AAAS), Nature Publishing Group, Proceedings of the National Academy of Sciences of the United States of America (PNAS), SAGE Publications, Wolters Kluwer Health, and the NIH.

A similar author disambiguation tool is Thompson Reuters' ResearcherID (researcherid.com). Like ORCID, ResearcherID not only allows authors to manage their publications but also allows for tracking of citation counts and *h*-index and includes a registry search function to identify potential research collaborators. It integrates with Web of Science resources and is ORCID compliant to allow uniform identity across multiple platforms. Many individual publication databases also have author profile functionalities, such as Google Scholar and Scopus author profiles. By registering for one or more of these author profiles, researchers can monitor that their publications are attributed accurately and make adjustments to attribution errors.

MAKE PUBLICATIONS DISCOVERABLE IN SEARCH ENGINES

While search engine optimization is beyond the scope of this chapter, there are a few basic principles authors can employ when preparing manuscripts for publication to improve article retrieval for humans and bibliographic systems, themselves. To be accurately indexed and retrieved in an academic search engine, article titles, abstracts, section headings, and image captions should be clear and descriptive. The International Committee of Journal Medical Editors (ICJME) is comprised of editorial membership from several journals, including Annals of Internal Medicine, British Medical Journal, Bulletin of the World Health Organization, Deutsches Ärzteblatt (German Medical Journal), Ethiopian Journal of Health Sciences, JAMA (Journal of the American Medical Association), Journal of Korean Medical Science, New England Journal of Medicine, New Zealand Medical Journal, PLoS Medicine, The Lancet, Revista Médica de Chile (Medical Journal of Chile), Ugeskrift for Laeger (Danish Medical Journal), the

US National Library of Medicine, and the World Association of Medical Editors [39]. The organization's Recommendations for the Conduct, Reporting, Editing, and Publication of Scholarly Work in Medical Journals (2017) recommends that titles provide a distilled description of the complete article and should include content that makes electronic retrieval sensitive and specific, including—in the case of clinical trials, systematic reviews, and metaanalyses—the study design type. The Journal of the American Medical Association (JAMA author guidelines (2017) further specify that study design type for clinical trials, systematic reviews, and metaanalyses be included as a subtitle (e.g., A Randomized Clinical Trial). Per the American Medical Association (AMA) Manual of Style [41], if a study is large and is best known by its group name or acronym, or if it is part of a series of reports from the same group, the research group name or series name may also be included as a subtitle (e.g., the Dietary Intervention Study in Children).

Because abstracts are the most substantive indexing source in many electronic databases (and the only portion many readers initially scan), researchers should ensure that abstracts accurately reflect the content of their articles [40]. This AMA Manual of Style [41] recommends including key descriptive terms, databases consulted, and/or study groups related to the article's subject in the abstract; however, individual journal guidelines for these elements may vary. It is also recommended to spell out abbreviations at first mention in the text.

Some medical journals solicit author-supplied keywords at the end of abstracts to assist the journal's indexer in categorizing manuscripts for retrieval or to help guide in the selection of peer reviewers; however, not all journals publish keywords. To optimize discoverability, researchers publishing in journals that do not use keywords should incorporate keywords within the text of the abstract as appropriate, provided doing so does not negatively affect readability.

In addition to optimizing manuscripts for accurate retrieval and comprehension, journal selection has also traditionally played a significant role in research impact. Waltman's [16] review of the literature on citation impact provides a helpful survey of the JIF [42] and various related journal-level metrics. It is important to stress avoiding publishing through predatory or unscrupulous publishers that deemphasize rigorous peer review and rigor in favor of maximizing profits from fees charged to individual scientists [43]. Recognizing the growing problem of predatory publishers, the ICJME recently strengthened its guideline language in this regard, stating that authors have a responsibility to evaluate the integrity, history, practices, and reputation of the journal to which they submit manuscripts [40].

When considering journals for article submission, researchers should not only consider metrics such as impact factor, which has been shown to affect citation metrics [44] but also consider other factors, such as a journal's ability to reach a target audience. For example, publishing in journal with a high impact factor and broad audience, such as *Nature* or *Science*, may not facilitate engaging with a specialized audience as would publishing in a specialty journal with a strong reputation, albeit with a lower impact factor. While the general convention is for researchers to pursue publication in journals with high impact factors in their disciplines, Lozano et al. [45] indicate that, since 1990, the relation between a journal's impact factor and individual articles' citation counts from within that journal is weakening.

There are several tools to assist researchers in choosing a journal, including Elsevier Journal Finder, EndNote Manuscript Matcher, and the Journal/Author Name Estimator (JANE) to name a few. These tools typically allow researchers to paste their manuscript title and abstract into a matching tool, chose disciplinary and/or open access options, and review potential journal matches. When searching for journals in which to publish, it can also be helpful to browse journal listings in bibliographic databases such as Scopus, Web of Science, the National Library of Medicine Catalog list of journals indexed in MEDLINE, the Directory of Open Access Journals, and the list of journals following ICJME Recommendations [46].

After identifying a potential journal for submission, the ICJME "Recommendations" (2017) are intended for use by authors who might submit their work for publication to ICJME member journals, but also work well for evaluating journal integrity, in general. In particular, the "Publishing and Editorial Issues" section is a useful reference point for several factors that authors should consider when exploring journals for potential submission. Similarly, the World Association of Medical Editors (WAME) [39] Principles of Transparency and Best Practice in Scholarly Publishing provides a useful framework for evaluating best practices in scholarly publications (2015).

SHARE PUBLICATIONS WIDELY USING MULTIPLE MEANS OF DISSEMINATION

Social media platforms provide almost limitless options for disseminating research and offer several strengths for connecting with research networks, practitioners, and the public: sharing and feedback are immediate, conversations are searchable on the open web, and researchers can provide additional perspective or context to their articles. In light of the possibilities and options related to scholarly social media, researchers should consider building a communication plan into projects from an early stage.

For many researchers, posting links to their research articles in social media (Twitter, LinkedIn, Academia.edu, etc.) or institutional blogs is common practice. Author agreements vary by publisher in relation to the transfer of copyright of research publications. However, when publisher policies allow, researchers should take advantage of posting (and linking to) prepublication manuscripts to university or disciplinary repositories (e.g., PubMed Central) as well as scholarly social media networks (e.g., Mendeley), and data repositories (e.g., Figshare).

With heavy use among researcher networks, publishers' use of Twitter is evolving to keep pace. For example, many journals are now experimenting with disseminating visual abstracts via Twitter. A visual abstract is simply a visual representation of research findings (similar to information presented in the abstract portion of an article) that is created by a journal publisher after an article is accepted [47]. More than 20 journals across medicine, basic sciences, and social science accept or produce visual abstracts as of 2016 [48]. In a study of visual abstract use by the *Annals of Surgery*, Ibrahim et al. [47] found that use of visual abstracts was associated with higher levels of dissemination as demonstrated by impressions, shares, and article visits on publisher websites.

However, it should be noted there is much debate about the potential complications and ethics of medical professionals engaging with patients and the public through social media, and institutional policies on this practice vary widely.

USE EVALUATION FRAMEWORKS BEYOND QUANTITATIVE CITATION METRICS ALONE

As previously mentioned in this chapter, Waltman [16] argues the importance of not relying on single citation metric to demonstrate research impact, but to instead look to several potential avenues, including peer review and various qualitative considerations, for a fuller measure of research impact. Sarli et al. [51] analyzed the research study life cycle and proposed a number of indicators (in addition to citation metrics) that can be documented for a more nuanced assessment of research impact.

The resulting evaluation framework developed at the Bernard Becker Medical Library (https://becker.wustl.edu/impact-assessment) calls these additional measurement areas "pathways of diffusion," consisting of the following categories: advancement of knowledge, clinical implementation, community benefit, legislation and policy, and economic benefit. Within each of the five pathways, the model offers researchers lists of indicators, or examples of instances to be recorded in various impact scenarios that can be used to document a broader measure of research impact.

Researchers should take time to familiarize themselves with the methods and tools used in impact evaluation by engaging with resources within their department or school, including librarians and bibliographic resources, open institutional repositories, blogs, media and social media training, and events with external stakeholders, to leverage the suggestions presented here to maximize research impact.

REFERENCES

[1] Greenhalgh T, Raftery J, Hanney S, Glover M. Research impact: a narrative review. BMC Med 2016;14. 78–7016-0620-8.

[2] Centers for Disease Control and Prevention. Program evaluation tip sheet: reach and impact. August 2011. Retrieved from: https://www.cdc.gov/dhdsp/programs/spha/docs/reach_impact_tip_sheet.pdf.

[3] National Institutes of Health. Overall impact versus significance. March 21, 2016. Retrieved from: https://grants.nih.gov/grants/peer/guidelines_general/impact_significance.pdf.
[4] National Science Foundation. Broader impacts review criterion. 2007, Retrieved from: https://www.nsf.gov/pubs/2007/nsf07046/nsf07046.jsp.
[5] Aggett S, Dunn A, Vincent R. Engaging with impact: how do we know if we have made a difference? International Public Engagement Workshop Report. London (United Kingdom): Wellcome Trust; 2012.
[6] Tuan MT. Impact capital measurement: approaches to measuring the social impact of program-related investments. Princeton (NJ): Robert Wood Johnson Foundation; 2011.
[7] Bill & Melinda Gates Foundation. Evaluation policy. 2017. Retrieved from: http://www.gatesfoundation.org/How-We-Work/General-Information/Evaluation-Policy.
[8] Office of Science and Technology Policy. About star metrics. 2017. Retrieved from: https://www.starmetrics.nih.gov/Star/About.
[9] Karras DJ, Kruus LK, Baumann BM, Cienki JJ, Blanda M, Stern SA, et al. Emergency medicine research directors and research programs: characteristics and factors associated with productivity. Acad Emerg Med 2006;13(6):637–44.
[10] Sarli CC, Carpenter CR. Measuring academic productivity and changing definitions of scientific impact. Mo Med 2014;111(5):399–403.
[11] Drew CH, Pettibone KG, Finch 3rd FO, Giles D, Jordan P. Automated research impact assessment: a new bibliometrics approach. Scientometrics 2016;106(3):987–1005.
[12] Declaration on Research Assessment. San francisco declaration on research assessment. 2012. Retrieved from: http://www.ascb.org/dora/.
[13] Centers for Disease Control and Prevention. Description of the science impact framework. 2016. Retrieved from: https://www.cdc.gov/od/science/impact/framework.html.
[14] Fineberg HV. The institute of medicine: what makes it great? president's address: institute of medicine annual meeting. 2013. Retrieved from: http://www.nationalacademies.org/hmd/~/media/Files/About%20the%20IOM/PresidentsAddress2013.pdf.
[15] Milat AJ, Bauman AE, Redman S. A narrative review of research impact assessment models and methods. Health Res Policy Syst 2015;13. http://dx.doi.org/10.1186/s12961-015-0003-1.
[16] Waltman L. A review of the literature on citation impact indicators. J Inf 2016;10(2):365–91.
[17] Leydesdorff L, Wouters P, Bornmann L. Professional and citizen bibliometrics: complementarities and ambivalences in the development and use of indicators – a state-of-the-art report. Scientometrics 2016;109(3):2129–50.
[18] Wilsdon J, Bar-Ilan J, Frodeman R, Lex E, Peters I, Wouters P. Next-generation metrics: responsible metrics and evaluation for open science. 2017.
[19] Colledge L. Snowball metrics recipe book. 2nd Amsterdam: Elsevier; 2014. [Snowball Metrics Program].
[20] Harzing A. Harzing.com. 2016. Retrieved from: https://www.harzing.com/resources/publish-or-perish.
[21] SPARC. SPARC article-level metrics primer. 2013. Retrieved from: http://www.sparc.arl.org/resource/sparc-article-level-metrics-primer.
[22] Wakefield AJ, Murch SH, Anthony A, Linnell J, Casson DM, Malik M, et al. RETRACTED: ileal-lymphoid-nodular hyperplasia, non-specific colitis, and pervasive developmental disorder in children. Lancet 1998;351(9103):637–41.
[23] Wilsdon J. The metric tide: independent review of the role of metrics in research assessment and management. 2017. 55 City Road, London.

[24] Costas R, Zahedi Z, Wouters P. Do "altmetrics" correlate with citations? extensive comparison of altmetric indicators with citations from a multidisciplinary perspective. J Assoc Inf Sci Technol 2015;66(10):2003–19.
[25] Rosenkrantz AB, Ayoola A, Singh K, Duszak Jr R. Alternative metrics ("Altmetrics") for assessing article impact in popular general radiology journals. Acad Radiol 2017. http://dx.doi.org/10.1016/j.acra.2016.11.019.
[26] Hutchins BI, Yuan X, Anderson JM, Santangelo GM. Relative citation ratio (RCR): a new metric that uses citation rates to measure influence at the article level. PLoS Biol 2016;14(9):e1002541.
[27] Hirsch JE. An index to quantify an individual's scientific research output. Proc Natl Acad Sci USA 2005;102(46):16569–72.
[28] Alonso S, Cabrerizo FJ, Herrera-Viedma E, Herrera F. h-index: a review focused in its variants, computation and standardization for different scientific fields. J Inf 2009;3(4):273–89.
[29] Casadevall A, Fang FC. Causes for the persistence of impact factor mania. Mbio 2014;5(2). e00064–14.
[30] da Silva JAT, Memon AR. CiteScore: a cite for sore eyes, or a valuable, transparent metric? Scientometrics 2017;111(1):553–6.
[31] Google. Google scholar metrics. 2017. Retrieved from: https://scholar.google.com/intl/en/scholar/metrics.html#metrics.
[32] Clarivate Analytics. The thomson reuters impact factor. 1994. Retrieved from: http://wokinfo.com/essays/impact-factor/.
[33] Scopus. What is CiteScore?. 2017. Retrieved from: http://help.elsevier.com/app/answers/detail/a_id/5221/p/8150.
[34] Arruda J, Champieux R, Cook C, Davis M, Gedye R, Goodman L, Jacobs N, Ross D, Taylor S. The journal impact factor and its discontents: steps toward responsible metrics and better research assessment. Open Scholarsh Initiat Proc 2016;1(0).
[35] Tahamtan I, Afshar AS, Ahamdzadeh K. Factors affecting number of citations: a comprehensive review of the literature. Scientometrics 2016;107(3):1195–225.
[36] Hicks D, Wouters P, Waltman L, de Rijcke S, Rafols I. Bibliometrics: the Leiden Manifesto for research metrics. Nature 2015;520(7548):429–31.
[37] Smalheiser NR, Torvik VI. Author name disambiguation. Annu Rev Inf Sci Technol 2009;43(1):1–43.
[38] ORCID. Orcid. 2017. Retrieved from: https://orcid.org/.
[39] World Association of Medical Editors. Principles of transparency and best practice in scholarly publishing. 2015. Retrieved from: http://www.wame.org/about/principles-of-transparency-and-best-practice.
[40] International Committee of Medical Journal Editors. ICMJE | recommendations. 2017b. Retrieved from: http://www.icmje.org/recommendations/.
[41] AMA manual of style. A guide for authors and editors. 10th ed. Oxford (New York): Oxford University Press; 2007.
[42] Garfield E. Citation analysis as a tool in journal evaluation. Sci (NY) 1972;178(4060):471–9.
[43] Beall J. Predatory publishers are corrupting open access. Nat News 2012;489(7415):179.
[44] Larivière V, Gingras Y. The impact factor's matthew effect: a natural experiment in bibliometrics. J Am Soc Inf Sci Technol 2009;61(2):424–7.
[45] Lozano GA, Larivèie V, Gingras Y. The weakening relationship between the impact factor and papers' citations in the digital age. J Am Soc Inf Sci Technol 2012;63(11):2140–5.
[46] International Committee of Medical Journal Editors. ICMJE | journals following the ICMJE recommendations. 2017a. Retrieved from: http://www.icmje.org/journals-following-the-icmje-recommendations/.

[47] Ibrahim AM, Lillemoe KD, Klingensmith ME, Dimick JB. Visual abstracts to disseminate research on social media: a prospective, case-control crossover study. Ann Surg 2017. doi:10.1097/SLA.0000000000002277.

[48] Institute for Healthcare Policy and Innovation. As scientists take to twitter, new study shows power of "visual abstract" graphics to share results. University of Michigan; April 2017. Retrieved from: http://ihpi.umich.edu/news/scientists-take-twitter-new-study-shows-power-%E2%80%9Cvisual-abstract%E2%80%9D-graphics-share-results.

[49] Bernard Becker Medical Library. Washington University School of Medicine. How to use the model. 2017. Retrieved from: https://becker.wustl.edu/impact-assessment.

[50] Sarli CC, Dubinsky EK, Holmes KL. Beyond citation analysis: a model for assessment of research impact. J Med Libr Assoc January 2010;98(1):17–23.

[51] OED Online. "impact, n." Oxford University Press: June 2017.

Chapter 23

Government Funded Research and Publishing

Susan E. Old, PhD, Neil Thakur, PhD
National Institutes of Health, Bethesda, MD, United States

> *I was an embarrassment to the department when they did research assessment exercises. A message would go round the department: 'Please give a list of your recent publications.' And I would send back a statement: 'None.'*
>
> Peter Higgs, British theoretical physicist (Higgs boson), Nobel Prize laureate for his work on the mass of subatomic particles.

WHAT IT MEANS TO AUTHOR A PAPER-BALANCING CREDIT AND RESPONSIBILITY

Scientists write peer-reviewed journal articles to disseminate their findings and earn recognition for their work. They primarily gain recognition through their journal choice and their role on the paper. A scientists' role is often conveyed through author position (e.g., first, second, last), and the meaning of author position is can vary by discipline. For example, in many fields, the last author position is thought of as the "senior author" and is almost as prestigious as being the first author. Boyak and Jordan [1] found if an National Institutes of Health (NIH) principal investigator was listed an author on a paper arising from their grant, it was 3.5 times more often as last author than first.

Authorship is also occasionally assigned as courtesy for sharing resources, even if the author has made no intellectual contribution to a paper and has not reviewed its contents. Note that the desire to use authorship as a means to gain credit, or acknowledge financial or other resource support, can be at odds with the notion that authors are responsible for the content. For example, the International Committee for Medical Journal Editors recommends that authorship should be based in part on an "agreement to be accountable for all aspects of the work in ensuring that questions related to the accuracy or integrity of any part of the work are appropriately investigated and resolved" http://www.icmje.org/recommendations/browse/roles-and-responsibilities/defining-the-role-of-authors-and-contributors.html.

Medical and Scientific Publishing. https://doi.org/10.1016/B978-0-12-809969-8.00023-1

Can authors be responsible for content if they are listed as a courtesy, or when there are dozens of authors, many of whom are listed as authors because they belong to a multisite data collection team or consortium? CASRAI has been working to more clearly identify specific author contributions through their CRediT program http://docs.casrai.org/CRediT. For example, an author may be noted for their role in developing methodology for a study or acquiring funding for a project.

When it comes to NIH funding, these subtle distinctions in credit are easily lost. If NIH awardees formally claim a paper is a product of their award, they assume responsibility for that paper. That means that they are responsible for ensuring that that paper complies with all NIH rules and conditions of their award.

Awardees can formally claim a paper by listing it in their annual progress report. It does not matter if their authorship is a courtesy, or even if they are not listed as an author. Boyak and Jordan found 27% of papers they surveyed acknowledged NIH grant support, but did not list the PI as an author. The awardees are still responsible for that paper.

In short, from a federal perspective, the primary consideration for publishing is not credit, but responsibility. If an awardee is not in a position to take responsibility for a paper, its contents, and its compliance with all applicable laws and policies, we advise him or her not to take credit for it. In cases where an awardee is tracking papers indirectly arising from his or her work (for example, papers from other labs arising from shared data), NIH has suggested (https://grants.nih.gov/grants/guide/notice-files/NOT-OD-16-079.html) awardees discuss those papers as indicators of impact, rather than direct products.

Case Study: A Research Center/Project Grant

A prestigious NIH-funded research center submitted a renewal application to NIH. Their center ran a popular local seminar series that they felt catalyzed many of the collaborations that made their center successful. They claimed every paper written by every regular member of that seminar series as a product of their award. In fact, most of the hundreds of papers from that center were linked to that seminar, and not arising from the direct spending on research. The reviewers, from that same discipline, agreed with that logic and gave them a perfect score.

In reality, the awardee had only a tenuous connection to most of those papers and, indeed, found many of them through online searches. For a steward of taxpayer funds, this center raises an interesting question since it is clear NIH could stop funding the research projects in that center and still maintain most of its "productivity."

The awardees were overly focused on credit and ignored the responsibility that came with it. The culture for their discipline supported their behavior. Our federal rules are less forgiving. Because the awardees had very little to do with hundreds of papers they had claimed credit for, they were unable to ensure compliance with NIH policies for many of them, and put their funding at risk (see the section on Rules, below). It took many months for the investigators to come into compliance for all the publications claimed. In other circumstances, it may have caused them not to be funded at all.

RULES AND HOW TO FIND THEM

When you get a new grant, you should check the terms and conditions of you in your Notice of Grant Award (https://grants.nih.gov/grants/policy/nihgps/HTML5/part_ii_subpart_b.htm). Not only may these terms change every year, but some awards have special conditions. NIH offers some broad guidance on managing awards at https://grants.nih.gov/grants/post-award-monitoring-and-reporting.htm. Your office of sponsored research also will have advice for you. In this section, we offer some advice and highlight some of the trickier aspects of publication management.

When it comes to publication, three practices are key: (1) correctly acknowledging federal funding, and (2) complying with public access; and (3) reporting papers back to NIH.

1. Acknowledging federal funding. Each paper that directly arises from NIH funds must:
   a. Offer an acknowledgment of NIH grant support in the text of the paper, such as:

   Research reported in this [publication, release] was supported by [name of the Institute, Center, or other funding component] of the National Institutes of Health under grant number [specific NIH grant number in this format: R01GM012345].

   b. Provide a disclaimer that says:

   The content is solely the responsibility of the authors and does not necessarily represent the official views of the National Institutes of Health.

   For the latest information on how to acknowledge funding see https://grants.nih.gov/grants/acknow.htm.
2. Complying with public access. Peer-reviewed journal articles arising from NIH funds need to be made available on PubMed Central, NIH's full text archive. This policy became a law in 2009 because so many NIH supported journal articles required subscription access or a significant payment to read, and limited the utility of NIH supported research (Fig. 23.1).
   Compliance with this policy is simple, but does take a bit of planning. Authors should determine how their paper will be posted to PubMed Central, secure any legal rights they need to make it happen, determine who will ensure

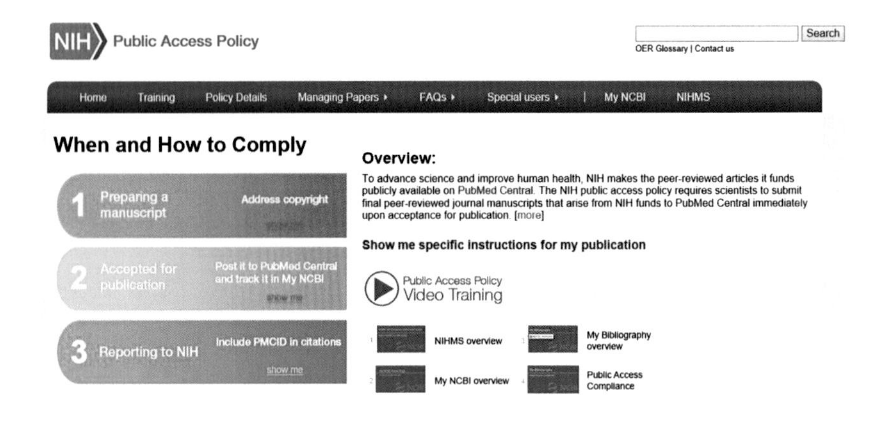

FIGURE 23.1 NIH Public Access Policy home page (https://publicaccess.nih.gov/).

that it happens, and then report that it happened. The specific steps vary by journal and level of access authors wish to offer the public. Fig. 23.2 shows PubMed Central submission methods.

If authors want to ensure a copy of their paper can be distributed for free to students, for example, they will need to secure a specific form of open access license from their publisher. Publishers often charge authors for open access licenses, but many (though not all) also simplify the NIH public access compliance process as well. If authors simply wish to ensure their paper is available without charge, under fair use, they can post their final peer-reviewed manuscript to PubMed Central themselves.

It can be a bit tricky to figure out what to do, and when. We recommend authors use this wizard, at https://publicaccess.nih.gov/determine-applicability.htm, to determine their compliance process. We suggest authors do so *before* they submit their manuscript to a journal.

3. Report the paper to NIH. We ask awardees to do so on annual progress reports, final reports, and competing renewal applications. We ask authors to include the PubMed Central identifier (PMCID) when they cite this paper so everyone knows that these papers are in compliance with the public access policy and can be easily found on PubMed Central. We also ask authors use include PMCIDs for their NIH supported papers on their application biosketches and in other parts of the publication, so reviewers can be sure that applicants comply with the public access policy.

The wizard mentioned above advises on citation formats. It also points to several electronic resources awardees and authors can use to collaborate on tracking, NIH reporting, and many other publication management tasks scientists face.

NIH Public Access Policy

Search

OER Glossary | Contact us

Home Training Policy Details Managing Papers ▸ FAQs ▸ Special users ▸ | My NCBI NIHMS

Submission Methods

There are four methods to ensure that an applicable paper is submitted to PubMed Central (PMC) in compliance with the NIH Public Access Policy. Authors may use whichever method is most appropriate for them and consistent with their publishing agreement. To identify the submission and reporting method for a specific journal, use the submission method identification wizard.

Overview

Details

Overview of Submission Methods

Version of Paper Submitted	Final Published Article	Final Peer-Reviewed Manuscript
Submission Process	Publisher posts the paper directly to PMC	Papers are ***required*** to be submitted via the NIHMS ***upon acceptance for publication***. Publishers, authors or their designee deposit files and the NIHMS converts them to the PMC native format.
Submission Method to Deposit Files	• **Method A:** Some Journals automatically post NIH supported papers directly to PMC • **Method B:** Authors must make special arrangements for some journals and publishers to post the paper directly to PMC	• **Method C**: Authors or their designee must submit manuscripts to the NIHMS • **Method D:** Some publishers will submit manuscripts to the NIHMS • Awardees are responsible for ensuring manuscripts are submitted to the NIHMS **upon acceptance for publication**
Approve Submission	Publisher	Author, via NIHMS
Approve PMC web version	Publisher	Author, via NIHMS
Responsible Party	NIH awardee	NIH awardee
To cite papers, from acceptance for publication to 3 months post publication	PMCID or "PMC Journal- In Process"	PMCID or NIHMSID
To cite papers, 3 months post publication and beyond	PMCID	PMCID

Details

Step-by-step Instructions and Best Practices	Participating Journals and Publishers
• Method A • Method B • Method C • Method D	• Method A Journals • Method B journals and publishers • Method D publishers

FIGURE 23.2 NIH Public Access Policy overview of PubMed Central submission methods (https://publicaccess.nih.gov/submit_process.htm).

Further, these electronic systems are integrated. Reporting a paper on an annual progress report will link that paper to the award electronically. Those associations will display in PubMed.

PUBLICATIONS IMPACT ON PEER REVIEW AND EVALUATION

Your publications inform NIH staff several things about you. Your publications describe your expertise and your scientific areas of interest. They show who your collaborators are and how you give acknowledgments to those whose published work you have built on. They say something about your thought process, your attention to detail, and your innovation and creativity.

NIH uses this information in a variety of ways.

When you apply for funding, NIH Scientific Review Officers (SRO) check your publication record and coauthorship for the last 3 years. They identify your collaborators and those that would be in conflict to review your application.

A collaboration creates a real, or perceived, conflict such that a collaborator may not be able to impartially judge the quality of your future work since he or she is vested in your current work.

SROs are also looking for experts in your field that can evaluate your application, but who do not have an establish research collaboration with you. They can also note that given your publication record, you would be a good reviewer for peer review meetings in the future.

NIH peer reviewers use your publications to evaluate your proposed grant application or contract proposal. Publications signal your independence if you are an early career investigator. Reviewers can also look at rigor and reproducibility of your work. In addition, they may also look at your publication references to see that you have acknowledged others in your field on which you are basing your work. Peer reviewers are generally familiar with the investigators in the field and will not only want to see that you have an independent research direction or concept but that you are well versed in the overall direction of the field. Be aware, however, that given the reviewers' workload, it is not expected that they will read or review the scientific citations you include in your application.

NIH Program Officers (PO) use your publications to evaluate your progress in your grant each year. POs also use your publications to stay current in the field, identify new areas of research to support, and identify gaps in knowledge that they could use to create a Funding Opportunity Announcement. Your publications can also lead to NIH press releases and research highlights for the public and congress. For example, the NIH Director highlights different scientific discoveries and advances on his blog: https://directorsblog.nih.gov/.

RESEARCH, REPRODUCIBILITY, AND INTEGRITY

NIH is very concerned about the quality of the research it funds and the integrity of the scientific record. NIH has enhanced its expectations on scientific rigor and reproducibility (https://www.nih.gov/research-training/rigor-reproducibility).

In concert with journal editors, NIH has set forward guidelines for rigor and reproducibility (https://www.nih.gov/research-training/rigor-reproducibility/principles-guidelines-reporting-preclinical-research) (Fig. 23.3). These include statistical analysis, transparency in reporting, and data and material sharing. In addition, several organizations have developed expanded guidelines to support research that is reproducible, rigorous, and transparent. An example is the Center of Open Science and their Transparency and Openness Promotion in Journal Policies and Practices Guidelines (https://osf.io/ud578/?_ga=1.211230620.829898984.1435325845).

The goal of all these guidelines is to create an expectation that scientific discovery is well founded and can be repeated by other scientists. When scientific discoveries cannot be validated, it brings into question the authenticity of the data. Nonreproducibility may be due to laboratory conditions (which determine

FIGURE 23.3 NIH rigor and reproducibility guidelines for preclinical research (https://www.nih.gov/research-training/rigor-reproducibility/principles-guidelines-reporting-preclinical-research).

why the methods sections need to be robust), lack of statistical rigor (assumptions, cohort size, power, etc.), or mistakes (intentional or nonintentional).

> *Research misconduct is defined as fabrication, falsification, or plagiarism in proposing, performing, or reviewing research, or in reporting research results, according to 42 CFR Part 93.*
>
> https://grants.nih.gov/grants/research_integrity/research_misconduct.htm

Your words reflect you, your colleagues, and your institution, and they have impact in any format (publications, grant applications, poster sessions, etc.). The scientific community examines your published work for rigor, reproducibility, and soundness of conclusion. There are several processes and rules guiding research integrity. You can find them at your institution, the journals, and NIH https://grants.nih.gov/grants/research_integrity/research_misconduct.htm

As mentioned previously, your name on a publication means you are responsible for its content. Your colleagues' reputation is also at stake when you copublish research findings. These relationships and reputation largely rely on trust. Once that trust is lost, or even questioned, things can get very difficult. An investigation of the researchers and the institution is costly and intrusive.

The best way to maintain trust is through transparency and scrupulous record keeping. Most scientists work long hours in extremely competitive environments and therefore could make the occasional error. You can help prevent stress and error from leading to misconduct by supporting an open culture. You should share your procedures, calculations, and assumptions with your collaborators and invite them to check your work.

It's also very helpful to try and continue this practice publicly, to the extent that human subject protections allow. There are a growing number of places to share data, protocols, and preprints (working drafts). These public products can serve as evidence of your productivity, as well as help others understand, replicate, and build on your work.

If you suspect research misconduct, report it to the NIH Research Integrity Office (Fig. 23.4) https://grants.nih.gov/grants/research_integrity/research_misconduct.htm.

Actions can be taken against the institution, as well as the researcher, when research is found to be false. Once a finding has been substantiated, it will be published at https://ori.hhs.gov/case_summary. A researcher can be barred from receiving future funding and from serving on government advisory and peer review committees. An institution may have to pay the government back for the dollars used to create the false data and publication.

Additional information can be found at the HHS Office of Research Integrity https://ori.hhs.gov/ and the NIH Office of Extramural Research https://grants.nih.gov/grants/research_integrity/research_misconduct.htm

TRICKS OF THE TRADE

Your publications are primarily a way to share and take credit for your research discoveries with the scientific community. However, they can have a broader impact if you share and discuss them with NIH. The NIH program and scientific review staff see a breadth of research, and your publications are an important resource for them. Often, they can put resources and investigators together to make a bigger impact in a field.

NIH excels as putting scientists of different fields together in a room and creating new research programs. By talking to NIH staff about your research and publications, you can put yourself in the "room." In addition, many NIH staff reach out to investigators about their publications to learn more about the science. This keeps them better informed and understanding the needs of the research field. NIH POs and SROs attend scientific meetings and site visits. This is an excellent time you share your finding and new directions you are interested in.

We are all working together to improve the health of the nation.

NIH must justify their budget every year to Congress. NIH staff are always on the lookout for publications that can demonstrate the importance of the government investment in biomedical research. These opportunities can arise

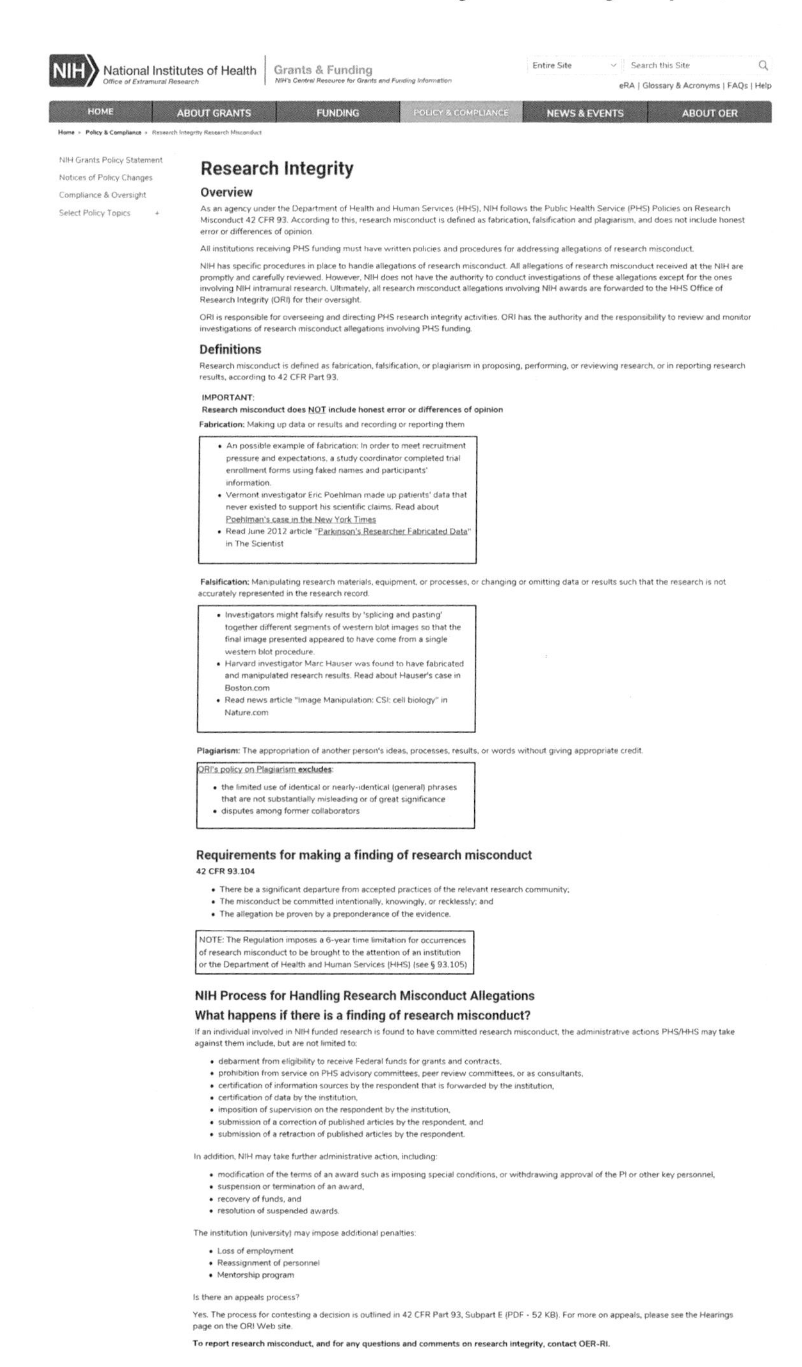

NIH National Institutes of Health
Office of Extramural Research
Grants & Funding
NIH's Central Resource for Grants and Funding Information

Entire Site | Search this Site
eRA | Glossary & Acronyms | FAQs | Help

HOME | ABOUT GRANTS | FUNDING | POLICY & COMPLIANCE | NEWS & EVENTS | ABOUT OER

Home » Policy & Compliance » Research Integrity Research Misconduct

NIH Grants Policy Statement
Notices of Policy Changes
Compliance & Oversight
Select Policy Topics +

Research Integrity

Overview

As an agency under the Department of Health and Human Services (HHS), NIH follows the Public Health Service (PHS) Policies on Research Misconduct 42 CFR 93. According to this, research misconduct is defined as fabrication, falsification and plagiarism, and does not include honest error or differences of opinion.

All institutions receiving PHS funding must have written policies and procedures for addressing allegations of research misconduct.

NIH has specific procedures in place to handle allegations of research misconduct. All allegations of research misconduct received at the NIH are promptly and carefully reviewed. However, NIH does not have the authority to conduct investigations of these allegations except for the ones involving NIH intramural research. Ultimately, all research misconduct allegations involving NIH awards are forwarded to the HHS Office of Research Integrity (ORI) for their oversight.

ORI is responsible for overseeing and directing PHS research integrity activities. ORI has the authority and the responsibility to review and monitor investigations of research misconduct allegations involving PHS funding.

Definitions

Research misconduct is defined as fabrication, falsification, or plagiarism in proposing, performing, or reviewing research, or in reporting research results, according to 42 CFR Part 93.

IMPORTANT:
Research misconduct does NOT include honest error or differences of opinion

Fabrication: Making up data or results and recording or reporting them

- An possible example of fabrication: In order to meet recruitment pressure and expectations, a study coordinator completed trial enrollment forms using faked names and participants' information.
- Vermont investigator Eric Poehlman made up patients' data that never existed to support his scientific claims. Read about Poehlman's case in the New York Times
- Read June 2012 article "Parkinson's Researcher Fabricated Data" in The Scientist

Falsification: Manipulating research materials, equipment, or processes, or changing or omitting data or results such that the research is not accurately represented in the research record.

- Investigators might falsify results by 'splicing and pasting' together different segments of western blot images so that the final image presented appeared to have come from a single western blot procedure.
- Harvard investigator Marc Hauser was found to have fabricated and manipulated research results. Read about Hauser's case in Boston.com
- Read news article "Image Manipulation: CSI: cell biology" in Nature.com

Plagiarism: The appropriation of another person's ideas, processes, results, or words without giving appropriate credit.

ORI's policy on Plagiarism **excludes**:

- the limited use of identical or nearly-identical (general) phrases that are not substantially misleading or of great significance
- disputes among former collaborators

Requirements for making a finding of research misconduct

42 CFR 93.104

- There be a significant departure from accepted practices of the relevant research community;
- The misconduct be committed intentionally, knowingly, or recklessly; and
- The allegation be proven by a preponderance of the evidence.

NOTE: The Regulation imposes a 6-year time limitation for occurrences of research misconduct to be brought to the attention of an institution or the Department of Health and Human Services (HHS) (see § 93.105)

NIH Process for Handling Research Misconduct Allegations

What happens if there is a finding of research misconduct?

If an individual involved in NIH funded research is found to have committed research misconduct, the administrative actions PHS/HHS may take against them include, but are not limited to:

- debarment from eligibility to receive Federal funds for grants and contracts,
- prohibition from service on PHS advisory committees, peer review committees, or as consultants,
- certification of information sources by the respondent that is forwarded by the institution,
- certification of data by the institution,
- imposition of supervision on the respondent by the institution,
- submission of a correction of published articles by the respondent, and
- submission of a retraction of published articles by the respondent.

In addition, NIH may take further administrative action, including:

- modification of the terms of an award such as imposing special conditions, or withdrawing approval of the PI or other key personnel,
- suspension or termination of an award,
- recovery of funds, and
- resolution of suspended awards.

The institution (university) may impose additional penalties:

- Loss of employment
- Reassignment of personnel
- Mentorship program

Is there an appeals process?

Yes. The process for contesting a decision is outlined in 42 CFR Part 93, Subpart E (PDF - 52 KB). For more on appeals, please see the Hearings page on the ORI Web site.

To report research misconduct, and for any questions and comments on research integrity, contact OER-RI.

FIGURE 23.4 NIH Research Integrity overview (https://grants.nih.gov/grants/research_integrity/research_misconduct.htm).

anytime you are at a professional meeting or other science-related setting. Be ready to describe the importance of your work for your colleagues, the press, and NIH in about 30 s and in plain language. Basic science, clinical science, and technology are all important to advance biomedical research.

Another important aspect of publishing is acquiring the skill to critique other scientific work and to accept critiques of your own work. Journal Clubs are a great way to learn how to critique, and consequently write, research papers. Serving on Peer Review panels is another.

Publications are the highlight to a series of experiments and hypothesis. It's critical that readers can understand your work. There are many ways to present data and conclusions. Experiment with reading scientific literature from different fields and see what resonates. Try presenting data in different ways to colleagues and mentors. Pay attention to what types of questions they ask.

Share draft publications and grant proposals with colleagues prior to submitting. It is much better to have critiques from those that can have a dialog with you about what you are trying to get across to the reader.

THINGS ON THE HORIZON

Scholarly publication has evolved over hundreds of years. It has been changing rapidly in the past few decades, mainly because of the Internet. In this section, we will describe three trends that appear to be gaining prominence today that most biomedical scientists need to be thinking about. In a few years, we may have a very different list.

Journal-, Article-, and Person-Level Metrics and ORCID

The Internet has enabled a whole new level of tracking and measurement of article impact. One class of metrics focus on the number of times a paper is cited by other scholarly papers. Some more traditional citation metrics, such as Journal Impact Factor, measure the citations for all the research articles in a journal. They are often used to judge the quality of an individual article, but articles vary in their citation rate even within a journal [2]. NIH is increasingly using the Relative Citation Ratio to evaluate awards, scientists, and portfolios. The Relative Citation Ratio compares the number of citations for a specific paper to the average number of citations of a similar set of papers [3].

Of course, academic citations are a poor measure of impact on clinical care, policy, human behavior, and other important benefits of research. Other metrics, often labeled "altmetrics," are rapidly evolving and it is difficult to tell which measures will be most meaningful to which audience (tenure promotion committees, grant reviewers, journal editors, etc.).

These new measures are possible because the web allows for relatively easy tracking of usage statics and other metadata. Science is also exploring similar kinds of metrics for people, as well as their papers. Companies provide these measures about you to the scientific community.

The best way to ensure these measures are even attributing the right scholarly products to you is to set up an ORCID profile (https://orcid.org/). It is free, and it will help you to link you to all your research products. ORCID is becoming increasingly integrated with many academic communities, including universities and NIH, and in the long run, will simplify tracking your research outputs.

For more information on these issues, please see Chapter 22 on Measuring Impact. Regardless of which measure becomes dominant, two things seem clear. (1) You need to have a coherent story for your work and its impacts, and not rely on numbers alone; and (2) you need to actively manage your scientific online presence, starting with your first public scientific product (poster, presentation, paper, patent, etc.) ORCID is a great place to start.

Interim Research Products—An Opportunity for New Investigators

Another trend, which is relatively new to biomedical research, is what NIH calls "Interim Research Products," defined as complete, public research products that are not final. A common form is the preprint, which is a complete and public draft of a scientific document. Preprints are typically unreviewed manuscripts written in the style of a peer-reviewed journal article. Scientists issue preprints to speed dissemination, establish priority, obtain feedback, and offset publication bias. Another common type of interim product is a preregistered protocol, where a scientist publicly declares key elements of his or her research protocol in advance. Preregistration can help scientists enhance the rigor of their work.

Preprints have been used in fields such as physics, mathematics, and economics for decades. Recently, psychology has been exploring preregistration as a way to reduce certain kinds of methodological bias.

NIH is encouraging the use of these interim products to speed dissemination and enhance rigor. They may be especially important if you are trainee or new investigator, who has not had a chance to finalize many of your research products. NIH offers guidance on how to use interim products in NIH applications and reports at https://grants.nih.gov/grants/guide/notice-files/NOT-OD-17-050.html.

Data Sharing—The Emerging Change in Science Culture

Biomedical scientists cannot be successful unless they publish their findings in peer-reviewed journals. That was not always true. It took many years and concerted effort to establish this practice as a standard.

We are in the midst of similar shift, in this case surrounding data. It is likely that future biomedical scientists will not be successful unless they share their data.

The biomedical community, including NIH, is carefully thinking through how to responsibly share research data. For many years, NIH has requested Data Sharing Plans as part of the NIH applications process for large projects (https://grants.nih.gov/grants/policy/data_sharing/data_sharing_guidance.htm). More recently NIH has published guidelines on Genomic Data Sharing (https://gds.nih.gov/).

As data become more accessible and transportable, new standards will probably emerge. It seems likely that the rigor and quality of publications will be increasingly linked to the quality of accompanying public data. It also seems likely that scientist's careers will be increasingly judged on their transparency, and their ability to support other scientific work through the data they share. For additional information, see Chapter 24 on Data Sharing.

REFERENCES

[1] Boyack KW, Jordan P. Metrics associated with NIH funding: a high-level view. J Am Med Inform Assoc July–August 2011;18(4):423–31. http://dx.doi.org/10.1136/amiajnl-2011-000213.

[2] Zhang L, Rousseau R, Sivertsen G. Science deserves to be judged by its contents, not by its wrapping: revisiting Seglen's work on journal impact and research evaluation. PLoS One March 28, 2017;12(3):e0174205. http://dx.doi.org/10.1371/journal.pone.0174205.

[3] Hutchins BI, Yuan X, Anderson JM, Santangelo GM. Relative citation ratio (RCR): a new metric that uses citation rates to measure influence at the article level. PLoS Biol September 6, 2016;14(9):e1002541. http://dx.doi.org/10.1371/journal.pbio.1002541. eCollection 2016 Sep. PubMed PMID: 27599104; PubMed Central PMCID: PMC5012559.

Chapter 24

Data Sharing

Marisa L. Conte, MLIS
Taubman Health Sciences Library, University of Michigan, Ann Arbor, MI, United States

INTRODUCTION

Sharing research data is becoming more commonplace for scientists in the biomedical and life sciences domains. This shift is due to several significant sea-changes, most notably: an increase in interdisciplinary research, requiring the sharing of research data among experts from different domains; an increase in pressure for scientific studies to be reproducible; and an increase in data sharing mandates from both federal and private funders. Data sharing can take many different forms, from publication of a dataset as an appendix to a scientific paper, to the deposit of digital datasets in open access online archives, to repositories where access is tightly controlled, to individual researchers sharing data with other researchers or groups. This chapter will provide an introduction to data sharing, including benefits and challenges, and an overview of funder mandates impacting American researchers. It will also address practical matters around data sharing that are most important to authors: repositories, metadata, standards, documentation, and other aspects that researchers should consider, beginning with the creation of a detailed data management plan. Finally, the concept of data sharing has led to the birth of new publications: data papers and data journals. This creates opportunities for researchers to disseminate their data for reuse or find new collaborators while simultaneously cashing in on the coin of the academic realm: publications leading to citations.

Mandates requiring authors to share the output or products of their research are not new in the health sciences domains. A number of prominent organizations, including government funders such as the National Institutes of Health (NIH) and Great Britain's Wellcome Trust, and private funders, including the Bill and Melinda Gates Foundation and the Howard Hughes Medical Institute (HHMI), have implemented policies to ensure open or public access to research publications funded by their respective organizations. US-based researchers in the biomedical and life sciences are probably most familiar with the NIH Public Access Policy, which has been in place since 2008 [1] and was made permanent in the Omnibus Appropriations Act of 2009 [2]. The Policy requires

Medical and Scientific Publishing. https://doi.org/10.1016/B978-0-12-809969-8.00024-3

authors of papers deriving from NIH-funded research to make their final, peer-reviewed manuscripts available via PubMed Central (PMC) on acceptance for publication. PMC, a free, full-text archive of literature in the biomedical and life sciences, is developed and maintained by the National Library of Medicine (NLM). In addition to archiving papers that fall under the Public Access Policy, PMC is the repository designated by many other funding agencies, including the Centers for Disease Control (CDC), Food and Drug Administration (FDA), National Institute of Standards and Technology (NIST), and US Department of Veterans Affairs (VA) [3]. As of June 2017, PMC contained over 4.3 million archived full-text papers [4]. Its impact is wide, and the Policy has increased public access to scientific research: "The significant usage of PMC underscores the value of public access to the peer-reviewed scientific literature. On a typical weekday, more than 1 million unduplicated users retrieve over 1.65 million articles. The number of articles retrieved has more than doubled in the past four years, from 17 million retrievals per month in 2009 to more than 40 million in 2013" [5]. Funding agencies such as NIH monitor author compliance with open or public access policies; noncompliance can affect grant distribution and endanger future funding opportunities [6].

DATA SHARING—ARGUMENTS FOR AND AGAINST

The argument for public access to the data underlying scientific research takes a slightly different line than the argument for public access to the published results of research (e.g., papers), primarily because the sharing of research data is less likely to be accessible or understandable to the general public, and is most likely to benefit researchers. One similarity between the two is that if public money (e.g., federal dollars allocated to the NIH) is used to support research, the public should be guaranteed the maximum return on investment. In this sense, maximum return on investment can be demonstrated by facilitating new research from existing datasets, and enhancing the capability of novel hypothesis generation. As Fecher et al. [7] state "The accessibility of research data has a vast potential for scientific progress. It facilitates the replication of research results and allows the application of old data in new contexts." Supporting reproducibility to enhance the credibility of scientific results is another use case for shared datasets; Poldrack et al. [8] note: "Although data sharing does not address the basic question of empirical replicability, it does help address the question of whether the same or a different set of analysis procedures can reproduce the original findings from the same data, and further, it allows for meta-analyses." Taken as a whole, data sharing is seen to have both scientific and societal value: "The sharing and preservation of data advances science by broadening the value of research data across disciplines and to society at large, protecting the integrity of science by facilitating the validation of results, and increasing the return on investment of scientific research" [5].

Data sharing can have a profound and immediate effect on human health. Modjarrad et al. state: "When a new or re-emergent pathogen causes a major outbreak, rapid access to both raw and analysed data or other pertinent research findings becomes critical to developing a rapid and effective public health response. Without the timely exchange of information on clinical, epidemiologic and molecular features of an infectious disease, informed decisions about appropriate responses cannot be made, particularly those that relate to fielding new interventions or adapting existing ones. Failure to share information in a timely manner can have disastrous public health consequences, leading to unnecessary suffering and death" [9]. Leaders in global and public health note that sharing data can help to stem or curtail disease outbreaks [9]. In the 2014–15 Ebola outbreak, sharing of genomic data helped researchers track the spread of the disease [10], and two novel vaccines were developed and trialled in a collaboration between researchers from Africa, Europe, and North America [9]. Inspired by genomics researchers who shared their Ebola data in real time, virologist David O'Connor followed their example. In 2016, in response to the Zika outbreak, O'Connor's lab began posting real-time results from its work in nonhuman primates, including viral load detection in the blood, saliva, and urine of infected macaques. As of 2016, the lab's data had been viewed by users in 98 countries [11]. Health organizations, research bodies, and publishers have all contributed to accelerating data sharing during global or public health emergencies. The World Health Organization (WHO) held a consultation in September 2015, to identify ways to facilitate data dissemination in real time (or as close as possible) and in advance of publication. A consensus statement published after the consultation "affirmed that timely and transparent prepublication sharing of data and results during public health emergencies must become the global norm" [12]. The consensus statement was made by the WHO and several journal publishers, including Nature Journal, the New England Journal of Medicine, and several PLoS Journals, with the journals stating they would not penalize authors who share data in advance of publication as warranted. In February 2016, funders and publishers released a joint statement on Data Sharing in Public Health Emergencies. In addition to making content related to these emergencies freely accessible, without endangering rights to future publication, the statement calls for the rapid and wide sharing of both interim and final data. The statement was signed by more than 30 journals, publishers, and organizations, including federal agencies, foundations, and international nongovernmental organizations [13].

Clinical trial data are another area increasingly identified by both funders and publishers as one in which data sharing is essential for the advancement of both research and human health [14]. Additionally, many see sharing data from clinical trials as a moral obligation. Hudson and Collins state: "When people enroll in clinical trials to test new drugs, devices, or other interventions, they're often informed that such research may not benefit them directly. But they're also told what's learned in those clinical trials may help others. ...

To honor these participants' selfless commitment to advancing biomedical science, researchers have an ethical obligation to share the results of clinical trials in a swift and transparent manner" [15]. Requirements for summary results reporting for sponsors and designated principle investigators were implemented as part of Title VIII of the Food and Drug Administration Amendments Act of 2007, but numerous sources [16,17] have pointed to low compliance numbers. A 2014 study [17] showed that, of 400 randomly selected clinical trials, nearly 30% did not publish the primary outcomes in either a paper or by reporting to Clinicaltrials.gov. A 2016 study by Zarin et al. states that of 224,000 studies registered in Clinicaltrials.gov, only 23,000 contain results information [16]. A regulation and policy change, both intended to increase results reporting and data sharing, were published concurrently in September 2016. First, the Final Rule clarifies the requirements for results reporting, including the relevant trial type, the type of information required, and timeframe, while also providing penalties for noncompliance [18]. Secondly, a new NIH Policy expands the Final Rule to include all NIH-funded clinical trials, including those not subject to regulation, and also provides penalties for noncompliance, including suspension of grant funding. Table 24.1 presents the Summary Table of HHS/NIH Initiatives to Enhance Availability of Clinical Trial Information, showing the new requirements and consequences of noncompliance. This has been both welcomed in many quarters and noted as long-overdue, with a *Nature* editorial stating that "For too long, researchers who don't like the results of a clinical trial have simply failed to publish them," and calling the concurrent Rule and Policy, "... a good first step, one that fosters scientific rigour and affords greater respect to patient commitments" [19].

Finally, there are new ways for researchers to benefit from shared data. Research datasets can be cited, just like papers, which can increase a researcher's impact metrics in systems from Altmetrics to h-index. Publishing datasets can also lead to new collaborations and research partnerships, which is increasingly important as many areas of research become more interdisciplinary. And new publication models, such as the data papers and data journals discussed at the end of this chapter, provide researchers with new opportunities to get scholarly credit for their work.

This is not to say that data sharing is without challenges. Many researchers have legitimate concerns about patient privacy, and others are concerned about yielding competitive advantage, or reluctant to allow others to benefit from their data, which has often taken years to collect, clean, and analyze [20]. Mbuagbaw et al. highlight a number of concerns of sharing clinical research data, including administrative and financial costs [21]. Gorgolewski et al. note barriers—real or perceived—within the neuroimaging community, including the time involved in properly curating a dataset for reuse, and state "Consciously or unconsciously, such logistical challenges are often compounded by fears of embarrassment related to exposition of data errors or poor handling practices" [22]. Farber [23]

TABLE 24.1 Summary Table of HHS/NIH Initiatives to Enhance Availability of Clinical Trial Information

Element	Final Rule	NIH Policy
Scope/ Applicability	Applicable clinical trials of FDA-regulated drug, biological, and device products and pediatric post-market surveillance studies of devices required by the FDA under the FD&C Act.	All clinical trials funded wholly or partially by NIH
	Does not apply to phase 1 trials or small feasibility device studies.	Includes phase 1 clinical trials and trials that do not involve any FDA regulated product such as trials involving only behavioral interventions
	Applicable clinical trials are (1) clinical trials of drug and biological products that are controlled, clinical investigations, other than phase 1 investigations, of a product subject to FDA regulation; and (2) prospective clinical studies of health outcomes comparing an intervention with a device product against a control m humans (other than small feasibility studies) or any pediatric post-market surveillance studies required by FDA under the FD&C Act.	Applies to NIH-funded clinical trials where applications or proposals are received by NIH on or after the policy's effective date
	Applies to public and private sector sponsors and other entities who meet the definition of a responsible party.	Applies to NIH-conducted clinical trials initiated on or after the policy's effective date.
Timeframe for registration on ClinicalTrials.gov	Not later than 21 days after enrollment of the first participant.	Same
Registration data elements to be submitted to ClinicalTrials.gov	Elements defined in the final rule. Consists of descriptive information, recruitment information, location and contact information, and administrative data.	Same

Continued

TABLE 24.1 Summary Table of HHS/NIH Initiatives to Enhance Availability of Clinical Trial Information—cont'd

Element	Final Rule	NIH Policy
Timeframe for results information submission to ClinicalTrials.gov	Not later than 12 months after primary completion date; possible delay of up to an additional 2 years for trials of unapproved products or of products for which initial FDA marketing approval or clearance is being sought, or approval or clearance of a new use is being sought.	Same
Results information data elements to be submitted to ClinicalTrials.gov	Elements defined in the final rule. Includes participant flow, demographic and baseline characteristics, outcomes and statistical analyses, adverse events, the protocol and statistical analysis plan, and administrative information.	Same
Potential consequences of noncompliance	• Identifying clinical trial record as non-compliant in ClinicalTrials.gov. • For federally funded trials, grant funding can be withheld if required reporting cannot be verified. • Civil monetary penalties of up to $10,000/day (amount to be adjusted going forward).	• May lead to suspension or termination of grant or contract funding • Can be considered in future funding decisions • Identifying clinical trial record as non-compliant in ClinicalTrials.gov
Effective date	January 18, 2017. Compliance date is 90 days from the effective date.	January 18, 2017

From: https://www.nih.gov/news-events/summary-table-hhs-nih-initiatives-enhance-availability-clinical-trial-information.

notes the challenges of working with diverse datasets given the inherent heterogeneity of biomedical data and produced a desiderata of sorts for a repository that could help address complex diseases, while noting the success of specific biological repositories (Protein Data Bank, GenBank, Gene Expression Omnibus) and lessons learned from looking at the challenges of aggregating data in other fields, e.g., neuroimaging.

It's also not true that the idea of sharing data has unanimous support among researchers or journal editors. Federer et al. surveyed NIH scientists and found that, among respondents who had shared data in the past, the most common motivation was to collaborate with another researcher or assist a known colleague, while researchers who had not or would not consider sharing data had a variety of reasons, including not believing anyone would find the data useful, concerns that other researchers might misinterpret data, and that current culture does not encourage data sharing [24]. In a 2016 editorial [25], *New England Journal of Medicine* editors famously alluded to "research parasites"—researchers who would go on to publish secondary studies using other peoples' data rather than generating novel ideas or meaningful work of their own. There are also legitimate concerns around privacy when looking at the sharing of protected health information (PHI) or patient-level data with the potential to be reidentified. Further to issues of privacy, researchers have also expressed concern that increased data sharing could lead to difficulties in obtaining informed consent from research participants who may not have a clear understanding of how and why data could be shared and their potential consequences. There is also the very real concern of time—with all of the pressures on researchers and authors to write successful grants, do impactful science and publish highly cited papers in high-profile journals, ensuring data are properly curated and shared is yet another thing to do, and something many researchers do not see any value in doing. Those who do value it are either frustrated by the lack of guidance or models or have difficulty finding help in navigating a complex and often poorly defined practice [26].

DATA SHARING MANDATES

Mandates for data sharing in the biomedical and life sciences have been taking shape over the last 15 years. Much of the impetus for data sharing on federally funded projects has risen since 2013. In February 2013, the Office of Science and Technology Policy (OSTP) issued a memorandum directing agencies with annual extramural research and development budgets exceeding $100 million to establish formal plans for providing public access to research products, including peer-reviewed publications and research datasets, in a manner consistent with existing laws and national security, among other factors. By February 2016, 16 agencies, representing over 98% of federal research spending, had developed these plans [27]. The following section will discuss the development of data sharing mandates for the primary federal funding agencies for US biomedical researchers, as well as several prominent foundations. Funder requirements for data sharing or data management plans are continually being updated; authors are advised to check with their funding agencies for specific data sharing mandates. For reference, the Scholarly Publishing and Academic Resources Coalition maintains a web resource of federal funding mandates: https://sparcopen.org/our-work/research-data-sharing-policy-initiative/funder-policies/.

National Institutes of Health

NIH has had a variety of data sharing requirements in place since the early 2000s. In 2002, NIH published the NIH *Draft Statement on Sharing Research Data*, which stated the importance of data sharing and stated that researchers would be expected to include a plan for data sharing or explain why project data could not be shared. 2002 also saw the release of the *NIH Intramural Policy on Large Database Sharing*, which applied principles of the *Draft Statement* to NIH researchers, including "facilitating the creation of new data sets when data from multiple sources are combined, avoiding duplication of expensive data collections, and expediting the translation of research results into knowledge and products to improve human health" [28]. In March 2003, the *NIH Data Sharing Policy and Implementation Guidance* was published [29]. This called for proposals requesting more than $500,000 in direct costs to include either a data management plan or justification for why research data could not be shared. Unsurprisingly, in the absence of a detailed policy, let alone instructions for compliance or enforcement, this was not generally followed. Additionally, individual Institutes and Centers (ICs) such as the National Institute on Aging (NIA), the National Heart Lung and Blood Institute, and the National Institute on Drug Abuse (NIDA) either expanded on the 2003 *Policy* or established their own data sharing policies; specific programs or projects also have their own data sharing policies, including the National Database for Autism Research, the ENCyclopedia Of DNA Elements (ENCODE) Consortia, and the Human Microbiome Project's Data Release and Resource Sharing Guidelines [5].

The NIH responded to the OSTP memorandum by directing each of the 27 ICs to generate IC-specific plans for compliance. In February 2015, the NIH published its *Plan for Increasing Access to Scientific Publications and Digital Scientific Data from NIH Funded Scientific Research.* This 44-page document details how NIH as a whole will comply with the directive related to public access to both peer-reviewed publications and data. Given the establishment of the NIH Public Access Policy, with PMC as the archive of record, the publications piece is already fairly well controlled. Data sharing is, comparatively, not as well planned for. The *Plan* states the intention of NIH to "make public access to digital scientific data the standard for all NIH-funded research," while taking into account mitigating factors, such as the privacy and confidentiality of research participants, and intellectual property rights or proprietary interests. Per the *Plan*, NIH will "Explore steps to require data sharing; ensure that all NIH-funded researchers prepare data management plans and that the plans are evaluated during peer review; develop additional data management policies to increase public access to designated types of biomedical research data; encourage the use of established public repositories and community-based standards; develop approaches to ensure the discoverability of data sets resulting from NIH-funded research to make them findable, accessible, and citable; promote interoperability and openness of digital scientific data generated or managed by NIH; explore the development of a data commons" [5].

This is a significant commitment and represents a significant departure from past efforts, including the 2003 *Policy* which merely "encourages researchers to share their final research data for use by other researchers in a timely way" [29] and calls for proposals requesting more than $500,000 in direct costs to include either a data management plan or justification for not sharing data. Notably, the *Plan* also calls for NIH support of and investment in not only more mature policies governing public access to data but also the significant infrastructure it will take to ensure that research data are appropriately curated and archived, including public repositories and standards to ensure that data are both findable and interoperable. To oversee planning and implementation, NIH Director Francis Collins established an Internal Steering Group on Public Access to Digital Scientific Data, and a new position, Associate Director for Data Science (ADDS). At the time of writing, the interim ADDS is Dr. Patricia Flatley Brennan, the Director of the NLM.

National Science Foundation

The National Science Foundation's (NSF) data sharing requires researchers to share primary data and other supporting materials within a reasonable timeframe and at cost (AAG). Since January 2011, researchers are required to include a formal data management plan with their proposals. Per the 2017 *Proposal and Award Policies and Procedure Guide*, the data management plan may include information about the types of data to be collected or produced, standards for data formats and metadata, policies for sharing data while assuring relevant protections, provisions for reuse, and plans to archive data to ensure preservation and access [30]. NSF Directorates, divisions or programs can specify additional requirements for data management plans: researchers should review requirements specific to their funding Directorate, division or program at https://www.nsf.gov/bfa/dias/policy/dmp.jsp.

Private Foundations

The Bill and Melinda Gates Foundation, which funds a significant amount of global and public health research, is establishing itself as a leader in open and public access. Established in 2015, the Foundation's open access policy calls for peer-reviewed papers resulting from Foundation funds to be made fully and freely available at the time of publication and also requires that "Data underlying published research results will be accessible and open immediately" [31].

The HHMI also supports open/public access to research outputs. Its 2015 Research Policy *Sharing published materials/responsibilities of HHMI authors (SC-300)* applies to all research publications submitted on or after January 1, 2016. The Policy requires funded investigators to provide sufficient detail about experimental procedures, including materials and methods, to allow other scientists to replicate the work described and also requires authors to make "data, software, and tangible work materials" available to researchers within 30 days of request, unless there are extenuating circumstances [32]. The Policy provides specific guidance on repositories and also refers to the complementary *Public Access to Publications* Policy, which requires investigators make their publications available via PMC within 12 months of publication.

In addition to tracking on funder requirements, authors also need to be aware of publisher requirements to share data related to manuscripts under consideration or accepted for publication. Publishers are requiring data be made available as a term of publication, though this practice is not yet widespread. A 2017 study by Vasilesky et al. reviewed 318 biomedical titles for data sharing policies and noted that, while only approximately 21% of the titles reviewed required data sharing, these titles published nearly 42% of all citable items during the study period and had higher impact factors than journals that did not require data sharing [33]. There are strong indications that publisher requirements for data sharing will become more commonplace. In January 2016, the International Committee of Medical Journal Editors (ICMJE) published a proposal and call for comment. The proposal stated that ICMJE would only accept publications of clinical trials if data were also made publicly available within 6 months of the journal article: "As a condition of consideration for publication of a clinical trial report in our member journals, the ICMJE proposes to require authors to share with others the deidentified individual-patient data (IPD) underlying the results presented in the article (including tables, figures, and appendices or supplementary material) no later than 6 months after publication" [34]. In June 2017, ICMJE published its final *Data Sharing Statements for Clinical Trials* that states "ICMJE will require the following as conditions of consideration for publication of a clinical trial report in our member journals: 1. As of July 1, 2018, manuscripts submitted to ICMJE journals that report the results of clinical trials must contain a data sharing statement as described below. 2. Clinical trials that begin enrolling participants on or after January 1, 2019, must include a data sharing plan in the trial's registration" [35]. The required data sharing statement must indicate whether individual deidentified participant data will be shared, what data will be shared, whether additional documentation will be shared, and when data will be available and under what conditions.

DATA SHARING FOR AUTHORS: PRACTICAL CONSIDERATIONS

Clearly, data sharing requirements are becoming a permanent part of both the research and publishing landscapes in the biomedical and life sciences. Researchers and authors should carefully review both funder mandates and journal contracts and facilitate compliance with all relevant requirements, while also taking steps to make their data findable and reusable. Van Tuyl et al. define data sharing as data that are Discoverable, with a URL or specific identifier provided to facilitate access; Accessible via an open platform; Transparent with documentation that lays out salient information about the dataset; and Actionable by researchers with a minimal amount of processing [36]. FORCE11, an international interest group focusing on issues of e-scholarship, has developed similar FAIR principles: data should be Findable, Accessible, Interoperable, and Reusable [37]. The following sections highlight specific areas for authors to consider both while initiating research projects and when writing and submitting data-based publications for which the sharing of complementary research data is either appropriate or required. For further assistance, researchers and authors, particularly at academic institutions, are encouraged to investigate whether research data management services are available through their institutions' libraries or research offices.

Data Management Plan

The initial requirement will likely take place during the grant-writing stage: the development of a detailed data management plan. Per the NIH *Plan*: "Data management planning should be an integral part of research planning, and planning for data collection or creation and management should take into account downstream data processing and dissemination" [5].

Depending on requirements of a funding organization, data management plans can be as brief as a few sentences or as long as several pages. In *10 Simple Rules for Creating a Good Data Management Plan*, Michener highlights common areas to address in a data management plan, including descriptions of data types, sources, and formats; organization of data, including file-naming conventions and versioning; a strong documentation strategy, including relevant descriptive metadata, provisions for quality assurance and quality control; storage and preservation strategies, including both working storage and a permanent repository; data policies for sharing, including licensing; procedures for dissemination; roles and responsibilities for project members; and development of a budget to cover costs associated with data management [38]. Questions to address include how will data be collected and how will information about the data collection methods (e.g., instrumentation) be collected and preserved? Are there concerns about privacy or confidentiality, and how will these be addressed? Where will working data be stored? What repositories may be appropriate for long-term preservation of final datasets? What common data elements (CDEs) or standards can be used to ensure that data are understandable and reusable outside of the context of the original research proposal? What metadata will be applied to the dataset to make it findable by other researchers?

Data management planning can be complex; many academic research libraries or university research offices offer consultations or services to help researchers develop robust data management plans. The Interuniversity Consortium for Political and Social Research (ICPSR) provides guidance on writing a data management plan (https://www.icpsr.umich.edu/icpsrweb/content/data-management/dmp/), Johns Hopkins has developed a questionnaire that helps researchers formulate data management plans (https://jh.app.box.com/v/questionnairejhudms), and DMPTool (https://dmptool.org/) provides templates, sample plans, and other resources to assist researchers writing data management plans.

Data Archives and Data Repositories

A primary concern of researchers and authors who are interested in or compelled to share their research data publically available is where to store it. What are appropriate places to store data for preservation and reuse? Many funding agencies and many publishers do not consider author-hosted data to be adequate. This has to do largely with concerns about an individual's ability to guarantee permanent preservation of and access to data. Most funders and publishers require data to be kept in a more permanent archive or repository.

The terms archive and repository are used somewhat interchangeably to refer to storage options for long-term preservation of data. For both researchers and authors, it's important to understand the difference between repositories or archives and storage. Generally speaking, storage is a local or cloud solution that is intended to hold either working data or data that are unlikely to be shared. Researchers may have access to personal account-based storage either via their institutions (e.g., institutional subscriptions to Box, Google Drive) or personal accounts (e.g., Amazon Web Services, Google Drive, DropBox) or institutions may have long-term storage solutions (e.g., tape storage) that are both cost-effective and convenient for researchers. Unfortunately, these solutions are generally not acceptable solutions to comply with funder or publisher mandates for data sharing as the data are generally not structured or documented for reuse and are generally impossible for external users to locate.

Data archives or data repositories are both more appropriate for data sharing and more likely to satisfy funder mandates and publisher requirements. Authors looking for an appropriate data repository have many options and should take the following factors into account: technical specifications (file size limits, dataset size limits), cost, level of assistance with deposit and curation, integration with other systems, available support, retention polices, security and backup procedures, licensing for reuse, and metadata requirements. Authors should also consider the reputation of the repository: how long the repository has been around and their level of confidence that the resource is stable and sustainable. Repositories can be hosted and maintained by a variety of sources, including federal agencies, universities, or research institutions. Several open-source research platforms also archive and host research data.

Federal agencies may have their own repositories for researcher-generated data. These repositories have several advantages: data deposit is generally free;

the deposit process can be mediated; and access to the data is often controlled through the program sponsor, so researchers do not have to worry about inappropriate sharing or reuse of data. NIH maintains a list of repositories that make research data available for sharing: https://www.nlm.nih.gov/NIHbmic/nih_data_sharing_repositories.html. The list includes the name, URL, and description of the repository and links to policies for data deposit (often governed by the specific funding mechanism or granting IC) and access to data for reuse.

In addition to federal repositories, generalist repositories are becoming more widely used. These repositories are generally account-based and do not require institutional subscriptions; several of the most common generalist repositories or research platforms are listed in Table 24.2. There are also a number of specialized or community-specific repositories, which archive data for specialized domains from sequencing data to model organisms.

One of the most well-known specialized or community-specific repositories is maintained by the ICPSR. Located at the University of Michigan in Ann Arbor, Michigan, ICPSR is an international consortium of hundreds of academic and research institutions and specializes in data curation and training in curation and analysis for the political and social sciences research communities. ICPSR's repository contains over a half million files from thousands of studies. While the repository focuses on social sciences or political sciences research, some of the domain areas intersect with health sciences research, and ICPSR archives research data from projects sponsored by several NIH ICs, including the NIA, National Institute of Mental Health, NIDA, and National Institute on Child Health and Human Development [39]. Many of the datasets archived in ICPSR are open for public use, but the repository also contains restricted data. There are two different deposit options: high-level curation of research data is provided for researchers affiliated with member institutions. Self-deposit for behavioral or social sciences data is also available via OpenICPSR (https://www.openicpsr.org/openicpsr/).

In addition to discipline-specific repositories, or repositories hosted by research institutes or federal agencies, many US research universities are also establishing data repositories to help their faculty and students comply with funder or publisher mandates to provide free, long-term access to their research data. Institutional repositories have a number of benefits, chief among them being cost (institutional repositories are generally free) and convenience. Like institutional publication repositories, institutional data repositories are often run through university libraries; librarians can provide researchers with expert assistance in depositing data as well as lending their expertise to curation services such as file conversion or creation of metadata and documentation, and other services, such as generating citations to datasets or minting a Digital Object Identifier (DOI) to facilitate both citation of and access to data. While there are benefits to institutional repositories, there are also potential disadvantages. As institutional repositories may lack the robust infrastructure

TABLE 24.2 Generalist (Nondiscipline-Specific) Data Repositories

Repository	URL	Description
Dataverse	https://dataverse.org/	Open-source repository; available to institutions or individual researchers through Harvard's instance. File size limits vary by instance.
Dryad	http://datadryad.org/	Curated repository suitable for data in medical or scientific disciplines. Partners with nearly 120 journals for integrated data submission; authors can also submit their own data after payment of a processing charge.
Figshare	https://figshare.com	Part of the Digital Science platform. With a free account, researchers can deposit data in a variety of formats, and individual files up to 5 GB. The free account allocates 20 GB of private space (suitable for datasets researchers may want to share with collaborators but not make publicly available) and provides unlimited space for unrestricted public data sharing in compliance with funder or other mandates.
Open Science Framework	https://osf.io/	An open-source research platform designed to facilitate collaboration, run by the Center for Open Science. Open Science Framework focuses on transparency throughout the research life cycle rather than only archiving final research outputs.
Zenodo	https://zenodo.org/	Open-source platform for deposit of data and related research outputs in the sciences; accepts files up to 50 GB; data stored in CERN Data Center

present in archives built to accommodate large-scale datasets, there may be limits on the size of data files and final datasets these repositories can accept. Additionally, most institutional repositories are not configured for sensitive data, including PHI, and do not meet Health Insurance Privacy and Portability Act (HIPPA) requirements for secure storage of sensitive data. Finally, many institutional repositories are configured for public access to all archived data and reuse under provisions of Creative Commons licenses and may not provide mechanisms for researchers to monitor who is accessing, downloading, and potentially making use of their data.

Find Trustworthy Repositories

Funders may state which repositories researchers must use to be in compliance with relevant mandates or may provide a list of acceptable repository solutions. Some journals or publishers also maintain lists of suggested repositories, including Public Library of Science (http://journals.plos.org/plosone/s/data-availability#loc-recommended-repositories) or the Nature Publishing Group's *Scientific Data* journal (https://www.nature.com/sdata/policies/repositories#healthsci). Finally, several web-based resources can help researchers or authors identify repositories in which to archive and share their research data. The Registry of Research Data Repositories (http://www.re3data.org/) lists over 1500 research data repositories in a variety of disciplines, and the Directory of Open Access Repositories, OpenDOAR (http://www.opendoar.org/) lists more than 2600 repositories.

Find Data for Reuse

Repository registries such as Re3data or OpenDOAR are also useful to researchers searching for relevant datasets to reuse in their own research. In addition to these registries, dataset catalogs can also help direct potential users to research data for reuse. Dataset catalogs are different than repositories; similar to a library catalog, dataset catalogs include descriptive information about a dataset but may not provide access to the dataset itself. Datasets included in these catalogs can be those generated by researchers on a specific topic or from a specific institution. Data.gov (https://www.data.gov/) is one of the most well-known dataset catalogs—it provides information about, and occasionally access to, nearly 195,000 datasets from various federal agencies, departments, and programs. Health data, including data from nearly 70 NIH repositories, are also available at HealthData.gov, a data catalog maintained by the Department of Health and Human Services (https://www.healthdata.gov/).

Ensure Shared Data Are Reusable—Common Data Elements, Standards, and Metadata

One of the biggest challenges researchers face in trying to reuse data is understanding the data itself. This applies not only to shared data but often to researchers attempting to reuse data generated in their own labs or collaborative groups. Poor versioning or file-naming conventions, ambiguous labeling of variable names, and inadequate documentation data collection methods or of processes used to clean and analyze data can render otherwise robust and potentially valuable datasets practically useless. Additionally, in many cases, data must be published or archived with the tools used to clean and analyze it for it to be useful for future use. The following section covers some of the things that should be considered when ensuring that shared data will indeed be useful to others.

Use Common Data Elements and Standards

The NLM defines a data element as "information that describes a piece of data to be collected in a study" and a CDE as "a data element that is common to multiple datasets across studies" [40]. Sheehan et al. expand on this definition, describing a CDE as "a combination of a precisely defined question (variable) paired with a specified set of responses to the question that is common to multiple datasets or used across different studies" [41]. CDEs provide a way to standardize data collection so that related data can be pooled and analyzed across multiple studies or to investigate relationships between data in unrelated datasets. Sheehan et al., in stating the value of CDEs, state "The use of CDEs, especially when they conform to accepted standards, can facilitate cross-study comparisons, data aggregation, and meta-analyses; simplify training and operations; improve overall efficiency; promote interoperability between different systems; and improve the quality of data collection" [41]. Simply put, researchers who employ CDEs can both streamline data collection and increase the potential for reuse of their shared data.

NIH encourages the use of CDEs, and a number of CDE collections have been developed by ICs or projects for specific diseases, conditions, or subject areas. NLM manages a CDE Resource Portal (https://www.nlm.nih.gov/cde/), which presents collections of CDEs by research area or subject, hosts a CDE Repository (https://cde.nlm.nih.gov/home), and provides access to tools and resources researchers can use to identify elements for their projects.

Standards can also facilitate data sharing and reuse by improving interoperability and allowing reuse of coded data. NLM coordinates clinical terminology standards for the Department of Health and Human Services, including management of standards for drug names, diseases, and laboratory measurements and values [42]. The Clinical Data Interchange Standards Consortium (CDISC) (https://www.cdisc.org/) is an international, nonprofit group working to develop consensus-based standards to enhance interoperability in clinical research. CDISC standards include the Biomedical Research Integrated Domain Group Model, as well as standards in more than 20 therapeutic areas for specific diseases or conditions [43]. To ensure research data have the widest possible use, researchers should investigate which standards may be available in their research domains and utilize them.

Invest Time in Documentation and Metadata

Sharing data involves more than just placing spreadsheets in a repository. Good documentation is essential to make data understandable for other researchers and reuse in different contexts. Documentation can take a variety of forms, from README files attached to datasets, to codebooks or data dictionaries, to metadata.

README documents are simple text files that provide additional information about and context for either specific data files or a whole dataset. Sample information provided in a README file can include a description of the data, the collection methodology, contact information for the creator, and details about any steps taken to clean or process the data for analysis. Cornell University's Research Data Management Services Group has published a guide to writing README documents at https://data.research.cornell.edu/content/readme. The Dryad repository also provides guidance in developing README files for datasets or individual files, with examples, at http://datadryad.org/pages/readme.

Codebooks or data dictionaries should also be shared with datasets whenever possible. These resources enable other researchers to understand a dataset at the variable level, which can help create more opportunities for reuse. A data dictionary includes specific information about each variable, including the format, units of measurement, and coded values (including null values). ICPSR is a great resource for more information on documentation and codebooks, including the 2011 *Guide to Codebooks* (http://www.icpsr.umich.edu/files/deposit/Guide-to-Codebooks_v1.pdf).

Metadata is succinctly defined as "data about data." Simply speaking, metadata is information that contextualizes data, making it useful and understandable not only to the original research team but also to other potential users. This may include experimental details such as instrumentation settings or timestamps of observations. In addition to providing information necessary to reuse data, metadata also improves the discoverability of publicly available datasets. By providing robust descriptive metadata, researchers can make their datasets more findable to other researchers. Repositories may require use of specific metadata standards, such as Dublin Core or Data Documentation Initiative. The Minimum Information for Biological and Biomedical Investigations began as a standard for The Center for Expanded Data and Retrieval is building a framework to help researchers create or identify useful metadata for experimental data in biomedical research using templates and other tools (https://metadatacenter.org/tools-training/cedar-metadata-tools). With formal training in information organization and retrieval, academic librarians are a wonderful resource for researchers looking for assistance with metadata.

Share Tools, Software, and Code

Sharing data is important for scientific discovery, novel hypothesis generation, reproducibility of results, and fostering collaboration. But in some cases, sharing only the data is not enough. Often, researchers must share not only well-documented datasets but also any related tools, code, or software they used to process or analyze these data, to make the data useful and useable. This is occasionally specifically required by funders or publishers but should be recognized as a best practice to maximize the utility of shared data. Researchers can use open-source repositories such as GitHub or SourceForge to make code and other related products available and link to these in dataset documentation and publications [44].

NEW PUBLICATION OPPORTUNITIES

The growing emphasis on and importance of data sharing has led to innovations in publication. Additionally, the growing emphasis on data sharing has led to the introduction of new publication formats: data papers and data journals.

Data papers are becoming more common in biomedical and life sciences publishing [45]. Rather than presenting the data in the context of a study, including analysis, discussion, and conclusions, data papers focus solely on the data itself. Data papers include information on how the data were acquired or collected [22]; they may also include the rationale for collecting the data, and suggestions for further use. Data papers do not provide any analysis or discussion of results or conclusions drawn from the data. In many ways, a data paper most closely resembles a very detailed "Methods" section. As data papers are published, they result in citations which researchers can use in CVs; data papers can also be cited by other researchers and can contribute to the original authors' h-index or other bibliometric research impact factors.

Data journals are also emerging as a publication venue for researchers seeking to formally publish datasets. Akers reports that there are currently over 200 data journals in health sciences disciplines; most of these journals are indexed for various databases, including PubMed, Scopus, and Web of Science. Nearly all data journals employ peer review for submissions and many are open access. Data journals take varying approaches to ensure readers have access to datasets being described in data papers: some allow authors to host the datasets on personal or institutional websites, while others require that datasets to be deposited in a publicly accessible repository [44].

CONCLUSION AND FINDING HELP

Sharing research data, either in conjunction with a formal publication or as a publication in its own right, is becoming not only increasingly common in the biomedical and life sciences but is also being mandated by funders and publishers. Researchers should inform themselves about mandates and opportunities for data sharing and follow best practices to make sure their research data are as open and useful to others as possible. For assistance with research data management, authors and researchers at academic instructions are encouraged to contact their libraries. Many university libraries now have research data services; these may include assistance with data management plans to the establishment and direction of an institutional data repository. With training in the organization and management of information, librarians are professionally equipped to partner with researchers in a number of issues around research data management, from planning for data use and reuse throughout a project life cycle to developing processes for documentation, curating data for discoverability and reuse including appropriate metadata, and identifying repositories and publication venues.

REFERENCES

[1] National Institutes of Health. Revised policy on enhancing public access to archived publications resulting from NIH-funded research. 2008.
[2] National Institutes of Health. The Omnibus Appropriations Act of 2009 makes the NIH public access policy permanent. 2009.
[3] National Library of Medicine. PubMed Central: Funders and PMC. https://www.ncbi.nlm.nih.gov/pmc/about/public-access/.
[4] National Library of Medicine. PubMed central. https://www.ncbi.nlm.nih.gov/pmc/.
[5] National Institutes of Health. National institutes of health plan for increasing access to scientific publications and digital scientific data from NIH funded scientific research. 2015.
[6] National Institutes of Health. Frequently asked questions about the NIH public access policy. 2014. https://publicaccess.nih.gov/faq.htm-6.
[7] Fecher B, Friesike S, Hebing M. What drives academic data sharing? PLoS One 2015;10(2):e0118053.
[8] Poldrack RA, Poline JB. The publication and reproducibility challenges of shared data. Trends Cogn Sci 2015;19(2):59–61.
[9] Modjarrad K, Moorthy VS, Millett P, Gsell PS, Roth C, Kieny MP. Developing global norms for sharing data and results during public health emergencies. PLoS Med 2016;13(1):e1001935.
[10] Pisani E, Aaby P, Breugelmans JG, et al. Beyond open data: realising the health benefits of sharing data. BMJ (Clin Res Ed) 2016;355:i5295.
[11] Kallas EG, O'Connor DH. Real-time sharing of Zika virus data in an interconnected world. JAMA Pediatr 2016;170(7):633–4.
[12] World Health Organization. Developing global norms for sharing data and results during public health emergencies. 2015. http://www.who.int/medicines/ebola-treatment/blueprint_phe_data-share-results/en/.
[13] Public Library of Science (PLoS). Statement on data sharing in public health emergencies. 2016.
[14] Hudson KL, Collins FS. Clinical trials: sharing of data and living up to our end of the bargain. NIH Director's Blog; 2016.
[15] Hudson KL, Collins FS. Honoring our promise: clinical trial data sharing. NIH Director's Blog; 2014.
[16] Zarin DA, Tse T, Williams RJ, Carr S. Trial reporting in ClinicalTrials.gov – the final rule. N Engl J Med 2016;375(20):1998–2004.
[17] Saito H, Gill CJ. How frequently do the results from completed US clinical trials enter the public domain? – A statistical analysis of the ClinicalTrials.gov database. PLoS One 2014;9(7):e101826.
[18] National Institutes of Health. Summary table of HHS/NIH initiatives to enhance availability of clinical trial information. https://www.nih.gov/news-events/summary-table-hhs-nih-initiatives-enhance-availability-clinical-trial-information.
[19] US clampdown on clinical-trial reporting is long overdue. Nature 2016;537(7621):450.
[20] Tenopir C, Allard S, Douglass K, et al. Data sharing by scientists: practices and perceptions. PLoS One 2011;6(6):e21101.
[21] Mbuagbaw L, Foster G, Cheng J, Thabane L. Challenges to complete and useful data sharing. Trials 2017;18(1):71.
[22] Gorgolewski KJ, Margulies DS, Milham MP. Making data sharing count: a publication-based solution. Front Neurosci 2013;7:9.
[23] Farber GK. Can data repositories help find effective treatments for complex diseases? Prog Neurobiol 2017;152:200–12.

[24] Federer LM, Lu YL, Joubert DJ, Welsh J, Brandys B. Biomedical data sharing and reuse: attitudes and practices of clinical and scientific research staff. PLoS One 2015;10(6):e0129506.
[25] Longo DL, Drazen JM. Data sharing. N Engl J Med 2016;374(3):276–7.
[26] Neylon C. As a researcher…I'm a bit bloody fed up with data management. 2017. Science in the open.
[27] Sheehan J. Increasing access to the results of federally funded science. 2016.
[28] National Institutes of Health. NIH announces draft statement on sharing research data. 2002.
[29] National Institutes of Health. NIH data sharing policy and implementation guidance. 2003. https://grants.nih.gov/grants/policy/data_sharing/data_sharing_guidance.htm.
[30] National Science Foundation. Proposal & award policies & procedures guide (PAPPG). 2017.
[31] Bill, Melinda Gates Foundation. Bill and Melinda Gates Foundation open access policy. http://www.gatesfoundation.org/How-We-Work/General-Information/Open-Access-Policy.
[32] Institute HHM. Research policies: sharing published materials/responsibilities of HHMI authors (SC-300). 2015.
[33] Vasilevsky NA, Minnier J, Haendel MA, Champieux RE. Reproducible and reusable research: are journal data sharing policies meeting the mark? PeerJ 2017;5:e3208.
[34] Taichman DB, Backus J, Baethge C, et al. Sharing clinical trial data–a proposal from the international committee of medical journal editors. N Engl J Med 2016;374(4):384–6.
[35] Taichman DB, Sahni P, Pinborg A, et al. Data sharing statements for clinical trials - a requirement of the international committee of medical journal editors. N Engl J Med 2017;376(23):2277–9.
[36] Van Tuyl S, Whitmire AL. Water, water, everywhere: defining and assessing data sharing in academia. PLoS One 2016;11(2):e0147942.
[37] FORCE11. FAIR principles. https://www.force11.org/group/fairgroup/fairprinciples.
[38] Michener WK. Ten simple rules for creating a good data management plan. PLoS Comput Biol 2015;11(10):e1004525.
[39] Inter-university consortium for political and social research. About ICPSR. https://www.icpsr.umich.edu/icpsrweb/content/membership/about.html.
[40] National Library of Medicine. Common data element resource portal: glossary. https://www.nlm.nih.gov/cde/glossary.html.
[41] Sheehan J, Hirschfeld S, Foster E, et al. Improving the value of clinical research through the use of common data elements. Clin Trials 2016;13(6):671–6.
[42] National Library of Medicine. Health information technology and health data standards at NLM. 2017. https://www.nlm.nih.gov/healthit/index.html.
[43] Clinical Data Interchange Standards Consortium. Standards. https://www.cdisc.org/standards/domain-information-module/bridg.
[44] Akers KG. Raising researchers' awareness of biomedical data journals to promote data sharing. In: Federer LM, editor. The medical library association guide to data management for librarians. Rowman & Littlefield Publishers; 2016. p. 49–68.
[45] Kratz J, Strasser C. Data publication consensus and controversies. F1000Res 2014;3:94.

Chapter 25

Open Educational Resources in the Health Sciences

Molly Kleinman, MSI, MA
Health Information Technology and Services, University of Michigan, Ann Arbor, MI, United States

INTRODUCTION AND BACKGROUND

Open Educational Resources (OER) are materials that are free to use for teaching and learning without asking for permission. For an educational resource to qualify as OER, it must be licensed not just for no-cost access but also for reuse, adaptation, and remixing. OER can encompass many kinds of educational materials, including lecture slides and notes, textbooks, images, videos, question banks, and interactive tutorials. In the health sciences, there have been a range of efforts to produce high-quality learning resources, especially in the area of global health.

Beginning in the 1980s, with the rise of personal computers and later the Internet, a constellation of open movements emerged that advocate for free, public access to various kinds of knowledge, information, and code. The open education movement argues that educational materials should be shared freely online in formats that are downloadable, shareable, revisable, and remixable. Other open movements include the open source movement, which promotes the development of software that is free for future developers to adapt and expand on and underpins several modern operating systems including Linux and Android; the open access movement, which advocates for free, online access to peer-reviewed scholarship such as journal articles and monographs; and the open data movement, which encourages researchers to share their data publicly to enable both replication studies and new research. All of these movements share as a premise the idea that the Internet makes possible free and instantaneous distribution of knowledge, and that we should harness the efficiencies of the network to promote the progress of society [1].

This chapter provides an explanation of OER and the copyright licenses that facilitate it, a brief history of OER in the health sciences, and a summary of some of the main benefits and challenges of creating and adopting OER.

Medical and Scientific Publishing. https://doi.org/10.1016/B978-0-12-809969-8.00025-5

THE FIVE Rs AND CREATIVE COMMONS

For an educational resource to qualify as "open," it must comply with what are known as "the five Rs of OER":

1. Retain—the right to make, own, and control copies of the content (e.g., download, duplicate, store, and manage)
2. Reuse—the right to use the content in a wide range of ways (e.g., in a class, in a study group, on a website, in a video)
3. Revise—the right to adapt, adjust, modify, or alter the content itself (e.g., translate the content into another language)
4. Remix—the right to combine the original or revised content with other material to create something new (e.g., incorporate the content into a mashup)
5. Redistribute—the right to share copies of the original content, your revisions, or your remixes with others (e.g., give a copy of the content to a friend) [2].

OER requires these five Rs to maximize the usefulness and reach of the content. It is also a reaction to trends in educational publishing that to allow students or institutions to "rent" electronic versions of textbooks or other resources that then expire at the end of the school year, rather than allowing them to own copies that they can refer to later [3].

The usual method for making content open is to apply a Creative Commons (CC) license. CC is a nonprofit organization that created a set of open licenses to offer a middle ground between the freedom of the public domain and the restricted "all rights reserved" of default copyrights [4]. CC licenses allow creators to mark their work with certain permissions, so that anyone who accesses that work will know they are free to use it in certain ways, without having to ask the copyright holder. The licenses are communicated in three ways: a brief, plain language deed that clearly states what is and is not permitted, a legal license, which communicates the same information in legally binding language that has survived multiple court challenges [5], and a computer code version, which allows search engines such as Google to find and return only results that are licensed for reuse. There are four different license features that can be combined to make a total of six different licenses:

- Attribution, which lets others copy, distribute, display, and perform a copyrighted work—and derivative works based on it—but only if they give the creator credit the way they request. This element is a part of all six licenses.
- Noncommercial, which lets others copy, distribute, display, and perform a work—and derivative works based on it—but for noncommercial purposes only.
- No Derivatives, which lets others copy, distribute, display, and perform only exact copies of a work, not derivative works based on it.
- Share Alike, which allows others to distribute derivative works only under a license identical to the license that governs the original work.

All of the licenses require Attribution as a bare minimum. The Share Alike component and the No Derivatives component are not compatible; either a user may make derivative works, such as translations, and must share them under the same license as the original or the user may not make any derivative works. For a license to qualify as an "open license," it must enable the five Rs. Not all CC licenses work for OER. No Derivatives licenses, which do not permit adaptation, are not suitable for OER.

BENEFITS OF OPEN EDUCATIONAL RESOURCES

Creating, sharing, and using OER can produce tangible benefits, both for individuals and institutions. While most of the research has studied general student populations, rather than health science students in particular, much of it is generalizable to the health sciences.

Cost Savings

Cost savings represents one of the most measurable institutional benefits of OER. When institutions make the switch to open resources as a replacement for textbooks, students often save millions of dollars. Relatively small investments can result in large savings. In Georgia, the public university system's Affordable Learning Georgia program awarded $2.1 million in "textbook transformation" grants; the estimated savings to students was $26.4 million [6]. A state-funded program in North Dakota parlayed a $110,000 appropriation into an estimated $2 million in savings for students [7]. In 2016, Rutgers University implemented the Open and Affordable Textbook Project providing grants to faculty who "chose to adopt, remix, or create a free or low-cost textbook." The program, which includes work in the health sciences, has a projected savings for students of $1.6 million dollars in its first year [8].

Improved Student Outcomes

Many students pay more for textbooks than they do for tuition, or choose not to purchase textbooks at all [9]. A 2012 survey of students in Florida found that when a course uses a traditional textbook, as many as 22% of students frequently do not purchase it, while another 26% only sometimes purchase it [10]. Students who do not have access to textbooks often do more poorly in their courses or drop them altogether. The timing of financial aid payments often means that even for students who plan to purchase the book, they have to wait to do so until several weeks into the semester. One study found that switching to OER had a statistically significant reduction in students who dropped the course [11]. In addition, several studies have found that students whose courses use OER do as well as or better than comparable students in a course with a traditional textbook [12–14].

Flexible Pedagogy

Adopting OER can enable instructors to tailor their courses more finely to the needs and interests of their students. Rather than hewing to a single textbook for an entire semester, instructors can build a syllabus from a range of open resources, try out with different media types such as video and audio, and adjust their course midway through the semester in response to student assessments and feedback. OER also permits students to interact more directly with their learning resources. Some instructors have even experimented with having students create their own OER as an assignment or as an integrated part of the course. In one example, students in a graduate-level chemistry class worked in groups to identify topics that were relevant to the course that were not adequately represented in Wikipedia and collaborated over the semester to develop and expand articles related to those topics [15].

CHALLENGES OF OPEN EDUCATIONAL RESOURCES

Accessibility

An ongoing for institutions engaged in the creation and dissemination of OER involves the accessibility of those resources for users with disabilities. The most dramatic case to date involves UC Berkeley. In August 2016 the US Department of Justice ruled that UC Berkley's OER were not accessible as required by the Americans with Disabilities Act [16]. UC Berkley, in addition to paying damages, would need to make all of their content accessible by meeting the benchmarks set in the WCAG 2.0 guidelines for accessibility at the AA level. Responding to the decision, the UC Berkeley Public Affairs office stated compliance would

> *...require the university to implement extremely expensive measures to continue to make these resources available to the public for free. We believe that in a time of substantial budget deficits and shrinking state financial support, our first obligation is to use our limited resources to support our enrolled students. Therefore, we must strongly consider the unenviable option of whether to remove content from public access [17].*

Beginning on March 1, 2017, Berkeley did in fact remove content from public access, a move that resulted in significant backlash from members of the Berkeley faculty community [18,19].

The events at Berkeley highlighted the importance of ensuring that OER are accessible to all users, both for legal liability reasons and ethical reasons. The OER movement aims to improve access to educational content to all users; this must truly mean all users, including those with visual, hearing, and learning impairments. It is much easier to create accessible OER at the outset than it is to attempt to remediate large bodies of inaccessible legacy OER, as would have been the case at Berkeley. OER initiatives must consider accessibility from the start to avoid legal challenges and to fulfill the values and mission of the OER movement.

Funding and Sustainability

Most OER initiatives begin with grants from philanthropic organizations, governments, or educational institutions. Ongoing funding is not guaranteed. This raises the challenge for OER initiatives of identifying a feasible long-term funding model. Some of the foundations that helped launch the OER movement continue to fund some OER initiatives, but they often demand plans for achieving "sustainability" as a condition of funding [20]. They may measure success for an initiative based on its ability to become self-sustaining after a certain period of time, often by having a business model that gives some content away for free and sells or licenses print copies or supporting content. While some projects, such as OpenStax, appear to have found a hybrid business model that works, many university-based OER initiatives are unlikely ever to bring in enough money to sustain their operations without support either from their home institutions or funding agencies. In some institutions, much of the energy that had previously gone into OER has shifted toward Massive Open Online Courses, which while they have "open" in their name rarely fulfill the definition of OER.

During the Obama administration, the federal government provided support for the creation of OER at the K-12 level through campaigns, such as the Department of Education's #GoOpen grants [21], and at the community college level through the Department of Labor's Trade Adjustment Assistance Community College and Career Training grant program [22]. However, traditional publishing companies have strongly argued against taxpayer-funded OER, and it is unclear if the current administration's Department of Education will continue to fund campaigns such as #GoOpen [23].

Copyright and Licensing

Despite the enormous quantity of CC licensed content currently available online (over 1.2 billion objects at last count) [24], there remain many copyright challenges related to the creation and distribution of OER (Fig. 25.1). On the creation side, producing educational content that is free from all third-party material, or that only uses appropriately licensed material, is harder than it looks. Many faculty who become interested in sharing their content wish to adapt slides, lectures, or videos that they already created for use in the classroom. However, the copyright laws governing what is permissible to use in the classroom or in a password-protected course website are very different from what an instructor may share on the open web. Furthermore, instructors may not attach CC licenses to material for which they do not hold the copyright. As a result, producing OER requires a certain level of copyright and licensing knowledge. Building that knowledge among faculty, or providing staff to offer consultations, can be prohibitive in terms of both cost and logistics.

On the distribution side, disagreement remains about which of the six CC licenses are ideal for OER. The license must permit adaptation and reuse, but there is ongoing debate about whether the noncommercial requirement, which

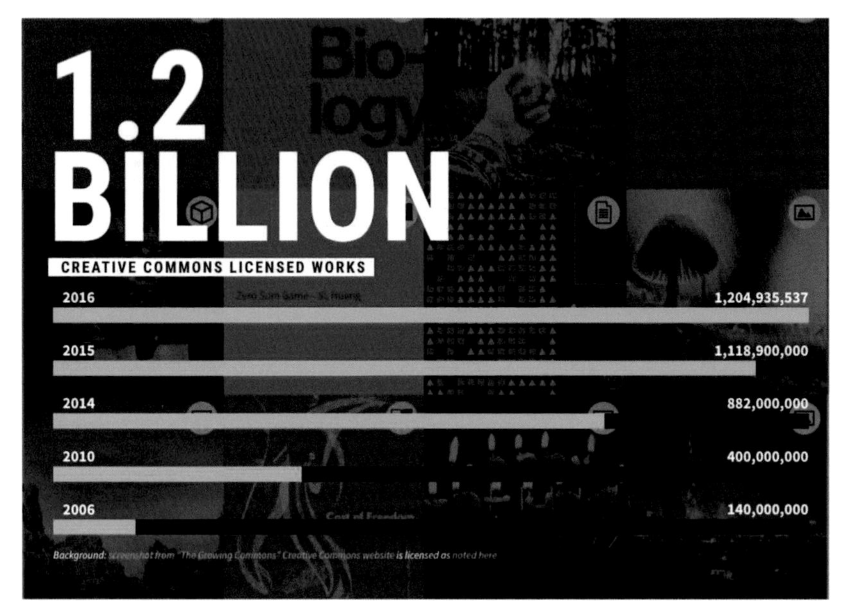

FIGURE 25.1 "State of the Commons" by Creative Commons, 2016. CC BY. (Courtesy of Creative Commons *https://stateof.creativecommons.org/.*)

is very popular among university faculty, is appropriate for OER or whether the share alike requirement is overly burdensome or confusing for lay users [25,26].

OPEN EDUCATIONAL RESOURCES IN THE HEALTH SCIENCES

A Brief History of Open Educational Resources

The first university-based OER initiative launched at Massachusetts Institute of Technology (MIT) in 2002. Called MIT OpenCourseWare (OCW), the initiative began as a proof of concept with 50 courses from across the university, with particular strengths in the sciences and engineering [27]. MIT OCW developed both a process and a platform for reviewing existing course materials, removing any material protected by third-party copyrights, and sharing the materials online under CC licenses. Most of the published courses included syllabi and lecture slides, while some also had quizzes and other assessments, or even lecture videos.

Major funding for the early stages of MIT OCW came from the William and Flora Hewlett Foundation, the Andrew W. Mellon foundation, and from MIT itself. The initiative grew out of a desire to take advantage of the Internet to expand the impact of the university. As Steven Lerman, a professor involved in the original proposal, put it to the New York Times in 2001, "Selling content for profit, or trying in some ways to commercialize one of the core intellectual activities of the university seemed less attractive to people at a deep level than finding ways to disseminate it as broadly as possible" [28].

In addition to providing support for MIT OCW, the Hewlett Foundation went on to make the production of and infrastructure for OER a major funding priority, and it was involved in helping to grow and shape the early OER movement. It funded CC, several university-based OER initiatives, the open textbook publisher Connexions, as well as technical and policy projects, with the aim of supporting "the creation of an ecosystem of OER groups" [29]. The Hewlett Foundation conceptualized its OER work as a "strategic international development initiative," using the Internet and high-quality educational content as tools to increase freedom around the world [30]. Its goal was "to use information technology to help equalize the distribution of high quality knowledge and educational opportunities for individuals, faculty, and institutions within the United States and throughout the world" [31]. Over time, Hewlett began to direct some of its OER funding toward projects specifically dedicated to health science education.

Early Health Science Open Educational Resources

Beginning with the earliest health science-specific OER initiatives, there has been a focus on global partnerships and global education. Tufts University's OCW initiative was announced as a campus-wide endeavor, but it was the first to focus on health science content when it launched in 2005. Tufts had a history of putting science content online with the Tufts University Science Knowledgebase (TUSK), "an integrated digital repository and curriculum knowledge management and delivery system" used by Tufts' for health sciences schools, as well as partner health sciences institutions in India, Tanzania, Uganda, Kenya, and the United States [32]. Faculty at Tufts who had been teaching and doing research in the developing world started using the TUSK system to support that work; the OCW initiative used the TUSK infrastructure to share openly licensed content with a global audience.

Other university-based OER initiatives with a focus on the health sciences included Johns Hopkins University, where the initiative was based in their School of Public Health, and the University of Michigan, where its Open Michigan was based in the Medical School. Each institution took a slightly different approach to its OER work. Open Michigan focused on developing legal, policy, and technical infrastructures to support the creation of OER. Johns Hopkins had a narrower focus on public health. Tufts used TUSK to build up external medical education and content delivery. All of these initiatives received funding from Hewlett; other major funders included the Andrew W. Mellon Foundation and the Bill and Melinda Gates Foundation.

In 2008, the Hewlett Foundation provided seed funding for OER Africa and its African Health OER Network, which support the development and use of OER in universities across Africa. The African Health OER Network consisted of a major partnership between several African universities, the University of Michigan, and global NGOs including UNESCO and Commonwealth of

FIGURE 25.2 Screenshot of Open Educational Resources (OER) Africa home page, 2017. *(http://www.oerafrica.org/.)*

Learning [33]. The vision for the Network was to support the development, adaptation, and sharing of health education resources to improve health-care provision across Africa [34]. The founders of the Network were also reacting to a tendency of OER to flow from wealthy countries to low-income countries, which served to reinforce historical dynamics of colonial power and control [35]. The project vision states "It is important to ensure that OER processes in health are driven from Africa, rather than imported to Africa, and see Africans as contributors to OER Networks not passive beneficiaries" [36]. By enabling faculty at African universities to create and share their own OER, they could address their most pressing local health-care subjects and reflect the resources and standards of care that trainees were likely to encounter in their local settings (Fig. 25.2).

MedEdPORTAL: Peer-Reviewed Open Educational Resources for the Health Sciences

OER initiatives today can loosely be categorized into two groups, textbook initiatives and repositories. Textbook initiatives focus on the "scaling up" of OER materials into larger-packaged products that are marketed as freely available alternatives to traditional course textbooks [37]. Prominent open textbook initiatives include OpenStax, a nonprofit open textbook publisher that makes its books freely available online while also selling a low-cost print option, and the Open Textbook Network, which trains faculty on the basics of creating and using OER and maintains a library of open textbooks across a wide range of disciplines [38]. Repositories focus on hosting OER materials in collections that are organized around courses or topics. One review in 2014 identified 80 repositories of varying quality in existence, although there are no doubt more [39]. Most OER repositories do not specialize in a single field, so there are few repositories that serve exclusively a health science audience. MedEdPORTAL is the notable exception.

The Association of American Medical Colleges funds the MedEdPORTAL with the mission to "promote educational scholarship and collaboration by facilitating the open exchange of peer-reviewed health education teaching and assessment resources." [40] Founded in 2005, MedEdPORTAL provides a platform for finding and publishing free, OER in the health sciences and accepts submissions from faculty and students around the world. Its primary focus is on medical education, but the repository includes content across disciplines including nursing, pharmacology, and dentistry and contains materials for levels from undergraduate to continuing professional development. Like other OER repositories, the educational materials in MedEdPORTAL not only include lectures, online modules, and assessments but it also includes content types specific to the health sciences, such as clinical cases and survey instruments.

The peer review aspect makes MedEdPORTAL unusual among OER repositories. MedEdPORTAL has an editor-in-chief and an editorial board, enforces a peer review policy, and uses expert reviewers, all of which makes it more like a peer-reviewed journal than an OER repository. While open textbooks often undergo the same rigorous review process as traditional textbooks, few OER repositories have undertaken to develop a peer review process for the course content they publish. One result of this process is that it can take some time for materials to become available in the collection; between the copyright screening process and peer review, it can take up to 3 months from submission to a final decision [41]. While this may put MedEdPORTAL at a disadvantage relative to other OER repositories that accept materials without peer review, the benefit is that a successful submission to MedEdPORTAL may be included as a peer-reviewed publication on a CV, and it is increasingly recognized as a desirable outlet for medical educators to share their work.

LOOKING AHEAD

Fifteen years after MIT launched the first large-scale OCW initiative, OER have become an accepted part of the landscape of content that supports teaching and learning, both in the health sciences and across disciplines and education levels. While early excitement about the mass democratization of knowledge has died down, OER continues to demonstrate advantages for instructors, for students, and for institutions. As the infrastructure for creating and distributing OER grows, so will the ease of use for instructors who wish to adopt or produce OER. In the health sciences, the potential impact for global health of making educational resources free to use and adapt for local contexts is enormous, and funders as Hewlett and Gates remain interested in promoting OER for that purpose. The existence of MedEdPORTAL, which accepts OER from faculty and students around the globe, means that health educators who wish to share their materials are not constrained by the lack of an OER initiative on their campus. This combination of funder support and a popular publishing platform provides the foundation for a vibrant ecosystem of open health education resources.

ACKNOWLEDGMENT

Jeff Bennett provided research support for this chapter.

REFERENCES

[1] U.S. Constitution, Article 1, Section 8, Clause 8.

[2] Wiley D. Defining the "Open" in Open Content and Open Educational Resources. Undated. http://opencontent.org/definition/.

[3] Thomassen E. Problem: overpriced textbooks, solution: opensource material. Pioneer Inst Public Policy Res 2015. http://pioneerinstitute.org/news/solution-to-overpriced-textbooks-opensource-material/.

[4] Creative Commons. What we do. Undated. https://creativecommons.org/about/.

[5] Creative Commons. Case law. Undated. https://wiki.creativecommons.org/wiki/Case_Law.

[6] Allen N. Turning point for OER use?. Inside Higher Ed.; April 19, 2017. https://www.insidehighered.com/digital-learning/article/2017/04/19/new-yorks-decision-spend-8-million-oer-turning-point.

[7] Allen N. Turning point for OER use?. Inside Higher Ed.; April 19, 2017.

[8] Open and affordable textbook project at Rutgers University. http://www.libraries.rutgers.edu/open-textbooks.

[9] Goodwin MAL. The open course library: using open educational resources to improve community college access (Unpublished doctoral dissertation). 2011.

[10] Donaldson RL, Nelson DW, Thomas E. Florida student textbook survey. Florida distance learning consortium. 2012. http://www.openaccesstextbooks.org/%5Cpdf%5C2012_Florida_Student_Textbook_Survey.pdf.

[11] Wiley D. Adopting OER is better for everyone involved. Open Content; January 22, 2015. https://opencontent.org/blog/archives/3743.

[12] Lovett M, Meyer O, Thille C. The open learning initiative: measuring the effectiveness of the OLI statistics course in accelerating student learning. J Interact Media Educ 2008;2008(1). Art-13.

[13] Bowen WG, Chingos MM, Lack KA, Nygren TI. Interactive learning online at public universities: evidence from a six-campus randomized trial. J Policy Anal Manag 2014;33(1):94–111.

[14] Hilton III JL, Laman C. One college's use of an open psychology textbook. Open Learn 2012;27(3):265–72.

[15] Moy CL, Locke JR, Coppola BP, McNeil AJ. Improving science education and understanding through editing wikipedia. J Chem Educ 2010;87(11):1159–63.

[16] U.S. Department of Justice. The United States' findings and conclusions based on its investigation under title II of the Americans with Disabilities Act of the University of California at Berkeley. 2016. DJ No. 204-11-309 https://news.berkeley.edu/wp-content/uploads/2016/09/2016-08-30-UC-Berkeley-LOF.pdf.

[17] Public Affairs, UC Berkeley. A statement on online course content and accessibility. 2016. http://news.berkeley.edu/2016/09/13/a-statement-on-online-course-content-and-accessibility/.

[18] Public Affairs, UC Berkeley. Campus message on course capture video, podcast changes. 2017. http://news.berkeley.edu/2017/03/01/course-capture/.

[19] Vogler C. Access denied. Inside Higher Ed.; April 18, 2017. https://www.insidehighered.com/views/2017/04/18/scholars-and-others-strongly-object-berkeleys-response-justice-department.
[20] McGill L. Sustainability. Open Educational Resources Infokit. https://openeducationalresources.pbworks.com/w/page/26789871/Sustainability.
[21] Press Office. U.S. Department of education launches campaign to encourage schools to #GoOpen with educational resources. October 29, 2015. https://www.ed.gov/news/press-releases/us-department-education-launches-campaign-encourage-schools-goopen-educational-resources.
[22] Trade Adjustment Assistance Community College and Career Training, 2011. https://doleta.gov/taaccct/.
[23] Herold B. Under Trump, Ed-Tech Leadership is a big question mark. Education Week; May 5, 2017. http://www.edweek.org/ew/articles/2017/05/10/under-trump-ed-tech-leadership-is-big-question.html.
[24] Creative Commons. State of the commons. 2016. https://stateof.creativecommons.org/.
[25] Pollock R. Why share alike licenses are open but non-commercial ones aren't. Open Knowledge International Blog; June 24, 2010. https://blog.okfn.org/2010/06/24/why-share-alike-licenses-are-open-but-non-commercial-ones-arent/.
[26] Wiley D. The SA fallacy: open knowledge foundation gets it wrong. Iterating Toward Openness; June 25, 2010. https://opencontent.org/blog/archives/1498.
[27] Massachusetts Institute of Technology. About OpenCourseware. Undated. https://ocw.mit.edu/about/milestones/.
[28] Goldberg C. Auditing classes at MIT, on the web and free. The New York Times; April 4, 2001. http://www.nytimes.com/2001/04/04/us/auditing-classes-at-mit-on-the-web-and-free.html.
[29] William + Flora Hewlett Foundation. Open Educational Resources. Undated. http://www.hewlett.org/strategy/open-educational-resources/.
[30] Atkins DE, Brown JS, Hammond AL. A review of the open educational resources (OER) movement: achievements, challenges, and new opportunities. 2007.
[31] Atkins, D.E., Brown, J.S., Hammond, A.L.. A review of the open educational resources (OER) movement: achievements, challenges, and new opportunities. 2007.
[32] Lee MY, Albright S, O'Leary L, Terkla DG, Wilson N. Expanding the reach of health sciences education and empowering others: the OpenCourseWare initiative at Tufts University. Med Teach 2008;30:159–63.
[33] OER Africa. African health OER network: about us. Undated. http://www.oerafrica.org/african-health-oer-network/about-us.
[34] OER Africa. A vision for the network. Undated. http://www.oerafrica.org/african-health-oer-network/vision-network.
[35] Luo A, Omollo KL. Lessons learned about coordinating academic partnerships from an international network for health education. Acad Med 2013;88:1658–64. http://dx.doi.org/10.1097/ACM.0b013e3182a7f815.
[36] OER Africa. A vision for the network. Undated. http://www.oerafrica.org/african-health-oer-network/vision-network.
[37] Mathewson TG. What's the state of OER and where do we go next?. Education Dive; March 11, 2016. http://www.educationdive.com/news/whats-the-state-of-oer-and-where-do-we-go-next/415429/.
[38] Open Textbook Network. Impact & benefits. https://research.cehd.umn.edu/otn/impact-and-benefits/.
[39] Motz R, Tansini L. Evaluating OER repositories. In: Interacción '14: proceedings of the XV international conference on human computer interaction. 2014.
[40] Meded Portal. Mission & vision. https://www.mededportal.org/about/missionandvision/.
[41] MedEdPORTAL. J Am Med Libr Assoc July 2016;104(3):250–2.

Part V

Case Study: Michigan Journal of Medicine

Chapter 26

From Concept to Publication: Laying the Groundwork for a Student-Run Medical Journal

Sagar Deshpande[1,2], **Spencer Lewis**[2]
[1]*John F. Kennedy School of Government, Harvard University, Cambridge, MA, United States;* [2]*University of Michigan Medical School, Ann Arbor, MI, United States*

INTRODUCTION

The Michigan Journal of Medicine (MJM) is a medical student-run peer review journal based at the University of Michigan. It was conceived, in large part, because of a story—a story we feel is ubiquitous (both at our institution and others), as well as unacceptable. It is a story we heard routinely when discussing extracurriculars with our classmates, who expressed shock and disbelief at the energy we devoted to research activities. It is the story of why our fellow students felt that research was a pointless and unproductive use of their time.

For us, this attitude was ceaselessly surprising. We were both involved in research from the most formative stages of our careers. For one of us, this experience started in high school, working in laboratories at Western Michigan University and Pfizer, and later in two research laboratories at the University of Michigan. For the other, it began as a fruitful collaboration with a University of Michigan neurosurgeon that whet his appetite for translational research. We both felt that our research efforts were the best investments we could make in our future careers. As one of our mentors was fond of saying, "your CV is forever." However, this attitude was not shared widely among our peers.

MOTIVATION AND JUSTIFICATION

At the University of Michigan, most students' first foray into medical research is through a 10-week summer intensive program, where students perform experiments and data analysis through one of the countless research groups at the University. After squandering Ann Arbor's most temperate months hunched

Medical and Scientific Publishing. https://doi.org/10.1016/B978-0-12-809969-8.00026-7

over a pipette or in front of a computer, our classmates would then crunch the numbers, with occasionally resoundingly disappointing results: a *P*-value greater than 0.05, a correlation that disappears on controlling for a confounder, an antibody that did not bind as expected. With no time to iterate their experimental design or refine their hypothesis before returning to class in the fall, their mentor deems their data "unpublishable" and the student is left with little to show but a sour taste in their mouth and the memories of their "last summer" being spent in fluorescently lit, windowless rooms. They felt that the unpredictable nature of research and the lack of payoffs to their hard work made it a pointless expenditure of time, and so they decided to just not "do" research. "I don't know how you can do it," they would say to us. "I tried it and I just couldn't keep doing it."

Unbeknownst to them, these early decisions can have substantial consequences on a medical student's career. Many medical students who decry the importance of research discover that publication is expected in residency or in practice, and find themselves lacking basic skills in statistics and scientific writing. Residents are frequently expected to assess primary literature, contribute to writing grants, and assist their seniors in completing manuscript reviews for journals. These are not innate skills, but ones that must be honed and practiced over time, putting residents who are research novices at a substantial disadvantage. These disadvantages can extend long beyond the confines of residency. In general, clinical track faculty are not explicitly expected to publish and are not represented on Faculty Senates. This does not affect the research-proficient clinical faculty members, who are typically in the instructional track for career advancement at the University.

The students' frustration was certainly understandable, but we felt their resentment was misplaced. Science does not fail students by disproving their hypotheses; rather, their mentors fail them by subscribing to the idea that only positive results are publishable. The academic community has warned of the dangers of positive publication bias—the tendency for experiments that fail to disprove the null hypothesis to go unpublished—since the 1950's. Karl Popper in 1963 described research as a "voyage of discovery," subject to unpredictability and fallibility [1]. Numerous papers have been published since then discussing the value of reporting negative results [2–4]. Yet trainees still face this pressure from both mentors and editors, with many rigidly insisting that negative studies are "unpublishable." Several top journals have made concerted efforts to correct this trend; PLoS One, for example, does not take into account whether a study is positive or negative when considering a manuscript for publication, and informal observation suggests that a substantial fraction of the articles published by the New England Journal of Medicine involve negative studies. However, these behaviors at top-ranked journals make little impact on the experience of most medical students, as many middle-tier journals remain focused on positive results as a driver of impact.

ENTER MICHIGAN JOURNAL OF MEDICINE

MJM was born out of a conversation between the two authors of this chapter to remedy these attitudes, and to provide a venue for exactly that type of student research. When we conceived of the journal, we had three priorities. First, we wanted to ensure that this journal would be a safe haven for all data that were rigorously and meticulously produced, without consideration of whether the study was "positive" or "negative." Second, we wanted to create a journal with enough legitimacy that medical students would actually list their publications in MJM on their CV's and residency applications. Third, we wanted medical students to be well trained in navigating the research process—submitting manuscripts, reviewing the manuscripts of others, and responding to reviewer concerns.

The conventional model of research is that medical students should feel grateful for the opportunity to publish, and for the unique learning experiences that opportunity provides. This new endeavor sought to disrupt that model because it carried the implicit assumption that students who perform quality research *deserve* to publish. As medicine is a notoriously lethargic field in exploring new attitudes, that had the potential to be a dangerous implication. But we felt, at its core, this project embodied a different truth: that the *information* deserved to be published. It was our stance in creating this journal that it is not our place to arbitrate the impact or value of our submissions beyond the validity of their science and the quality of their writing. Indeed, both of our formative scientific careers had already encountered situations where our laboratories would have saved countless months and tens of thousands of dollars if another investigator had published that a certain idea, or application of an experimental technique, did not work. In the current scientific model, it was up to each research group not to waste precious resources discovering these unfortunate truths individually and in parallel.

With our idea in hand, we decided that the next step was finding a faculty mentor for our project. The natural choice for us was to pitch it to the leadership of our medical school curriculum. The University of Michigan Medical School was in the process of transforming the way medical students are taught and we felt that this new journal could be a part of that initiative. We were pleased with the positive response and the high level of commitment we received. Our faculty sponsor, in addition to having a stereotypically busy clinical practice, is a well-established researcher, overseeing numerous studies and student research projects. Moreover, he is incredibly passionate about student education at the undergraduate, medical student, and resident levels, was a familiar research mentor to countless medical students, and was overseeing the creation and implementation of the new medical school curriculum (Fig. 26.1).

We discovered that our idea serendipitously intersected with what is referred to as a *window of opportunity* in political theory—the confluence of events and circumstances that allow a motivated actor to achieve an otherwise unlikely

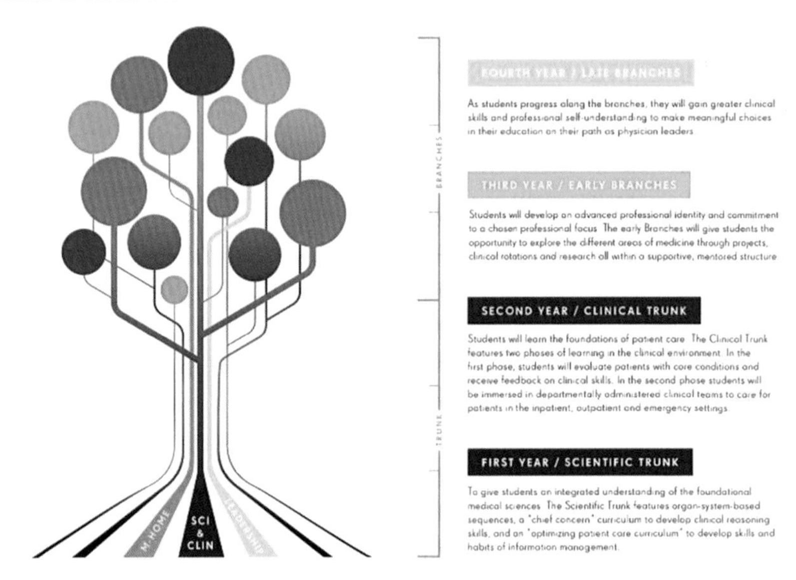

FIGURE 26.1 "Tree" schematic depicting the University of Michigan Medical School curriculum. *(From https://medicine.umich.edu/medschool/about/key-initiatives/curriculum-innovation.)*

outcome. With the medical school curriculum revision, the journal could be incorporated as a longitudinal course rather than an extracurricular activity, facilitating accountability on the part of both students and faculty mentors. This simultaneously addressed several barriers to success for MJM. First was legitimacy; instead of being simply an initiative of a collection of students at a university, the Journal would be officially sanctioned and promoted by the medical school administration. Second was longitudinal leadership, preservation of institutional memory, and the ability to survive editorial transitions of power. Student publications frequently fail when the initial cohort of extremely motivated students graduate and move on, and their replacements lack the enthusiasm to ensure the initiative continues to succeed. By being incorporated into the curriculum, a prescribed pathway for MJM was developed that trained and groomed students for leadership positions, ensuring a smooth and sustainable handoff of responsibilities. And lastly, integration into the curriculum came with a built-in intrainstitutional reputation. For this project to achieve its potential, we needed faculty at Michigan to perceive this as a valuable resource. Research investigators would have to allow their students to publish their data in our journal, and faculty who served as external journal editors would need to volunteer their time to mentor students learning the editing and review process.

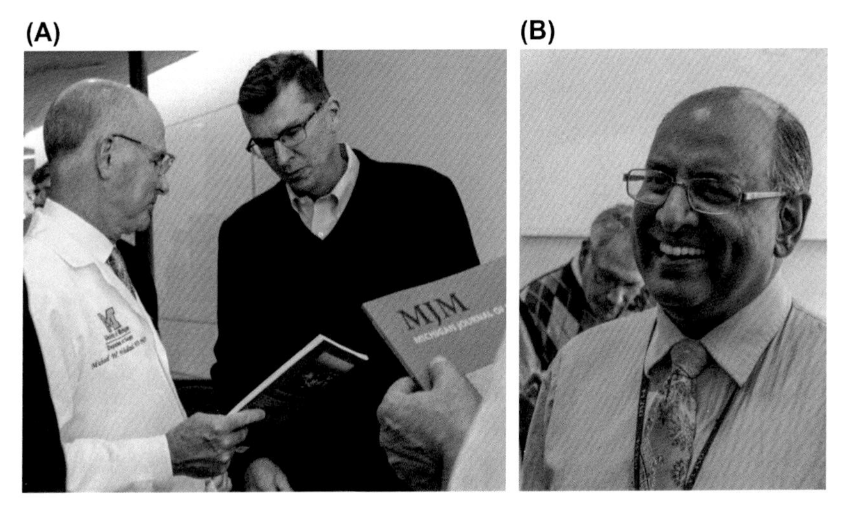

FIGURE 26.2 (A). Dr. Michael Mulholland, Chair of Surgery at Michigan, and Dr. Michael Englesbe, Michigan Journal of Medicine (MJM) Faculty Advisor, discussing MJM at the inaugural issue launch party, May, 2016. (B). Dr. Rajesh Mangrulkar, Associate Dean for Medical Student Education, at MJM launch party. *(Photos courtesy Marc Stephens, Ann Arbor, MI.)*

Our faculty sponsor quickly gave his enthusiastic endorsement for MJM and recruited funding and support from other faculty members at the medical school, including division chiefs and chairs of departments (Fig. 26.2). He also put us in touch with a medical school colleague with journal publishing experience who ran a learning design and publishing unit and provided publishing services and consulting for faculty, staff, and students. The two became the coadvisors for the journal, forming an incredible base of support and advice, and advocating tirelessly on the organization's behalf to the deans of the medical school as well as other adjunct organizations, such as the University Library, who were integral in making MJM a reality.

However, to truly cement the Journal as a functional and sustainable initiative, we needed the financial support of more than just discretional departmental funding. An early long-term goal of the journal was to eventually be indexed in PubMed, and establishing independent funding was an important step toward achieving that goal. Furthermore, an investment through a central university funding body would both functionally advertise MJM and give it additional reputational legitimacy. We found an excellent opportunity for such funding with the Transforming Learning for the Third Century (TLTC) at the University of Michigan (Fig. 26.3).

The TLTC initiative is a large, $50 million dollar educational initiative established at the University to commemorate its bicentennial and the transition into the "third century" of the University by investing in innovative, multidisciplinary teaching and scholarship ideas [5]. We applied for a "Quick Wins" grant, which was directed at "relatively small-scale, shovel-ready" projects that

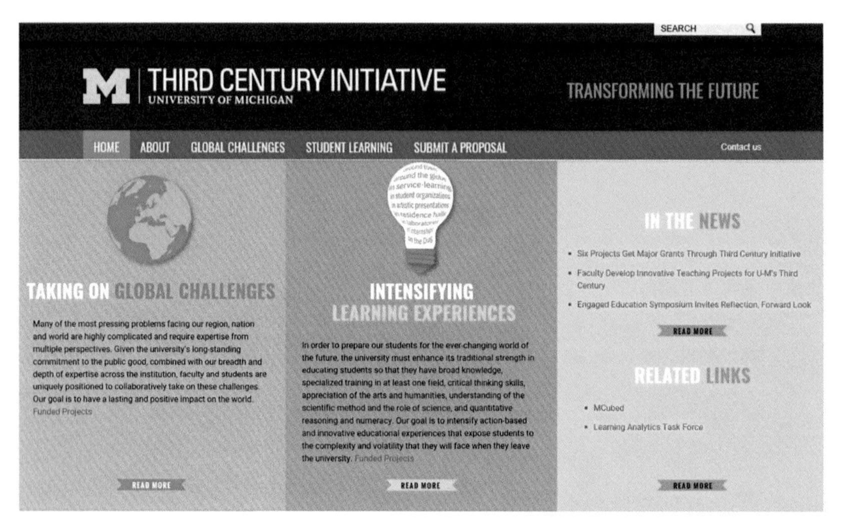

FIGURE 26.3 University of Michigan Third Century Initiative. *(From http://thirdcentury.umich.edu/.)*

have transformative potential for curriculum, pedagogy, and student learning, and/or projects that embrace risk, discovery, and experimentation, empowering faculty members and staff to explore opportunities beyond the traditional. We felt that our project, which paired an innovative "learning through doing" approach to student authorship and editorship with faculty membership, fit both clauses of the grant description well.

Although the maximum funding for the grant was substantial ($50,000), we applied for a more modest sum of funding ($15,500). This decision was strategic. Our slim budget was generously accommodated, keeping our expenses relatively low, as all the incredibly costly backend computer services required for running and maintaining a journal were provided through the University's Health Information Technology and Services team and the University Library through its Michigan Publishing unit. We hoped that the comparatively small investment would allow for easier access to seed funding that could quickly build the necessary legitimacy to access larger funding pools. In this case, our strategy was successful; we were funded by the TLTC committee. In addition to providing financial support, the TLTC grant also gave us very valuable exposure and publicity university-wide. MJM was a real initiative. We were ready to go… but we needed editorial leadership.

This was the last potential major barrier to the success of MJM. We had approached our faculty sponsor about the Journal in the fall of our second year of medical school. By the time the journal organization was forming, we were well into the school year. During our time at Michigan, medical students started their clinical responsibilities in their third year of medical school, and we were rapidly approaching that threshold. We and our faculty advisor knew that we could not reasonably expect to successfully transition into this new stage of

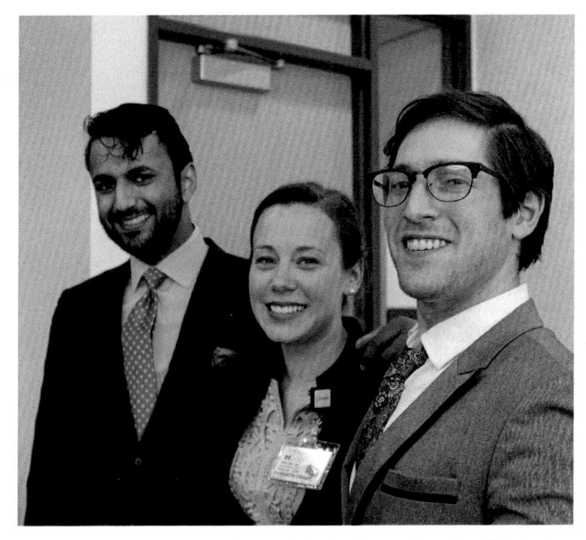

FIGURE 26.4 The founders of MJM along with the first Editor-in-Chief, photo taken at May 2016 Launch party for the journal. On the left, Sagar Deshpande, on the right, Spencer Lewis, coauthors of this chapter and cofounders of MJM. In the center, Shannon Cramm, the first Editor-in-Chief. *(Photo courtesy Marc Stephens, Ann Arbor, MI.)*

our education while developing the journal and recruiting and managing the student authors and editors. Therefore, we had to find a student to take charge of the project and accomplish the incredibly difficult task of turning someone else's idea into a reality. Our faculty advisor had the ideal candidate in mind: Shannon Cramm, then a fourth year medical student, became the inaugural editor-in-chief for the Michigan Journal of Medicine (Fig. 26.4). In Chapter 27 she describes that exciting first year.

REFERENCES

[1] Conjectures and refutations: the growth of scientific knowledge. London: Routledge; 1963.

[2] Matosin N, et al. Negativity towards negative results: a discussion of the disconnect between scientific worth and scientific culture. Dis Model Mech February 2014;7(2):171–3.

[3] Scargle JD. Publication bias (The "File-Drawer Problem") in scientific inference. ArXiv:physics 9909033. 1999.

[4] Fanelli D. Do pressures to publish increase scientists' bias? An empirical support from US States data. PLoS One 2010;5:e10271.

[5] http://thirdcentury.umich.edu/.

Chapter 27

Michigan Journal of Medicine as a Learning Tool: Establishing a Student-Run Journal and Editorial Team

Shannon Cramm, MD
Department of Surgery, Massachusetts General Hospital, Boston, MA, United States

To create a student-run journal at the University of Michigan Medical School, a team that could complete this task had to be established. The success of the first year of the Michigan Journal of Medicine (MJM) was due to the extraordinary team of students, faculty, staff, and collaborators who brought enthusiasm, expertise, and commitment to this unique endeavor. In Part 1 of this chapter, I will discuss some of the team members who each played a key role in creating the journal, how team members were recruited, and what qualities made for successful as part of the MJM team. In Part 2 of this chapter, I will discuss how the journal itself was established, including finding a need in the curriculum to fill; getting investment from the medical school, faculty, and students; and lessons learned from our first year.

PART 1: ESTABLISHING A TEAM

To establish a student-run journal, we needed a team that included not only students but also those with expertise in scientific publishing, website design, publicity, and more. For MJM to be both educational and successful as a journal, it was important that we had enthusiastic students interested in learning and experienced professionals with a desire to pass on their expertise and skills. Here I will detail some of the key members of our team and their roles.

Medical School Faculty Advisor

The first step for medical students with a new idea for an academic endeavor such as a student-run journal is to find faculty members who will invest in

Medical and Scientific Publishing. https://doi.org/10.1016/B978-0-12-809969-8.00027-9

the idea; further, identifying specific faculty as designated faculty advisors is essential to create clearly defined expectations for busy faculty members. While this may seem counterintuitive for an idea to create something that is *student* run, finding a faculty advisor who will provide mentorship is a critical step for several reasons. Faculty investment is important in order to guide students who may have little to no experience. For example, MJM was unique because it would offer new opportunities for medical students who often have little to no experience as reviewers, editors, or publishers. Thus, it was essential that we brought in mentorship with experience in these areas. Additionally, faculty often have broader connections within a university as well as credibility and leverage that are useful in bringing in other experts to the team. Finally, as I will discuss in Part 2 of this chapter, when creating a scientific journal it is essential that you have a level of credibility to encourage authors to submit their work to your publication. Having sponsorship from a faculty member, or even better from the medical school itself, elevates the level of credibility of a new journal.

Our founding editors (Fig. 27.1) approached one of the surgeons who was also interested in medical education and was an ideal candidate for a faculty mentor. As director of the evolving clinical curriculum, he had investment in creating and finding innovative ways to enhance opportunities for student education. This role also meant that his mentorship would add to the credibility of the journal when trying to get investment from the medical school. He was a respected researcher who had published extensively. He also had a long history of mentoring medical students and advocating for medical student ideas. He had several important connections to the publishing services offered at the University, including Michigan Publishing Services [1], the unit of the University Library that would ultimately serve as our publisher and was able to

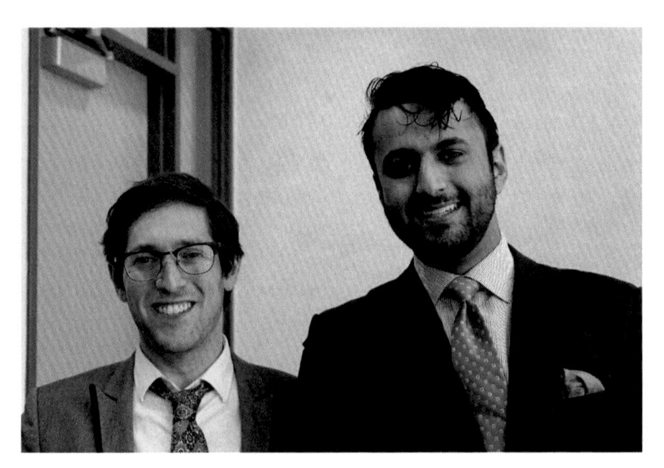

FIGURE 27.1 Founding editors, Spencer Lewis (left) and Sagar Deshpande (right).

recruit future team members that were integral to MJM's success. Lastly, he was excited by the idea. Finding a mentor who is energized and enthusiastic about a proposal is a key part to its success.

Our mentor's role as faculty advisor was to provide guidance to the student editorial staff, to create an actual course for the senior editors, to network to provide other key team members to MJM, and to ensure that MJM was a quality educational and professional experience for all involved. Ultimately, he also served as a liaison to bring in other important team members from across the university, augmented the educational aspect of MJM by creating a course with guest lecturers who were knowledgeable in publishing for participating students, provided the authority figure when issues of professionalism with medical students arose, and was instrumental in getting buy-in from the medical school itself.

For more on MJM as a concept and how it evolved into a real publishing initiative, please see Chapter 26.

Editor-in-Chief

As a then fourth-year medical student at the University of Michigan, I was brought on as the inaugural Editor-in-Chief (EiC) for the first year of MJM in part because the founding editors were just beginning their clinical training. Given the time constraints and importance of their first-clinical year in medical school, the founding editors and faculty advisor agreed that another student should be brought on to lead the journal in its first year. I was recruited by our faculty advisor because of my previous leadership roles in student-run organizations, research experience, and thus some familiarity with scientific publishing, and the fact that I doing a full-time research fellowship for the year and would not have clinical duties. My leadership and research experience in combination with my less demanding schedule gave me the skills and time needed to take on this challenging role.

The EiC role in the first year of MJM was multifaceted. Over the year, I had many jobs. In the creation of the journal, I recruited student editors and reviewers, worked with the publisher to develop a strategy and timeline for the creation of our journal, designed the work flow for the journal for the first year and for future publication years, and worked closely with the other team members to create the resources that the journal would ultimately rely on including our website and our reviewer tools. The various components of the creation of the journal will be outlined in Part 2 of this chapter. Once we had submissions for the journal, I oversaw our editorial team and reviewers and ultimately was the final quality check in the publishing process. The ability to plan ahead, delegate, communicate, and be flexible were all essential to my role as EiC. This role was incredibly challenging and rewarding; over the course of the year I was able to develop my ability to plan and strategize, to manage my peers, to communicate with professionals, and to review scientific work. In this way, MJM provides a unique opportunity for students to hone their leadership and managerial skills under the guidance of skilled mentors.

Publishing Advisor

In addition to having a faculty advisor with educational and scientific experience, it was crucial for MJM to have an advisor with expertise and experience in scientific publishing from the publication perspective. Through the faculty advisor's connections, a senior level director who ran a publishing and curriculum design team at the Medical School. Her role as our publishing advisor was critical to MJM in the first year because of her experience with scientific and medical publishing and publication ethics. She was engaged and enthusiastic about the project and became a key advocate who, with the faculty advisor, brought in our publishing partner, Michigan Publishing Services. To create a new scientific journal, MJM needed someone who had successfully established journals before and a team that knew the ins and outs of publishing. Our publishing advisor worked hand in hand with the faculty advisor to create a syllabus for the course that our senior editors would take, which ensured that the educational component would be robust and have practical use. As publishing advisor, she took on a large part of the managerial work as well—ensuring the various collaborators were completing their assignments, checking in with our EiC to make sure deadlines were met, and helping to move things along when necessary.

Michigan Publishing Services

Our publishing advisor then brought on the staff from Michigan Publishing to work through the publication aspects of MJM. Having the Michigan Publishing staff on our team at MJM provided detailed knowledge and experience of the individual tasks that would need to be completed to reach our goal of publishing the inaugural issue of MJM in May 2016, in time for medical school graduation. Beyond publication experience, Michigan Publishing also had experience working with student groups on publications. Thus, their knowledge and experience was particularly tailored to our needs. They helped our editorial team to create a timeline and deadlines, a marketing plan, a website, and a submission/reviewer tool. Their involvement with MJM not only made the journal's publication timely, successful, and of professional quality but also they allowed us to distill the educational value from the experience so that difficult tasks that might not have been valuable for editors to learn from were done efficiently by experienced staff at Michigan Publishing. The recruitment of collaborators at Michigan Publishing was key to the success of MJM in the first year. However, without the knowledge and connections of our publishing advisor, I would have never known who or how to reach out to this resource. This highlights the crucial point that recruiting experienced advisors and mentors to our team truly allowed us to leverage all that the University of Michigan had to offer for MJM to be successful.

Senior Editors

Senior editors were senior medical students who signed up to an editor in the first year of MJM and take a longitudinal course called "Medical Editing." For more on the course, please see Chapter 1. The senior editors were the members of our editorial team, took part in monthly meetings, and oversaw a team of reviewers to review submitted manuscripts and then worked with authors on resubmissions. They were key leadership within MJM and their input on our vision, timeline, and tasks was essential. As part of the course they helped to create author guidelines and reviewer worksheets. In Chapter 28, the role of the senior editors within the organization will be further detailed. Having the requirement for students to be involved in a course was a particularly useful aspect for our editorial team. This ensured that students signing up to be editors were committed to the 1-year commitment required. It also meant that these were students with a considerable amount of interest in learning about medical editing and publishing. Lastly, having an accompanying course allowed our editors to receive training prior to having to review submissions; this allowed them to be more successful in leading a team of inexperienced reviewers through the process of reviewing a manuscript and working with authors to improve the quality of their work.

Reviewers

Reviewers were medical students of any year who were interested in participating with MJM but were not a part of the course. They were invited to come to the Medical Editing lectures and encouraged to attend monthly meetings; however, these were not required for reviewers. Chapter 28 will detail more specifically their role in MJM. We recruited reviewers by sending email requests to the medical school, which generated a lot of interest in the journal. The reviewers were excited about the opportunity to participate in scientific publishing, learn more about scientific writing, and how to work in teams. As MJM continues on, future members of our editorial board will hopefully have started as reviewers for MJM. This will augment the richness of the education experience and improve our ability to provide quality feedback to our authors. The strong support and interest we got from the medical school is a significant indicator that there is a real need in the curriculum for hands on experience in scientific writing and publishing. As discussed later in the chapter, by fulfilling that niche MJM is able to help ensure its continued success.

Lessons Learned From Our Team

MJM had an incredible team that allowed our first year to be successful. From our advisors to staff collaborators to students, we had a dedicated group that was enthusiastic about our vision. There were several key takeaways of our team that I believe contributed to our success:

- *Everyone had a voice.* Whether it was a reviewer who only attended one meeting and reviewed one article or our publishing advisor, our team had the mentality that all opinions should be heard and valued. Recognizing that each member brings different experiences to the team allowed us to maximize contributions from all involved. While ultimate decisions fell on editorial board and advisors, our culture of open communication and valuing all contributions helped every team member to feel invested in our vision.

- *Simple decisions were made quickly and with appropriate team members, while impactful decisions were made after thoughtful deliberation.* There were many small decisions that needed to be made during MJM's first year. Because we had such a diverse team of students, staff, and faculty, not every team member was involved in every decision. However, large decisions that would shape the vision of MJM were made as a team and only after all team members had the opportunity to provide input. This allowed MJM to be efficient with our collaborators' time, while making informed decisions that the team felt invested in.
- *All team members were enthusiastic about the creation of MJM.* This is a point that I have mentioned several times, but I think it is important. Our team was composed of individuals who were excited about MJM and its possibilities; thus, they were committed to putting in the time to make their contributions great. All team members and their contributions were valued and respected. With a culture of enthusiasm and respect, MJM was able to have a diverse team come together to make a great product.
- *We were open and responsive to feedback.* Throughout the year we had many ideas that were ultimately changed or scrapped based on the real-time feedback of various team members. It was important that everyone in our team remained flexible and open to the expertise of our collaborators and the suggestions of our team. Additionally, at the end of our first year, we completed a survey of all of our student members. We asked about work flow, the vision of MJM, the areas of need in the curriculum, and the lessons we learned as a team throughout the year. By eliciting and being receptive to feedback, both on a short-term and long-term scale, our team was able to put the best ideas into effect.

PART 2: ESTABLISHING A JOURNAL

There were four key components to establishing MJM: (1) adding value to medical student participants, (2) having credibility to encourage submissions and involvement, (3) publicity and visibility for the new journal, and (4) creating a sustainable work flow that would enable the journal to continue to exist beyond the individual graduating classes.

Adding Value

From the editorial and review side, MJM provided a unique educational and professional opportunity for students to participate in scientific publishing. For students interested in an academic career, learning how to review scientific papers, how to write scientific papers, and how scientific publishing works is an invaluable experience that will serve them throughout their career. While students may be able to experience this to some extent by participating in research and writing/submitting manuscripts, opportunities to be a part of the other side of the process are almost nonexistent because students generally are not qualified to review others' work or to serve on a journal's editorial board. In this way, MJM filled a serious vacancy in the education of students at our medical school.

However, to be successful, MJM also needed to add value for the students who would be submitting their work for review. After all, without submissions to review and publish there would be no learning experience. As an editorial board, we inherited the vision from the founding editors that MJM would be a peer-reviewed, student-led journal for other students to publish their work that they otherwise would not be able to publish. It was up to us to determine how that vision fit in with the realities of the student researchers' needs at our medical school. To understand where a need might exist, we turned our student reviewers and editors into focus groups. After all, they were our target audience: students who were motivated to be involved in scientific research and publication. We used our first few meetings to discuss where students felt MJM could fill a need for student authors. We acknowledged that students with manuscripts they would be able to publish in established, PubMed indexed journals would not have a need to submit to MJM, but that there were many times students participated on a research project that never resulted in publication. It was this latter group we were targeting.

From our discussions, we devised three areas where students might benefit from an opportunity to publish in a student-run journal: (1) quality research with negative results that a research mentor might otherwise deem unworthy of publication, (2) the opportunity to publish a smaller contribution of a larger study as first author, and (3) topic reviews that students may create as background for research projects that otherwise would not be published. From these areas, we created our three submission types: (1) Original Research Article, (2) Brief Communication of Original Research, and (3) Review Articles. We also agreed as a team that we would not outright reject any submissions. We felt that MJM had the opportunity to add value to our authors by providing more thorough feedback to authors of papers that were lower quality. We felt a commitment to augment all of our submissions to a quality worthy of publication and thus enhance the educational value for authors, reviewers, and editors. In our first issue, published in May 2016 (Fig. 27.2), we had five Brief Communications of Original Research, two Original Research Articles, and four Review Articles. In the future, MJM hopes to partner more closely with the medical school to be a productive outlet for student work that would not otherwise be published to increase the educational value of student work for authors, editors, and reviewers as well as to publicize the excellent work being done by our medical students to a larger community.

FIGURE 27.2 First issue of *Michigan Journal of Medicine* on display at a medical school event.

Credibility

In addition to adding value, it is important that a student-led journal has credibility among our faculty and peers to be successful. In the future, the credibility of MJM will derive from the journal's history of quality publications and professionalism. However, in the inaugural year, we had to establish credibility with potential authors and readers in other ways. First, we had the benefit of being part of an institution with a long history of student-led initiatives. Second, our team included experts in scientific publishing and established faculty members. Third, we had a robust educational component for the student editors and reviewers prior to their reviewing the work of our authors. These pieces allowed us to have substantial engagement from the medical school during a time that the school was looking to innovate and expand its curriculum. Having the backing of the medical school was a huge benefit to our credibility with faculty and students. Our first email calling for submissions was sent from the Office of Medical Student Education by our Associate Dean for Medical Student Education.

Publicity and Visibility

To submit manuscripts for review, potential authors had to know that MJM existed, that MJM was looking for submissions, and what the deadline for our first issue would be. To spread the word and increase interest around our first issue, we created a publicity strategy with the help of our team members at Michigan Publishing. Our strategy included an online presence, social media presence, email reminders, fliers posted throughout the medical school, and a lunch talk for students.

- *Website*: With the help of Michigan Publishing and collaborators in Information Services at the Medical School, we created a website at www.michjmed.org. The website would be a central place for information about MJM, including submission deadlines, author guidelines, and the vision of MJM. It would also be the place where authors would submit their manuscripts for review and where MJM would ultimately publish and publicize our authors' work. The URL was short and memorable and was included on all of our fliers, in our emails, and easily accessible from our social media accounts. The MJM website continues to evolve to meet the needs of its users. For more information about the web design process, please see Chapter 29.

- *Social Media*: We created a Facebook (www.facebook.com/michjmed) and Twitter (@umich_MJM) for MJM to better reach our audience. There is precedence for this as many established medical and scientific journals take to social media to promote their publications. Through social media we were able to further create value and establish credibility by promoting useful information regarding scientific publishing and research. We are also able to better publicize the work of our authors to a much broader audience.
- *Email campaigns*: Our first call for submissions was sent out by our Associate Dean for Medical Student Education. This was a powerful and very effective way to generate interest in MJM. In addition to letting students know about the new journal it also signaled the strong support MJM had from the medical school. Subsequent reminders and notifications came from either our EiC or Faculty Advisor. This was an easy way to reach medical students, publicize our journal, and distribute our website and social media accounts.
- *Fliers*: As was common practice of student groups at University of Michigan Medical School, we posted fliers (Fig. 27.3) with our submission deadline, website, and social media accounts in high traffic areas throughout the medical school. These semipermanent reminders allowed us to maximize the visibility of MJM and also increase web traffic to our site and social media accounts.
- *Lunch Talks*: Additionally, we held a talk for medical students with free lunch where we discussed the value of submitted to MJM, the process of our review, and answered questions. By providing free lunch, we were able to reach a large number of students and provide detailed, meaningful information. We also had members of our editorial team attend required lectures for medical students of all years where they were able to make announcements, quickly answer questions, and provide our website and social media accounts for further interest.

The publicity and visibility of MJM will continue to be a two-pronged effort: aimed at both recruiting submissions and at publicizing the excellent work that we publish. In this way, both efforts help the other. Having quality publications that are visible to the medical school will help to recruit authors and allow MJM to fill a need at UMMS to publicize the work of our students. Additionally, as students are interested in submitting their work, they will spend more time on our website and social media accounts where the work we have published will gain further attention.

FIGURE 27.3 Call for submissions flier for MJM.

Lessons Learned From Creating a Review Process

The work flow and review process utilized in MJM's second year will be detailed in Chapter 28. Here I will discuss how our team approached creating a review process from scratch and the lessons we learned throughout the year. As an editorial team, in collaboration with our partners at Michigan

Publishing, we created a 1-year timeline with the goal of our first issue published in May 2016. Based on this end date, our colleagues at Michigan Publishing estimated that we would need final version of articles to be ready for copy editing and publishing by early January 2016. This highlights the importance of having team members with experience and expertise. Without prior publication experience, it would have been impossible for our editorial team to create a reasonable timeline for the copy editing, printing, and publication tasks for our journal. Because Michigan Publishing has had many years of experience working with student publications, they were able to provide useful guidance that allowed us to create a workable timeline. With feedback from Michigan Publishing, we set a submission deadline of October 15, 2015. We felt this date allows the editorial staff time for adequate review and resubmission, as well as allowed students time to put together meaningful projects for submission. The key to our successful timeline is that our editorial board would be working toward several goals simultaneously. At our monthly meetings we would check in regarding our progress toward these goals. Things that we had to complete prior to our October 15 submission deadline for our first volume included the following:

- Publicity for MJM (as discussed above) including creation of website, Twitter, and Facebook accounts
- Centralized submission tool for authors to submit and editors/reviewers to use as they reviewed manuscripts
- Training for the submission tool for editors and reviewers
- Standardized review forms to allow the review of manuscripts to be more uniform across a variety of students with different knowledge and skill levels
- Creation of guidelines for submission for authors for each manuscript type (Original Research, Brief Communication, and Review Articles)

Many of these tasks were delegated to the editorial board as part of their coursework, including creation of author guidelines and the standardized review forms; others were completed with our partners at Michigan Publishing. We learned many lessons about strategy, teamwork, communication, and planning through this yearlong endeavor. There were several tenets that we found to be specifically critical to completing a long-term task, such as publishing a journal, on time:

- Clear and specific long-term goals (e.g., final publication of printed and online version by May 2016)
- Clear and specific short-term goals (e.g., reviewer worksheets will be ready for approval in 1 month)
- Accountability to editorial staff, faculty mentors, and other collaborators
- Open and honest communication between all team members
- Group and individual check ins between EiC and all team members
- EiC kept in loop by all team members, serving as point person to ensure all parts were being completed successfully

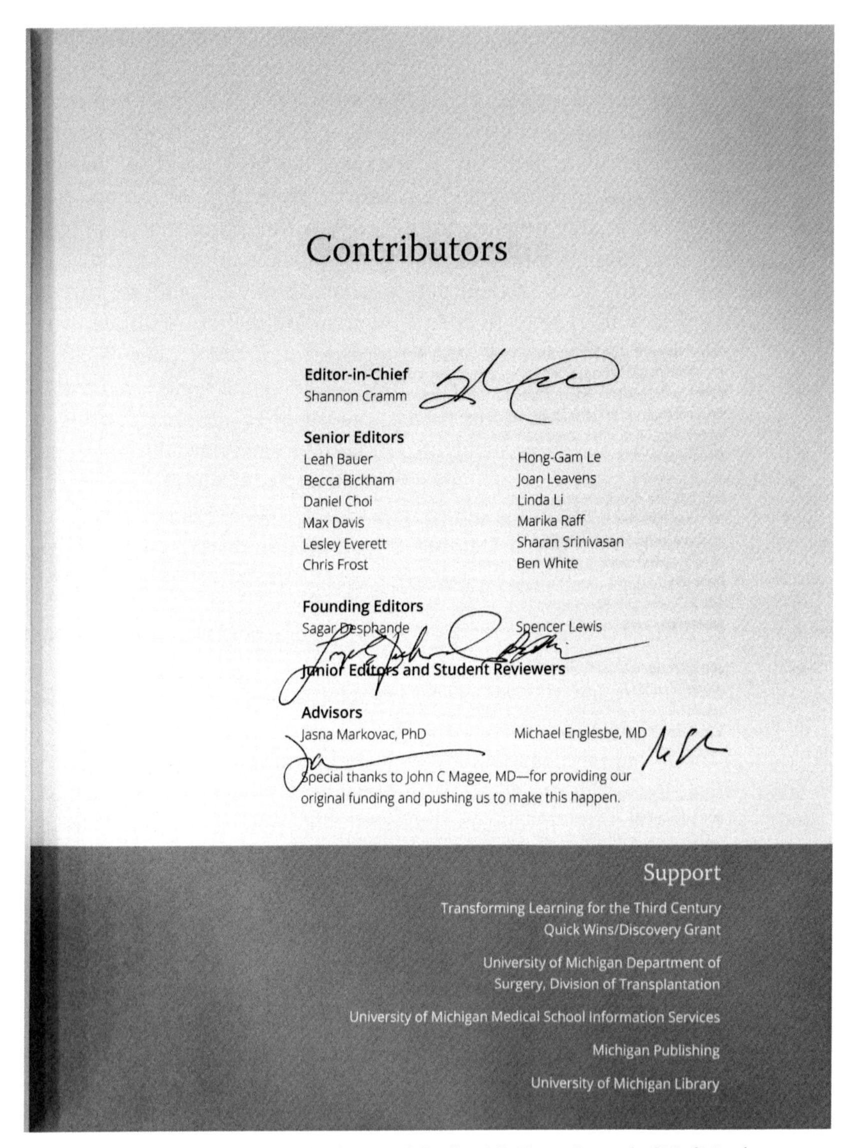

FIGURE 27.4 Signed masthead page of the first *Michigan Journal of Medicine* issue.

Having an EiC who is able to see all the moving pieces at once and make sure all the parts of individuals' work fit nicely together was an important component of the success of MJM in its first year. Whether it was an author who was late on resubmission or an editor who was not finalizing their review quickly enough, the EiC having an idea of what was going on with every aspect of

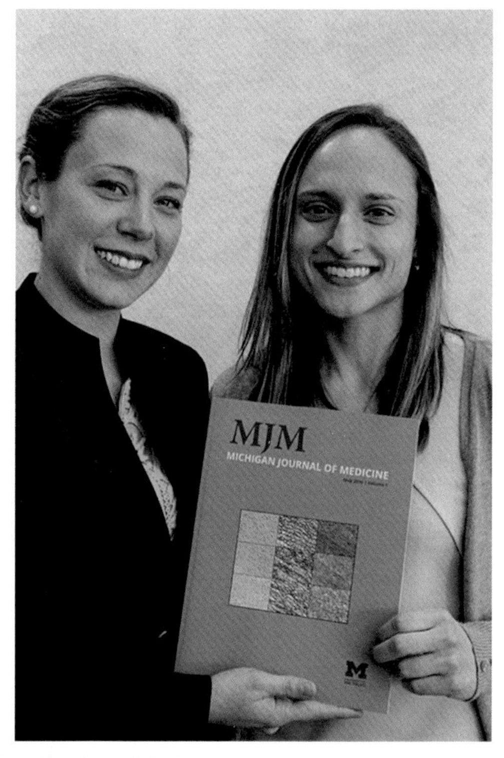

FIGURE 27.5 First Editor-in-Chief (EiC), Shannon Cramm (left) and second EiC, Alisha Lussiez (right).

the journal ensured a timely delivery of our final product. Ultimately, it was the incredible work of a diverse team of students, staff, and faculty that made MJM's first year a success [2]. Our team dynamic that valued people and their contributions fostered an environment where everyone took pride in the quality and timeliness of their work.

In the words of Bo Schembechler (a well-known and very accomplished former University of Michigan football coach): "The team, the team, the team." At MJM, we had established "the team," a talented and passionate group that allowed us to take the idea of a student-run journal and make it a reality (Figs. 27.4 and 27.5).

In May, 2016, the first issue of MJM was officially celebrated at the journal launch party held at the medical school (Fig. 27.6).

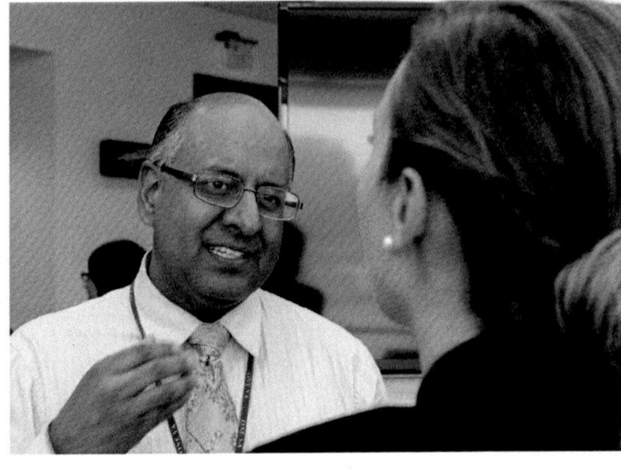

FIGURE 27.6 Medical school leadership show support for *Michigan Journal of Medicine* at May, 2016, launch party. Dr. Michael Mulholland, Chair of Surgery, left; Dr. Rajesh Mangrulkar, Associate Dean for Medical Student Education, near right, with Shannon Cramm, far right. *(Photos courtesy Marc Stephens.)*

REFERENCES

[1] https://www.publishing.umich.edu/services/.

[2] Smith TM. Michigan students launch peer-reviewed medical journal. AMA Wire September 19, 2016. https://wire.ama-assn.org/education/michigan-students-launch-peer-reviewed-medical-journal.

[3] Relevant sites. www.michjmed.org. www.facebook.com/michjmed.@umich_MJM.

Chapter 28

Michigan Journal of Medicine as a Learning Tool: Perspectives From the Editor-in-Chief

Alisha Lussiez, MD
Department of Surgery, University of Michigan, Ann Arbor, MI, United States

MICHIGAN JOURNAL OF MEDICINE AS A LEARNING TOOL

The *Michigan Journal of Medicine* (MJM) is a unique learning opportunity for all medical students involved [1]. At the University of Michigan, there are ample opportunities to conduct research, but there are fewer opportunities to critique manuscripts and practice giving authors feedback. As a completely student-run journal, MJM provides a rare opportunity to practice the skills many future academic physicians will need—both as Editors/Reviewers and as authors [2]. By getting involved with MJM, student authors are exposed to the entirety of the research life cycle from initial submission to resubmission(s) to approving copy-edited text and page proofs for the print journal. On the other side, the student Editorial Staff helps guide authors with intensive feedback from their first submission to finer word editing in later submissions.

In its second year, the MJM Editorial staff was set up with a three-level hierarchy (Fig. 28.1). The Editorial Staff was made up of 42 reviewers, 12 editors, and one Editor-in-Chief (EiC). The student Reviewers were mostly second- and third-year medical students, while the Editors and EiC were fourth-year medical students. This large group was further broken down into 12 Editorial Teams made up of one Editor and three to four Reviewers. Once an author submitted a manuscript, the manuscript was blinded by the EiC and assigned to one of the 12 Editors. From there, the Editor assigns the manuscript to the Reviewers. The Reviewers reviewed the manuscript and then the Editor compiled the reviews into one succinct document for the authors. In addition, Editors, with help from the EiC, were in charge of giving Reviewers feedback on their reviews. The EiC oversaw the work from the editors and made the final pass through each paper, modifying the feedback to the author as needed. The EiC was also in charge of communicating with authors. These Editorial Teams

Medical and Scientific Publishing. https://doi.org/10.1016/B978-0-12-809969-8.00028-0

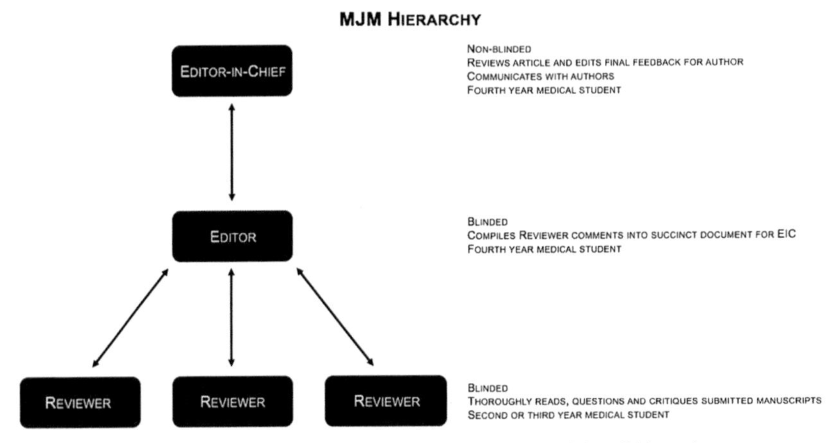

FIGURE 28.1 *Michigan Journal of Medicine* editorial staff hierarchy.

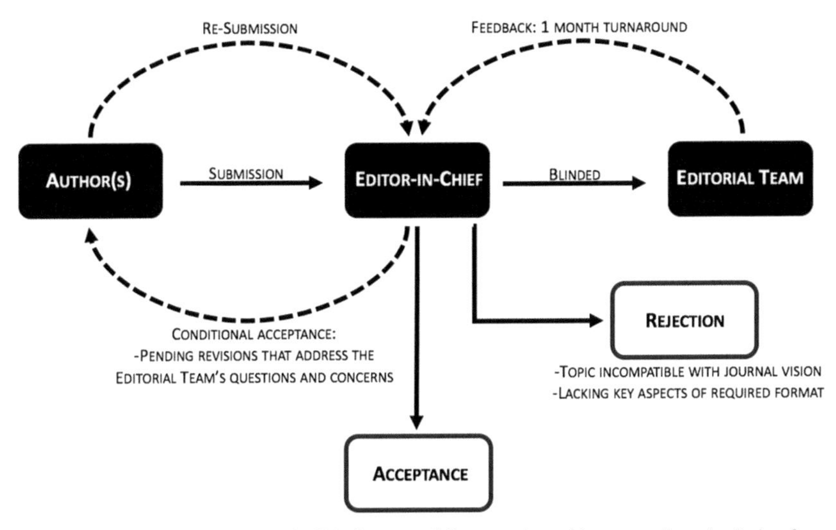

FIGURE 28.2 *Michigan Journal of Medicine* workflow starting with manuscript submission from medical student authors. *(With permission from Ania Owczarczyk, MD.)*

did not change throughout the year to foster mentor–mentee relationships. The workflow can be seen in Fig. 28.2.

Learning Opportunities for Editors and Reviewers

Editors for MJM were fourth-year students, most of whom had prior research experience as scientists and authors. However, few of them had experience on the other side of the research life cycle as an Editor/Reviewer and none of them had experience in writing all the types of articles published in the MJM (Case Report, Perspective, Brief Communication of Original Research,

Original Research and Review Article). Therefore, these fourth-year students received training to prepare them for their work as Editors. They received a week of Medical Editing didactics in which they improved their understanding of quality scientific writing and data display, learned about the ethics and editorial process and standards in scientific publishing, and familiarized themselves with establishing and improving their online publishing profile. For more information about each of these topics, please see the relevant chapters in this book. To familiarize themselves with the different manuscript types MJM accepts for submission, Editors had access to resources such as "how-to guides" for authors and examples of published articles of different types. The work in developing these didactics and resources was essential to give authors useful feedback, as well as to give the Reviewers quality feedback on their reviews.

The infrastructure of having the Editorial Teams, composed of the EiC, one Editor and three to four Reviewers, enabled the editorial team to build confidence making editorial decisions. For more information about Journal Editorial Teams, please see Chapter 27. Each manuscript had at least five different people look at it and contribute to the manuscript feedback. While it is true that editing is subjective and the same point can be made in multiple different ways, there are some hard rules that need to be followed for each article type. For example, all Original Research manuscripts should have a clear hypothesis and sound methods. The learning objectives at MJM were to become familiar with the necessities for each article type.

In addition to reviewing manuscripts, one of the unique opportunities for Editors was the chance to mentor younger medical students—both in growing their editing skills and in navigating medical school. This year's Editors at MJM were applying into a broad range of fields including Neurology, General Surgery, Interventional Radiology, Psychiatry, and OB/GYN among others and reviewers were paired with editors based on their future career interests. By providing feedback to Reviewers, Editors developed their skills on delivering constructive criticism and professional communication. For more information about Peer Review, please see Chapter 7.

The structure of MJM was designed to give learning opportunities to Reviewers, as well. Reviewers typically consisted of second- and third-year medical students, and no research or editing experience was required. As reviewers for MJM, students had the opportunity to develop skills related to analyzing submitted manuscripts. To help Reviewers critique an article, worksheets for each article type were created to help guide the editorial staff (Fig. 28.3). These worksheets contained questions that helped shape the review and highlight what Reviewers should be thinking about. Like Editors, Reviewers were often assigned to an article type they were not familiar with, in a field they may have or may not have known anything about. This required familiarizing themselves with the unknown to provide authors feedback.

Original Research Review Worksheet

1. (CHECKBOXES) Does this submission meet the following requirements?
   a. Abstract < 500 words with appropriate subdivisions
   b. No missing sections in the article body (introduction, methods, results, discussion, conclusion, citations)
   c. All tables and figures have captions
   d. Not previously published
2. (SHORT ANSWER) What problem are the authors addressing?
3. (SHORT ANSWER) Please paraphrase what the research question and hypothesis are. Are they clearly identifiable?
4. (Y/N) Does the introduction share relevant background and make it clear for someone without any expertise in this field?
5. (Y/N) Are the methods, including materials, study population, study design, and analysis, clearly and concisely described?
6. (Y/N) Is the study design adequate to test the research question?
7. (MULTIPLE CHOICE) On a scale of 1-5, how sound are the methods employed in this paper?
   a. 1 (extremely poor with major flaws)
   b. 2
   c. 3 (minor flaws requiring explanation)
   d. 4
   e. 5 (exemplary)
8. (SHORT ANSWER) Please provide an explanation for the grade you assigned to the methods section in the prior question.
9. (SHORT ANSWER) What are the limitations to the author's methods of addressing this problem?
10. (SHORT ANSWER) How would you improve this study design?
11. (Y/N) Are the results presented in an organized and understandable manner (and can you obtain clear information from the tables and figures in less than 1 minute)?
12. (Y/N) Does the discussion summarize the main findings and place them in context of current literature?
13. (Y/N) Are the conclusions supported by the results?

FIGURE 28.3 Excerpt from a reviewer worksheet for an Original Research manuscript.

Learning Opportunities for Authors

Authors submitting manuscripts to MJM benefitted from careful peer review of their work and thoughtful feedback. Feedback from the editorial team was detailed and offered suggestions on how to fix problems (when applicable) instead of just pointing them out. In addition to writing the manuscript, authors gained experience in formatting their work to fit submission guidelines, responding to Reviewer comments and working with publishers to create a final version.

Articles that were well written went through the review process much like at other journals. However, articles that may have otherwise been quickly rejected at other journals were not immediately rejected at MJM. Instead, the editorial staff would do a full review of the manuscript and suggest "Revision and Resubmission." When this was the case, the author was emailed by the EiC to set up a meeting to go over their manuscript and the feedback. The goal was to ensure quality feedback because it is a challenge trying to give authors feedback when it was unclear what they were trying to say.

As EiC, I found these meetings one of the most rewarding parts of the position. I have been fortunate to have received high-quality mentorship through each of my research experiences from conducting the research to analyzing the data to writing the manuscript. I have drafted sections only to have almost every word rewritten. Though I still have a lot to learn as a researcher, I learned an incredible amount from these experiences and was able to pass some of what I have learned on to authors either writing an article on their own or to authors who did not receive helpful feedback on their work before submitting. It was incredibly rewarding to read a well-written resubmission—it was an indication that feedback had been clear and well delivered.

LEADERSHIP LESSONS

As EiC, I am thrilled with the success and production of the second issue. It was incredibly satisfying to have lead a team of 70–80 medical students through that year, and it was a rewarding experience for all those involved. It is the team's hope that other schools and organizations will soon offer similar experiences. I have shared below some of the struggles and difficulties from this year.

Submission Volume

As a young journal in its second year, MJM struggled obtaining manuscript submissions for various reasons. There were 14 total submissions for the second issue; 11 were being published and the other 3 were reviewed but not resubmitted. With a team of 12 Editors and >40 Reviewers, low submission volume left many without an avenue to contribute. As a partial remedy, I assigned multiple Editorial Teams to review and edit the article at the beginning of the year when there were only a couple submissions. However, only one Editorial Team was assigned as the "Official Editorial Team." The "Unofficial Editorial Teams" were not required to review the article; it was an option for students interested in the learning experience. About half of the Unofficial Editorial Teams reviewed the article, and those who did found the learning experience beneficial.

The Editorial Staff as a whole made an effort to understand from faculty and classmates why there were not more submissions. The most prominent reason from both students and faculty was the hesitation to submit work to a journal that was not yet indexed. Authors will continue to initially shoot for top-tier journals with high impact factors given all the hard work that goes into the conduction, analysis, and write-up of research; the fact that lab funding is partially dependent on publications; and the need to grow one's resume in anticipation of a future job search. This will be hard to compete with for many years to come. That said, MJM is an incredible learning opportunity for those who take part in any part of the submission-editing process. It will continue to be a great avenue for sharing medical student research with the community, for publishing Original Research that is not yet complete in the form of a Brief Communication, for work yielding negative results, or for work that was high quality but lacked one of several key factors (favorable response rate for surveys, large sample size, significant *P*-values, etc.).

Leading Your Peers

The EiC position at MJM was a unique opportunity because I was in charge of leading a group made up of my peers, some of whom had more experience in research and editing than I did. In other leadership positions I have held, hierarchy and prior experience had played to my advantaged (captaining a soccer team as an upperclassman, for example). Furthermore, this was leading a group of my friends—people who knew me more casually and not in the context of working toward the completion of a project.

The yearlong experience was a constant and conscious balance between playing it "chill" and being formal with more rigidity. For the most part, I was lucky to work with motivated, organized, and efficient medical students. The tricky part came with imposing deadlines and trying to coordinate the busy and unaligned schedules of fourth-year medical students. Fourth year of medical school is the year when everyone is doing their own thing and trying to enjoy life outside of school and the hospital as much as they can.

I found setting deadlines that were well ahead of schedule worked well. This was possible because MJM was only being produced once a year and the submission volume was low. Early deadlines left room for extensions as well as minimal consequences if deadlines were missed. This allowed me to maintain relaxed and friendly relationships while still getting the work done. Separately, I planned ahead by asking all Editors for their yearlong schedule to know when he or she would have a lighter rotation, and I avoided assigning work when an Editor was on a more time-intensive rotation. Editors appreciated this and, in turn, took advantage of the free time they had.

Delegating

An important lesson I learned this year is figuring out when and, equally importantly, when *not* to delegate. It is important to make sure team members feel fulfilled, productive, and that they are making a contribution. It was helpful to remember this when delegating tasks that seemed faster to complete myself. Creating roles at the beginning of the year allowed for a clear division of labor and an opportunity for each Editor to be an expert in that area. These were roles for fourth-year medical student Editors that involved duties outside of editorial tasks. Some examples include publicity, meeting secretary, summer research program outreach, and journal design.

When being involved in a young initiative, name and brand consistency is key in building a picture our medical student peers recognize. For example, having the same person contact the school via email every time allows the school to start to pair the sender with the topic. This way, it also becomes clear whom to contact when questions arise.

As a leader and team member, it was an excellent opportunity for me to learn how to step back, as I have a tendency to be fairly hands-on. I learned a lot from delegating and then seeing what my classmates came up with on their own. Overall, delegating made the year a lot easier and though, at times, some of the work had to be revised due to miscommunications or disagreements, it was useful to view this as an opportunity for collaboration and teaching.

Two things I would have done differently include improving role definition and expectations as well as not delegating certain key tasks. Part of the reason I frequently had to follow-up on tasks was because I had not made my expectations clear. All the MJM roles were being piloted and therefore were necessarily fluid. However, it would have been helpful to share a sort of rubric or outline of what was expected of each role at the beginning of the year, and then edit them as necessary. Secondly, I learned there are some things that are best done by the person in charge. As EiC, it is important to be the one to communicate with faculty and other organizations, especially when trying to build an official presence for one's organization. This way, one has a good handle on the direction of the initiative; and if questions come up, they can be more readily answered. If delegated inappropriately, it can also be much harder to recover from these situations if there was a miscommunication or time-sensitive issue.

BEYOND THE EDITING

Templates

Investing the time into making all tasks templated (e.g., forms with clear directions on how to fill them out instead of a prompt such as author feedback forms, reviewer worksheets, classroom didactics, etc.) and available remotely was worth the effort. Both Editors and authors could then operate completely on their own schedule and could finish tasks quickly, even if working remotely. Learning to communicate clearly and succinctly over email is essential, and email templates can also help save time. Templates also make directions very clear on what needs to be completed so that minimal time is spent on answering questions and communicating individual instructions.

Cover Art

The cover of a paper journal is extremely important and is often the sole image that comes to mind when thinking of the journal. It creates an opportunity to involve students in the design and construct of a visual to represent the work of the team. This is empowering and a unique chance to let the creative juices flow. This can potentially also be an area for conflict since so many people contributed to the journal and have different esthetic tastes. This year, there was one Editor who designed the cover art. In the future, it could be useful to hold a submission contest in which students vote for the winning design. Submitters and voters can include a body as broad as the medical school or as narrow as just the Editorial Staff. For more on medical illustrations, please see Chapter 14.

Website

Like cover art, a journal's website is also instrumental in building a brand identity for a young initiative. Among the important features include ease of use, esthetics, and transparency. Website visitors will often come across the page not knowing anything about the initiative. Given this, a reader should immediately see the purpose of the website and find visible links to things such as the journal issues, submission instructions, news updates, and contact information. In addition, as a student-run journal, information about the process and editorial team should be readily available for students and faculty. For more on website development as it relates to MJM, please see Chapter 29.

IN CONCLUSION

Now that my tenure as the EiC for MJM is over and I have turned over its reigns to the next student editorial team, I am very interested to see the progression and development of MJM, both as an educational tool for future physicians and as a viable publication vehicle for their research. Fig. 28.4 shows the front cover of the second issue of MJM.

FIGURE 28.4 Front cover of the second issue of the *Michigan Journal of Medicine*, published in April, 2017.

The support that MJM continues to receive at the University of Michigan gives me great optimism for its success. I am looking forward to seeing the publication of more papers, more frequent issues, and ultimately even submissions from beyond our school.

REFERENCES

[1] http://www.michjmed.org/.

[2] Smith TM. Michigan students launch peer-reviewed medical journal. AMA Wire September 19, 2016. https://wire.ama-assn.org/education/michigan-students-launch-peer-reviewed-medical-journal.

Chapter 29

Applying Design Thinking to the Design of an Online Electronic Journal

Chris Chapman, MA, Aki Yao, Jason Engling, MA
Health Information Technology and Services, University of Michigan, Ann Arbor, MI, United States

INTRODUCTION

Design Thinking is an approach used by designers to systematically design products ranging from buildings to appliances to websites [1]. Design Thinking allows designers to collaborate with clients, users, and nondesigners in the design process [2,3]. In the case of the medical student-run journal launched at the University of Michigan, ***Michigan Journal of Medicine (MJM)***, we employed a design thinking process to work closely with the MJM faculty subject matter experts and student editors to design the journal website in an efficient and effective manner. This chapter presents our overall design process and the ways we specifically applied it to the design of the MJM web presence. This same process can easily be adapted to laboratory websites, patient portals, and other products for the dissemination of medical and scholarly content.

The Design Thinking Process—Overview

The process we used to design educational products and applications is a hybrid derived, combined, and customized from other sources to meet the unique needs and constraints at our institution.

Specifically, the approach we use is adopted from a process entitled "High-Tech Anthropology" from Menlo Innovations (https://www.menloinnovations.com/) (a software development company based in Ann Arbor, Michigan) and design skills taught at Skillcrush (https://skillcrush.com/) (an interactive online learning community specializing in digital design instruction). We also applied some of the design and development methodologies taught in two texts: "Contextual Design: Defining Customer-Centered Systems (Interactive Technologies)" [4] and "User Story Mapping: Discover the Whole Story, Build the Right Product" [5]. Our team combined

Medical and Scientific Publishing. https://doi.org/10.1016/B978-0-12-809969-8.00029-2

the knowledge learned from these resources with existing internal processes to develop a Design Thinking approach that we are now implementing for various online publications.

The Design Thinking approach described here has been developed further for the design of webpages and websites for medical publications. This approach can be modified for other purposes such as the design of laboratory websites.

Application: Michigan Journal of Medicine

Our team had been working with faculty instructors on a then new Medical Editing course for medical students. A student-run medical journal (Michigan Journal of Medicine (MJM)) was launched as part of that course and published by the University Library. For more information about the MJM journal concept and about Library Publishing please see Chapters 26 and 4.

Our team worked with faculty subject matter experts (SME) to create a new top-level home page for MJM. This website was to provide a front-end to the journal with seamless linking to the published journal articles housed on the publisher's system, as well as features such as editorial interviews, student editor bios, social media feeds, and other nonpeer-reviewed content. To develop a website that would provide what the readers, authors, editors, and educators needed and wanted to use, the design team engaged the SME in the Design Thinking process.

The following section describes each of the Design Thinking steps used for the MJM project, along with more general methods for other applications. Documents, designs, and layouts (with annotations) generated during the MJM design process will be used to illustrate the specifics of each of the steps.

THE STEPS

Project Intake

The project team meets with the client to learn more about their goals for their site. A website questionnaire is used to obtain specific information about the project and the website (e.g., budget, launch date, reasons for wanting a site, etc.), the target audience, site perception (e.g., adjectives to describe how the audience should perceive the site, similar sites, etc.), and the types of content that will be accessible from the site (e.g., text, video, existing, to-be-created, categories of content, etc.). The resulting information provides the designers with a general sense of how the site should look, who will be using it, the scope of the project, the timeline, and budget. This first step is critical for the success of the project—the designer must understand the purpose of the site, how it will be used, and what it aims to achieve for each of the user groups. For the MJM project, the intake session included input from faculty, staff, and students associated with the journal as well as an investigation of other medical journal home pages. The questionnaire used for MJM is shown in Fig. 29.1.

1	**Project Title:**	MJM Website Redesign
2	**Today's Date:**	1/26/2017
3	**Project Champion(s):**	Dr. J.M., Dr. M. E.
4	**Project Stakeholder(s):**	Dr. E, Dr. M, Dr. W.
5	**Department/Affiliation:**	H.I.T.S., Library (Michigan Publishing Services), Department of Surgery
6	**Consultant(s):**	J.E., A.Y.
7	**What problem or opportunity is this project intended to address?**	Current Site: http://www.michjmed.org/ • Home page needs to be be more inviting. • Ready for more national exposure
8	**Who is the intended audience of this project?**	• Medical Educators • Medical student • Physician • Professionals
9	**What are the goals and/or learning objectives of this project?**	• Gateway to an educational tool • Raise awareness on a more national level
10	**If this is a learning program, what prerequisite knowledge do you expect your learners/users to have?**	The site is the gateway to the work that is created during the course.
11	**Learning context: where does this program fit in terms of the overall learning context (i.e., supplements a lecture, an overview that is elaborated on in other experiences, etc.)?**	• Medical students opportunity to learn about the publishing process
12	**Delivery method to audience?**	Online
13	**How long will the final program be from a user perspective (i.e., x number of minutes for a video, y number of pages for a website, etc.)?**	N/A
14	**Who are the other stakeholders, departments, etc. that need to be involved to ensure success of**	• MJM Editors in Chief • Michigan Publishing

FIGURE 29.1 Intake questionnaire used for *Michigan Journal of Medicine*.

Persona Creation and Selection

Personas are fictionalized individuals who represent the people who will be using the product being designed in some capacity [6]. In the case of a website, personas may represent users, content creators, administrators, etc. The personas are based on observations [7] and/or research on the types of people who will be using, creating, and/or supporting the site and are created by the design team, giving each the project-relevant identifying attributes, such as age, demographics, and reasons

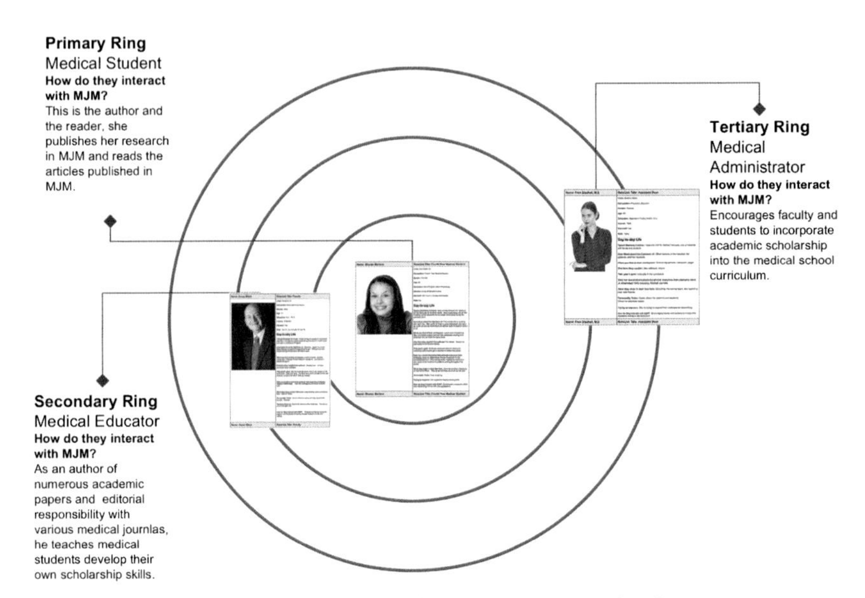

FIGURE 29.2 Persona map for *Michigan Journal of Medicine.*

they would use the site. For a complex project with numerous categories of users, the design team can create as many 15–20 personas, organized by importance. The primary persona represents the main person who will be using the site, so all design decisions must support and not interfere with their use of the site. For example, if the primary persona only has a desktop computer from which to use the site, then the first priority would be to make the site work on this type of a machine and hence, designing for a smartphone would not necessarily be a priority and, if the site were to be smartphone friendly, it would still need to function as intended on a desktop computer. The secondary and tertiary personas are also used to inform design decisions (e.g., maybe the secondary persona only uses a smartphone, so the site also needs to function on such a device), but the priority for design decisions starts with the primary persona, follows to the secondary personas, and concludes with the tertiary personas. Outside of technical and compliance standards, design decisions normally are made by consulting the personas (e.g., "Would Persona X want this feature?" or "What does Persona X need?"). Features not required by the personas are normally not designed or implemented.

For the MJM project, because of its relative simplicity, we only created three personas, the medical student (primary), the medical educator (secondary), and the medical administrator (tertiary) (Fig. 29.2). These personas then became proxies for people using the site.

Inventory

At this stage, the team works with the client to identify all the possible components and features that the site will have. Once all of the components are identified, they are grouped into relevant categories to form an initial organizational representation

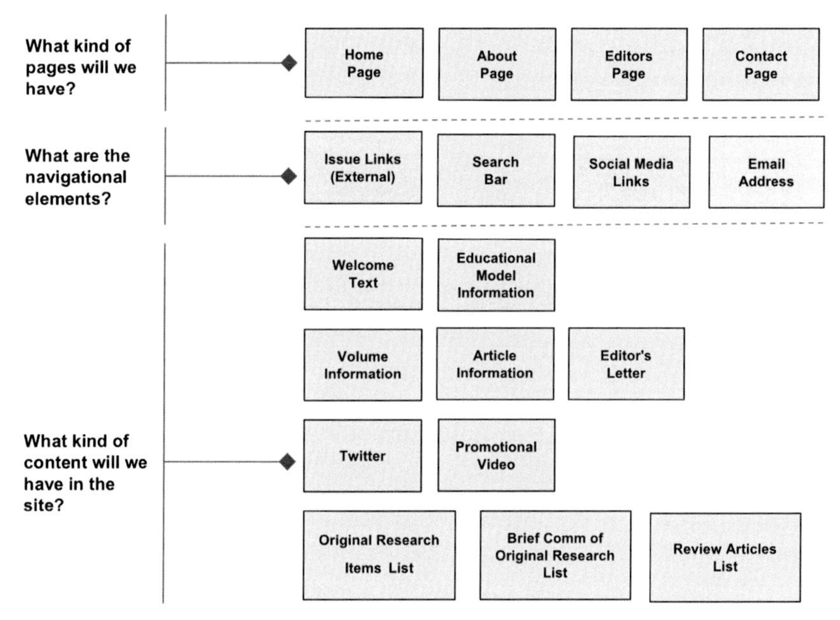

FIGURE 29.3 Inventory organization of *Michigan Journal of Medicine* components.

of the site. This allows the client and the design team to see the basic structure of the site, to develop a sense of the scope of the site, to ensure that all of the site components are accounted for, and to remove redundancies and unneeded elements. This process can be done with sticky notes on a board if all those involved happen to be located in the same place. But it can just as easily be done electronically to accommodate multiple geographical locations. For MJM we were able to look at the publisher's journal site and break it down into its components as the basis for the new website, adding additional components and features as determined by the persona needs. Fig. 29.3 illustrates the organization of the MJM components.

User Goals and Entry Points

Now the team works with the client to identify the main goals and objectives the client has for the users (using the personas as proxies for the users) of the site. For MJM, goals include having the user view a newly published article, sign up for a newsletter, or contact the site owner. Goals are established for each persona, and for each goal a corresponding entry point(s) is established. Goals are the objectives the users want to accomplish when using a site. For MJM we created a mix of goals that included, for example, submitting a paper, contacting the journal editor, and how learning about how MJM is used as a learning tool.

Entry points are a way to explore the variety of ways a user will access a site or product. For the MJM site we needed to think about how someone would come across the URL or learn about the site. Since entry points can be varied and somewhat arbitrary, it is important to stretch your thinking, use your imagination, and consider things you might not have thought of before.

Potential Entry Point:	Goal Action:
Link from email	Submit student research paper
See a tweet while they are scanning their twitter feed	Want to explore the journal.
A friend forwards them a link from a news article	Want to find out about more about MJM as an educational model.
While reviewing residency applications, reviewer sees that applicant listed they published in MJM.	Want to explore the journal.
Home Page	Want to contact the editor to learn more.

FIGURE 29.4 Selection of entry points and goals for *Michigan Journal of Medicine*.

For example, if the goal for a medical student is to view a specific video on the MJM home page, the entry point may be that the medical student receives an email from a colleague suggesting they go to the site and watch the video. The entry point would then be the email they received from the colleague with a link to the site embedded in the message; the goal would be that they view the video. Fig. 29.4 illustrates a selection of entry points and goals for MJM.

User Flows

At this point the design team works with the client to determine how the personas will achieve the goals that have been set for them [8]. For each goal, the entry point is recorded, and each step or click the persona would take to reach the goal is documented (e.g., receive email about a video to view, click on link in the email, go to the home page of the site, click on the "videos" page, and finally view the video and achieve their "goal").

This user flow is mapped out for each of the persona goals and entry points (Fig. 29.5). During this step, it is important to keep in mind the user behaviors of the people represented by the personas and, as such, to minimize the number of clicks required to reach the desired goal. Medical students are busy; they want to be able to find what they are looking for with a minimum number of clicks. The navigation needs to be intuitive, and the content they need must be easily accessed. The user flow visualizes the user's experience from the entry point to the goal and one can see how easy or difficult it is for a user to reach the goal. In designing an intuitive navigation, it is advantageous to consider the various

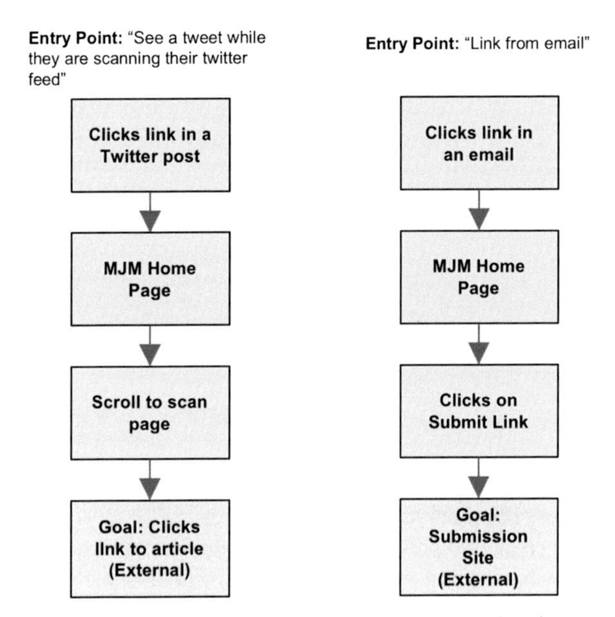

FIGURE 29.5 Example user flow map for *Michigan Journal of Medicine* persona.

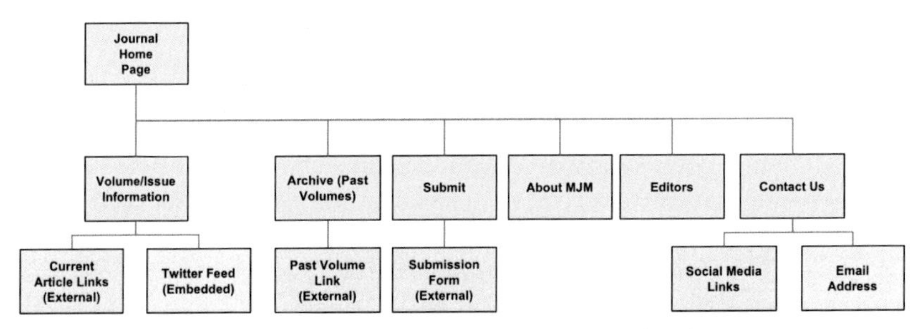

FIGURE 29.6 Site map for *Michigan Journal of Medicine*.

locations users may go through and the actions they will take in those locations. Based on this information, the design can be adjusted for maximal efficiency.

Site Mapping

Based on the content inventory and the user flows, the design team then maps out the architectural structure of the site. The site map takes the items from the inventory and organizes them in a way that they relate to one another. This becomes important for the next step, the development of mock-ups. Most of the content on the MJM website is settled on the home page with external links to the publisher site for submission information and published articles. The site map for MJM is depicted in Fig. 29.6.

Sites maps can range from complex structures to simple, linear progressions depending on the project. If the site is very large and multilayered, the design team may work with the client to establish a staged development approach in which sections of the site map are prioritized for phased development and release [5]. Using the site map in this way allows the design team and the client to plan and track the process and to make adjustments based on user behavior along the way.

Text Content

Once the site has been mapped and the development priorities have been established, the client is asked to provide text copy that will be used on the site. Although most sites feature visual elements, clear concise text is essential for optimal audience use (e.g., the site has a title, contact information, headlines, and content). For the MJM home page, we were able to use some of the text found on the publisher's site and supplement that with additional content (videos, photos, and other elements).

Mock-Ups

At this point mock-ups are created for the site based on the information obtained in the previous stages. The mock-ups (also called wireframes) are sketches of the site pages [9] and are created to present possible layouts. The mock-up process begins the move toward what the user will see and interact with. For MJM we started with quick drawings on a white board (Fig. 29.7A). This way we could easily try out different ideas and layouts within a short period of time. If all parties involved at this stage are not in the same location, this can easily be accomplished electronically using various share-screen features. Mock-ups can be further refined with digital tools such as Balsamiq (https://balsamiq.com/), as shown for MJM in Fig. 29.7B.

Here it is important to develop mock-ups quickly and efficiently so that ideas and multiple options can be rapidly explored and iterated. Based on the feedback we received from medical students as well as from our SME, several rounds of mock-ups were created to further refine the website design (Fig. 29.7C). Photos and images can be included at this stage to better visualize the end product. Text is added to help refine the layout.

(A)

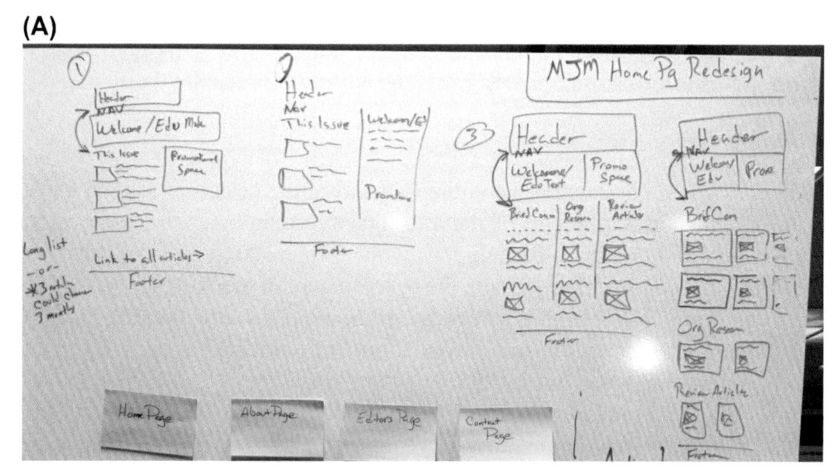

FIGURE 29.7 (A) Michigan Journal of Medicine (MJM) mock-ups (white board). (B) MJM mock-ups (digital). (C) MJM mock-ups (with content details).

(B)

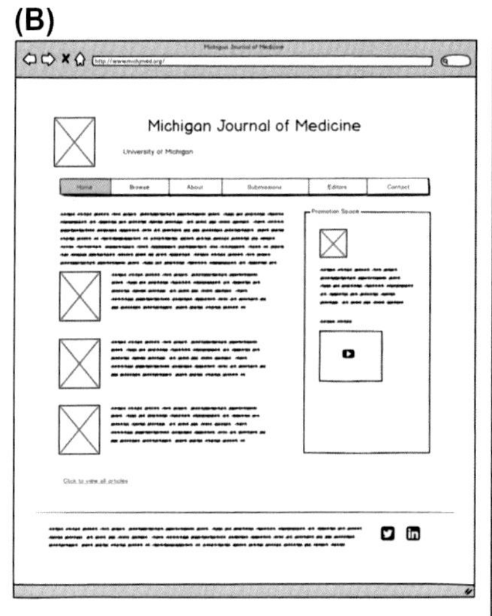

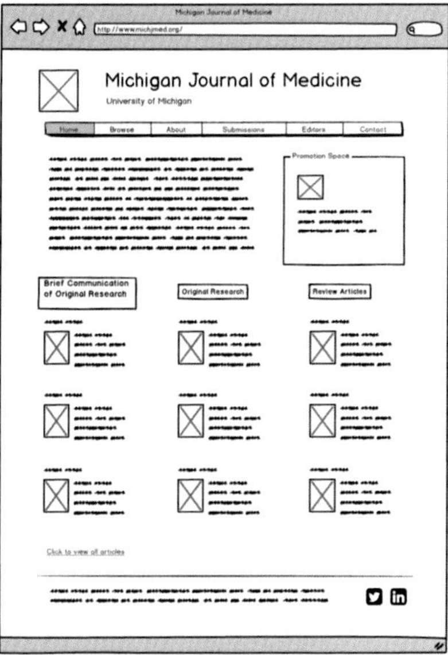

(C)

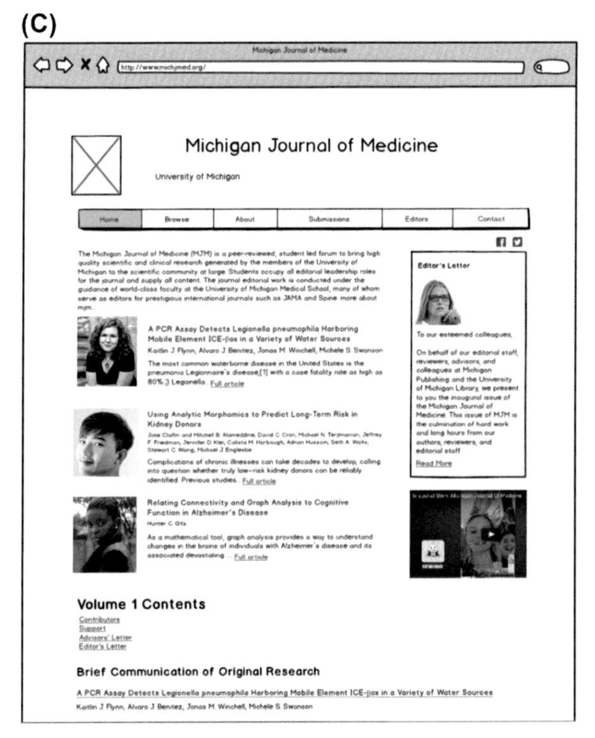

FIGURE 29.7 Cont'd.

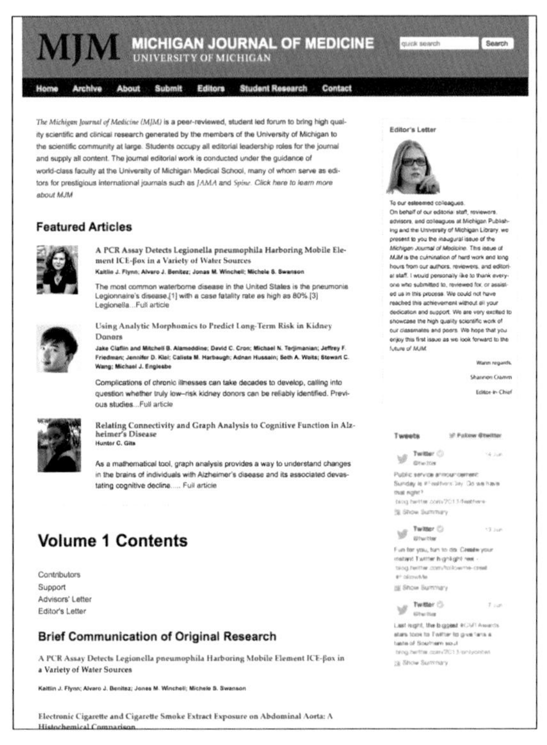

FIGURE 29.8 *Michigan Journal of Medicine* composite.

Composites

Using the mock-ups as a guide, along with information gathered from the website questionnaire (e.g., descriptive adjectives describing the site, color preferences, examples of other similar sites), the graphic designer creates full color, full fidelity layouts of the final wireframe illustrations [9] in a graphics program such as Photoshop or Illustrator. At this stage, the journal logo, branding elements, correct fonts, real text and images, graphics, and other elements were incorporated. Fonts, font sizes, and the visual relationship between text elements are determined at this stage. The composites may go through several rounds of iteration until a final working design is developed. Fig. 29.8 illustrates the composite we created for MJM.

The Final Product

After the composites are finalized, they are translated into a functional website using the technology selected for the specific project. This can be a particular content management system such as Drupal or custom coding using HTML and/or other technologies. During the production stage, slight modifications may be made to the designs based on feedback received from the various stakeholders. The MJM home page is shown in Fig. 29.9A and B. Because the MJM user groups access the journal in a variety of ways using a variety of devices, it

(A)

(B)

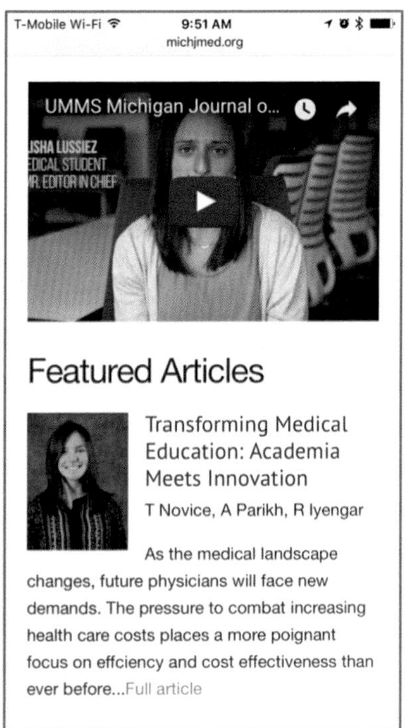

FIGURE 29.9 (A) The final product—Michigan Journal of Medicine (MJM) website—desktop version. (B) The final product—MJM website—mobile version.

was important that the site be fully functional (and visually pleasing) on desktops and laptops as well as on smartphones and tablets. To develop the latter, we used responsive technology to allow the site layout to adjust for different display sizes.

SUMMARY

In developing websites, it is essential to consider user expectation and user behavior as the primary driving force throughout the process. We used a Design Thinking approach to design the MJM website. This process allowed our team to rapidly create an appealing and user friendly site taking into account what medical students, educators, and administrators wanted and needed. We were able to collaborate closely on the design with the SME and the student Editors-in-Chief while being respectful of their time and availability, which is vitally important especially when dealing with busy physicians and other professionals. The Design Thinking process described in this chapter can be used and adapted to design simple and complex websites, and other electronic (and non-electronic) programs, applications, and products. This rigorous design process allows for collaboration and exploration of ideas, takes into account user needs from the outset, and leads to the efficient creation of quality products.

REFERENCES

[1] Beckman SL, Barry M. Innovation as a learning process: embedding design thinking. Calif Manag Rev 2007;50(1):25–56.

[2] Bjögvinsson E, Ehn P, Hillgren PA. Design things and design thinking: contemporary participatory design challenges. Des Issues 2012;28(3):101–16.

[3] Seidel VP, Fixson SK. Adopting design thinking in novice multidisciplinary teams: the application and limits of design methods and reflexive practices. J Prod Innov Manag 2013;30(S1):19–33.

[4] Beyer H, Holtzblatt K. Contextual design: defining customer-centered systems. San Francisco (CA): Morgan Kaufmann Press; 1997.

[5] Patton J, Economy P. User story mapping: discover the whole story, build the right product. Sebastopol (CA): O'Reilly Media, Inc; 2014.

[6] Pruitt J, Adlin T. The persona lifecycle: keeping people in mind throughout product design. San Francisco (CA): Morgan Kaufmann; 2010.

[7] Goodwin K. Getting from research to personas: harnessing the power of data. Cooper Newsletter; 2002. Retrieved from: https://www.cooper.com/journal/2002/11/getting_from_research_to_perso.

[8] What is a user flow? 2017. Retrieved from: https://learn.skillcrush.com/skill-resource-flow/.

[9] Garrett JJ. The elements of user experience: user-centered design for the web and beyond. 2nd ed. Berkeley (CA): New Riders; 2011.

Index

Note: Page numbers followed by "f" indicate figures, "t" indicate tables.

F

G

H

I

J

N

R

S